Embedded Systems Building Blocks,

Second Edition

Complete and Ready-to-Use Modules in C

Jean J. Labrosse

R&D Books
Lawrence, KS 66046

R&D Books
1601 West 23rd Street, Suite 200
Lawrence, Kansas 66046
USA

Cover art created by: Robert Ward.

Distributed in the U.S. and Canada by:
Publishers Group West
1700 Fourth Street
Berkeley, CA 94710
1-800-788-3123

ISBN 0-87930-604-1

A United News & Media publication

*To my loving and caring wife and best friend, Manon,
and to our two lovely children,
James and Sabrina.*

Table of Contents

Preface

This is the second edition of *Embedded Systems Building Blocks, Complete and Ready-to-Use Modules in C*. This is a book of software modules that you can use to design embedded systems. The modules are some of the most common *building blocks* of embedded systems: keyboard scanners, display interfaces, timers, and I/Os. Most of the code is written in highly portable C.

Managers will like this book because it can reduce the amount of time, and thus money, required for some of the more repetitive aspects of embedded systems design. Each chapter is independent of the others, allowing you to use only the module(s) you need. Each chapter describes what the module does, how it works and, what services it provides. This information will help you estimate the resources you'll need to implement your product.

What's new in the Second Edition?

I made a number of changes from the first edition. The most notable one is, of course, the hard cover which makes the book more durable. The second major change is that all of the code and examples have been revised to use µC/OS-II. µC/OS-II is a Real-Time Operating System that I wrote and is fully described in my other book, *MicroC/OS-II, The Real-Time Kernel* (ISBN 0-87930-543-6), R&D Books. A scaled down version of µC/OS-II is provided in object form to allow you to run and change the sample code.

I decided to use the Borland C/C++ compiler V4.51 instead of V3.1 because some of you had indicated that the version 3 tools are no longer available. I also included a `makefile` to build the sample code instead of relying on the IDE (*Integrated Development Environment*). The `makefile` can easily be changed so the code can be compiled for just about any other target processor.

Chapter 1, "Sample Code", has been completely revised. Chapter 2, "Real-Time Systems Concepts", now contains over 10 new pages. For all the building blocks, I now have a section that presents the APIs (*Application Programming Interfaces*) in a standard format. This allows you to better use the interface functions of each building block. In the first edition, Appendix F contained all the data sheets of electronic components I used. I decided to move the data sheets to the companion CD-ROM in PDF form to reduce the book size by about 100 pages and save a few trees in the process.

In the first edition, I included the execution times of each of the building block interface functions provided in the book. This process was quite tedious and so I decided to drop this in the second edition. Also, the 80386 computer I had used to come up with the execution times was retired a few years ago.

Goals

This book is designed to aid embedded systems programmers by providing ready-to-use modules. If the code in this book doesn't match your exact requirements, you can use the code as a starting point. In other words, it is a lot easier to modify code than to start from scratch. The main objective of this book is to save you time.

Intended Audience

This book is for embedded system programmers, consultants, and students interested in embedded systems. I assume you know C and have a minimal knowledge of assembly language. You should also understand microprocessors and have a basic electronics background. The hardware presented in this book is, however, fairly easy to understand. Because the code is written in C, you can apply the concepts presented in this book to a much broader range of microprocessors (assembly language would not be portable).

If you are a student interested in embedded systems, this book will take some of the mysteries out of the unique requirements of embedded system software design by providing you with concrete programming examples. This book will also allow students to build much more complex embedded systems than would otherwise be possible in the classroom.

Portability

The code presented in this book is written in ANSI C and is highly portable. C has been the language of choice for embedded system designs because C has the following features:

- C code is easier to write and understand than code in assembly language.
- The code generated by some C compilers approaches assembly language in efficiency.
- Once written, C code often can be used on different processors. This is not the case for assembly language code.

In many cases, less than 10% of the code uses more than 90% of the CPU time. You can always optimize this time-critical code by using assembly language. The non-time critical code (90% of the code), can still be written in C. If you are still using assembly language to design embedded systems, you should consider obtaining a C compiler and writing portions of your code in C.

Hardware interface functions have been carefully isolated to minimize the amount of work required to adapt the module to your own hardware environment. I have kept the assembly language to a minimum, and in the places where I have used assembly language, I have kept the code as clear and simple as possible.

What Will You Need to Use this Book?

The code supplied with this book assumes you will be using a PC (80486 minimum) computer running under either Windows 95/98/NT or DOS v4.x and higher. The code was compiled with Borland International's (now called Inprise) C++ v4.51 (see www.borland.com). You should have about 5 Mbytes of free disk space on your hard drive.

Acknowledgments

First and foremost, I would like to thank my wife for her encouragement, understanding, and patience. This book would never have been possible without her. I would also like to thank my children James (age 9) and Sabrina (age 6) for putting up with having just a mom for a few months while I was 'hiding' in my office working on this new edition. I hope one day they will understand. Special thanks to Dr. Bernard Williams and all the fine people at R&D Books for their help in making this book a reality. Finally, I would like to thank you for buying this book and I hope it will live up to your expectations.

Introduction

I've been designing embedded systems for more than 17 years. During that time, I've noticed that some of the pieces always seem to keep coming back. I have concluded that 80+ percent of the code for an embedded product seems to be similar to the previous product. I always seem to need to read analog and discrete inputs, output control signals on analog and discrete outputs, provide some form of user interface and thus, I need to read/scan keys on a keyboard and put information on a display device of some sort (7-segment numeric and/or to an LCD module). Most embedded controllers seem to have an asynchronous serial port (i.e., UART, *Universal Asynchronous Receiver Transmitter*) and interfacing to a laptop seems like a natural thing to do. I also find myself needing to trigger events when a certain amount of time expires, and to keep track of the date and time. Although it was fun and challenging to develop some of these modules at one point in my career, having to do the same thing over again for each new project has become mundane and even unpleasant. I find that the real challenge is to develop application code that makes my products unique. Over the years, I've written fairly generic modules to accomplish some of the functions mentioned above. As I used these modules, I optimized and enhanced them, giving me a good collection of Embedded Systems Building Blocks.

As Steve McConnell mentions in his book, *Code Complete*, "The single biggest way to improve both the quality of your code and your productivity is to reuse good code." In his fine book, *The Art of Programming Embedded Systems*, Jack Ganssle states that, "It's ludicrous that we software people reinvent the wheel with every project. ... Wise programmers make an ongoing effort to build an arsenal of tools for current and future projects ... Collect algorithms!"

If you already write software for embedded systems, this book will provide you with portable, ready-to-use code so that you can save time with your next embedded system design. Time to market is becoming just as important (and in some instances, more important) than the cost of the product itself. Reduced time-to-market provides a competitive advantage.

If I can save you days or even weeks of programming time on one of your products, I will have met my objectives. You might decide to use the code provided in this book for rapid prototyping or as a permanent addition to your final product. All of the modules presented in this book most likely have nothing to do with what makes your product unique. In other words, your *application* code is what makes your product different. For example, you may need a keyboard scanning routine and an LCD display module in a FAX machine. What you provide in this product is your FAX machine expertise and you shouldn't have to spend time with keyboard scanning and LCD display details.

It is very difficult to write 100% reusable code. This is especially true for embedded systems because most embedded systems have very unique requirements and most likely limited memory to hold both the executable portion of your code and its data. The code presented in this book is not intended for embedded systems that will be sold in very large volume. This is because large volume applications are very cost sensitive which means you must typically account for just about every single byte of memory (ROM and RAM); my focus was not to save every single byte.

Figure, Listing, and Table Conventions

You will notice that when I reference a specific element in a figure, I use the letter 'F' followed by the figure number. A number in parentheses following the figure number represents a *specific* element in the figure that I am trying to bring your attention to. 'F1.2(3)' thus means look at the third item in Figure 1.2.

Listings and tables work exactly the same way except that a listing starts with the letter 'L' and a table starts with the letter 'T'.

Source Code Conventions

All of the building block objects (functions, variables, #define constants and macros) start with a prefix indicating that they are related to the specific building block. For example, all clock module functions and variables start with Clk. Similarly, all timer manager functions and variables start with Tmr.

Functions are found in alphabetical order in all the source code files. This allows you to quickly locate any function.

You will find the coding style I use is very consistent. I have been adopting the K&R style for many years. However, I did add some of my own enhancements to make the code (I believe) easier to read and maintain. Indention is always 4 spaces, tabs are never used, always at least one space around an operator, comments are always to the right of code, comment blocks are used to describe functions, etc.

I also use and combine acronyms, abbreviations, and mnemonics (AAMs) to make function, variable, and #define names in a hierarchical way (see Appendix C).

Figure I.1   A block diagram representing the key areas covered in this book.

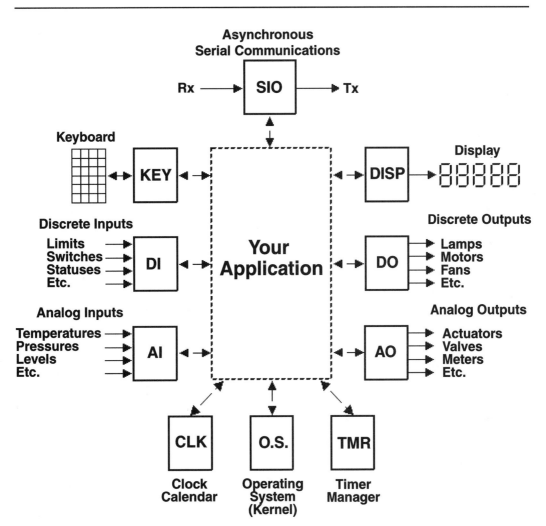

Figure I.1 is a block diagram representing the key areas covered by this book. Even though the building blocks shown in the figure interact mostly with hardware, I have carefully isolated hardware-dependent code to a few easy-to-change functions or constants. This makes the code easy to port to your own environment. Also, I avoided using assembly language except when absolutely necessary.

Chapter Contents

Each chapter describes one or more of the building blocks shown in the figure. The building blocks are mostly independent of one another, so you can jump to any chapter you need. However, you should read at least Chapter 1 to familiarize yourself with some of my conventions. You will also need to understand the material presented in Chapter 9 in order to understand Chapter 10.

Chapter 1   tells you how to install the software provided on the CD-ROM. The chapter also tells you about some of the conventions I use and then provides you with an example on how to use some of the modules presented in this book. I decided to include this information early in the book to allow you to start using the code as soon as possible.

Chapter 2   introduces real-time systems concepts such as foreground/background systems, critical sections, resources, multitasking, context switching, scheduling, reentrancy, task priorities, mutual exclusion, semaphores, intertask communications, task synchronization, task coordination, interrupts, clock ticks, etc.

Chapter 3   describes one of the building blocks shown in Figure I.1, keyboards. Chapter 3 describes keyboard basics and provides you with a general purpose module that can scan and decode any keyboard matrix from a 3x3 to an 8x8 key arrangement. The keyboard module can buffer keystrokes, repeat the same key if the key is held down for a certain length of time, keep track of how long the key has been pressed, and allow you to define multiple scan codes for each key. The code can be easily expanded to support larger keyboards.

Chapter 4   will show you how to control LED (Light Emitting Diode) displays. LED displays can consist of discrete LEDs, seven-segment modules, or any combination of both. Chapter 4 provides you with a module that can multiplex LEDs from a 3x3 to an 8x8 arrangement. The code can easily be changed to accommodate larger displays.

Chapter 5   provides you with a software module that will control Character LCD Modules which are based on the Hitachi HD44780 Dot Matrix LCD Controller & Driver chip. Character LCD (Liquid Crystal Display) modules are display devices that can display alphanumeric data.

Chapter 6   describes a software-driven clock/calendar module that keeps track of hours, minutes, seconds, days, months, years (including leap years) and day-of-week. The code also provides you with a 32-bit timestamp which can be used to mark the occurrence of events.

Chapter 7   describes a module that manages up to 250 countdown timers. Each timer can be preset to timeout after up to 100 hours with 0.1 second resolution. You can define a function that will be executed when the timer expires (one for each timer).

Chapter 8   provides a module that can read discrete inputs and control discrete outputs (up to 250 each). For discrete inputs, the module will tell you whether the input is high, low, transitioned from low to high, high to low or both. When a transition is detected, a user-definable function can be executed (one for each input). Each discrete input can also simulate a toggle action (push-ON, push-OFF). Each discrete output can be turned ON, turned OFF, or made to blink at a user-definable rate.

Chapter 9   will give you tools to improve the efficiency of mathematical calculations in embedded processors. The concepts presented in this chapter will be used in Chapter 10.

Chapter 10 describes how to read and scale analog inputs and how to scale and control analog outputs. This chapter also provides you with code that will read and scale up to 250 analog inputs and scale and update up to 250 analog outputs.

Chapter 11 discusses asynchronous serial communications and specifically provides you with code that performs buffered serial I/O on a PC. There are actually two versions of this code. One version can be used by a DOS application while the other assumes the presence of a real-time kernel.

Appendix A describes how to use MicroC/OS-II, The Real-Time Kernel. µC/OS-II (for short) is a portable, ROM-able, preemptive, real-time, multitasking kernel. The internals of µC/OS-II are fully described in my other book, *MicroC/OS-II, The Real-Time Kernel*, which is also available (along with a diskette containing the source code) from R&D Books (see the ad at the back of the book). Most of the code presented in *Embedded Systems Building Blocks* assumes the presence of a real-time kernel. Specifically, I make use of semaphores and time delays which are available on most (if not all) commercially-available real-time kernels. To allow you to use the code in this book, I have included a compiled version of µC/OS-II (compiled using a Borland C++ v4.51 compiler for an Intel 80x86 Large Model).

Appendix B describes some of my programming conventions. Specifically, I describe my directory structures and C programming style.

Appendix C lists the acronyms, abbreviations, and mnemonics that I used in the code presented in this book.

Appendix D presents two DOS utilities that I use: TO and HPLISTC. TO is a utility that I use to quickly move between MS-DOS directories without having to type the CD (change directory) command. HPLISTC is a utility to print C source code in compressed mode (i.e., 17 CPI) and allows you to specify page breaks. The printout is assumed to be to a Hewlett Packard (HP) Laserjet type printer.

 Appendix E describes how to install the source code provided on the companion CD-ROM included with this book and describes the licensing policy with regards to using the code in commercial applications.

Web Site

To provide better support to you, I created the µC/OS-II web site (www.uCOS-II.com). You can obtain information about:

- news on µC/OS, µC/OS-II, and *Embedded Systems Building Blocks*,
- upgrades,
- bug fixes,
- answers to frequently asked questions (FAQs),
- application notes,
- books,
- classes,
- links to other web sites, and more.

Bibliography

Ganssle, Jack G.
The Art of Programming Embedded Systems
San Diego, California
Academic Press, Inc.
ISBN 0-12-274880-8

McConnell, Steve
Code Complete, A Practical Handbook of Software Construction
Redmond, Washington
Microsoft Press
ISBN 1-55615-484-4

## Chapter 1

Sample Code

This chapter provides you with an example on how to use some of the embedded systems building blocks described in this book. I decided to include this chapter early in the book to allow you to start using the code as soon as possible. Before getting into the sample code, I will describe some of the conventions I use throughout the book.

The sample code was compiled using the Borland International (now called Inprise) C/C++ compiler V4.51 and options were selected to generate code for an Intel/AMD 80186 processor (large memory model) although the compiler was also instructed to generate floating-point instructions. I realize that the 80186 doesn't have hardware assisted but most PCs nowadays contain at least a 80486 processor which has floating-point hardware. The code was actually run and tested on a 300 MHz Intel Pentium-II based PC which can be viewed as a super fast 80186 processor (at least for my purpose). I chose a PC as my target system for a number of reasons. First and foremost, it's a lot easier to test code on a PC than on any other embedded environment (i.e., evaluation board, emulator etc.) — there are no EPROMs to burn, no downloads to EPROM emulators, CPU emulators, etc. You simply compile, link, and run. Second, the 80186 object code (Real Mode, Large Model) generated using the Borland C/C++ compiler is compatible with all 80x86 derivative processors from Intel or AMD.

Embedded Systems Building Blocks assumes the presence of a real-time kernel. For your convenience, I included a copy (in object form) of *µC/OS-II, The Real-Time Kernel* (see Appendix A for details).

1.00 Installing Embedded Systems Building Blocks

R&D Books has included a companion CD-ROM to *Embedded Systems Building Blocks* (ESBB). The CD-ROM is in MS-DOS format and contains all the source code provided in this book. It is assumed that you have a DOS, Windows 95, Windows 98, or Windows NT-based computer system running on an 80x86, Pentium, or Pentium-II processor. You will need less than about 10 Mbytes of free disk space to install ESBB and its source files on your system.

Before starting the installation, make a backup copy of the files found on the companion CD-ROM. To install the code provided on the CD-ROM, follow these steps:

1. Load DOS (or open a DOS box in Windows 95/98/NT) and specify the C: drive as the default drive
2. Insert the companion CD-ROM in your CD drive
3. Enter `<cddrive>:INSTALL <cddrive>[drive]`

Note that `<cddrive>` is the drive letter where your CD is found and, `[drive]` is an optional drive letter indicating the destination disk on which the source code provided in this book will be installed. If you do not specify a drive, the source code will be installed on the current drive.

INSTALL is a DOS batch file called `INSTALL.BAT` and is found in the root directory of the companion CD-ROM. `INSTALL.BAT` will create a `\SOFTWARE` directory on the specified destination drive. `INSTALL.BAT` will then change the directory to `\SOFTWARE` and copy the file `ESBB.EXE` from the `A:` drive to this directory. `INSTALL.BAT` will then execute `ESBB.EXE`, which will create all other directories under `\SOFTWARE` and transfer all source and executable files provided in this book. Upon completion, `INSTALL.BAT` will delete `ESBB.EXE` and change the directory to `\SOFTWARE\BLOCKS\SAMPLE\TEST` where the example code executable is found.

Make sure you read the `READ.ME` file on the companion CD-ROM for last minute changes and notes.

Also see Appendix E for a list of files and directories created.

1.01 How Each Chapter Is Organized

Each chapter in this book briefly introduces and describes the features of the "Embedded Systems Building Block" provided in the chapter. A more detailed description generally follows the introduction. Next, I describe the internals of the module. You will find:

- the name of the directory where the module's files are located,
- the name of the files for the building block,
- the naming conventions related to the module, and
- the step-by-step description of how the module works.

Your application interfaces with each module through functions. *Interface functions* allow the details of the module to be hidden from your own code. This is called *data abstraction*. If done properly, data abstraction allows you to change the implementation details of the module without affecting your application code. In other words, your application always sees the same module even though you may change the internals of the module. Each interface function is presented along with a description of how to use the function and what arguments are expected.

The modules provided in this book have been developed for use on fairly low-end 8-bit processors. I purchased an IBM PC/AT compatible breadboard to test some of the hardware aspects of the modules presented in this book. This breadboard made testing a breeze. The breadboard I used was the JDR Microdevices (see bibliography) PDS-601 which cost only $80. The PDS-601 contains an ISA bus interface, decoding logic, an Intel 8255A chip, an Intel 8253 (similar to an 82C54), and a large breadboard area.

In every building block, I tried to isolate target-specific code into a few functions and configuration constants, i.e., `#defines`. This allows you to easily adapt the code to your own environment. Thus, each chapter has a *configuration* section which describes how to change the code so that it can work in your target system.

Some of the chapters, specifically Chapters 3, 4, 8, 10 and 11, include a section called, "How to Use the ??? Module." This section provides an example on how you can actually use the module in an application. The example describes how to properly initialize the code and how to invoke some of its services.

Each chapter ends with a bibliography, source code listings, and pointers to one or more data sheets (stored on the CD-ROM) of an electronic components mentioned in the chapter.

1.02 **INCLUDES.H**

You will notice that every .C file in this book contains the following declaration:

Listing 1.1   Master **INCLUDE** *file*

```
#include "includes.h"
```

INCLUDES.H allows every .C file in your project to be written without concern about which header file will actually be included. In other words, INCLUDES.H is a Master include file. The only drawback is that INCLUDES.H includes header files that are not pertinent to some of the .C file being compiled. This means that each file will require extra time to compile. This inconvenience is offset by code portability. You can certainly edit INCLUDES.H to add your own header files. The actual INCLUDES.H I used is found in Listing 1.24 at the end of this chapter.

1.03 Compiler Independent Data Types

Because different microprocessors have different word lengths, I have created a number of type definitions that ensures portability (see \SOFTWARE\uCOS-II\Ix86L-FP\OS_CPU.H (see Appendix A, Listing A.1) for the 80x86 real-mode, large model). Specifically, ESBB and µC/OS-II code never make use of C's short, int, and long data types because they are inherently non-portable. Instead, I defined integer data types that are both portable and intuitive as shown below.

Listing 1.2   Compiler independent data types

```
typedef unsigned char   BOOLEAN;
typedef unsigned char   INT8U;
typedef signed   char   INT8S;
typedef unsigned int    INT16U;
typedef signed   int    INT16S;
typedef unsigned long   INT32U;
typedef signed   long   INT32S;
typedef float           FP32;
typedef double          FP64;
```

The INT16U data type, for example, always represents a 16-bit unsigned integer. ESBB, µC/OS-II, and your application code can now assume that the range of values for variables declared with this type is from 0 to 65535. A compiler for a 32-bit processor could specify that an INT16U would be declared as an unsigned short instead of an unsigned int. Where the code is concerned, however, it still deals with an `INT16U`. The above code fragment provide the declarations for the 80x86 and the Borland C/C++ compiler as an example.

1.04   *CFG.C and CFG.H*

To allow you to easily adapt the code in this book to your environment, I created two user-configurable files called `CFG.C` and `CFG.H`. All the target-specific code has been conveniently located for you in `CFG.C` and `CFG.H`. You don't have to edit every `.C` and `.H` file to use the code in this book. If you adapt `CFG.C` and `CFG.H` to your environment, you can use every module 'as is'.

`CFG.C` (Listing 1.22) contains the hardware-specific functions of the modules presented in this book. `CFG.H` (Listing 1.23) contains the configuration `#defines` for each module. `CFG.C` and `CFG.H` are found in the `\SOFTWARE\BLOCKS\SAMPLE\SOURCE` directory. In order to use `CFG.C` and `CFG.H`, you must 'tell' the compiler to ignore the same declarations in the code for the modules. You accomplish this by defining the constants `CFG_C` and `CFG_H` in `INCLUDES.H`.

1.05   *Global Variables*

The following is a technique that I use to declare global variables. As you know, a global variable needs to be allocated storage space in RAM and must be referenced by other modules using the C keyword `extern`. Declarations must thus be placed in both the `.C` and the `.H` files. This duplication of declarations, however, can lead to mistakes. The technique described in this section only requires a single declaration in the header file, but is a little tricky to understand. However, once you know how this technique works, you will apply it mechanically.

In all `.H` files that define global variables, you will find the following declaration:

Listing 1.3   External references

```
#ifdef    xxx_GLOBALS
#define   xxx_EXT
#else
#define   xxx_EXT extern
#endif
```

Each variable that needs to be declared global will be prefixed with `xxx_EXT` in the `.H` file. 'xxx' represents a prefix identifying the module name. The module's `.C` file will contain the following declaration:

Listing 1.4   .C file declarations of global variables

```
#define   xxx_GLOBALS
#include "includes.h"
```

When the compiler processes the `.C` file it forces `xxx_EXT` (found in the corresponding `.H` file) to "nothing" (because `XXX_GLOBALS` is defined) and thus each global variable will be allocated storage space. When the compiler processes the other .C files, `xxx_GLOBALS` will not be defined and thus

XXX_EXT will be set to extern, allowing you to reference the global variable. To illustrate the concept, let's look at DIO.H (from Chapter 8) which contains the following declarations:

Listing 1.5   *Example using* **DIO.H**

```
#ifdef    DIO_GLOBALS
#define   DIO_EXT
#else
#define   DIO_EXT extern
#endif


DIO_EXT  DIO_DI       DITbl[DIO_MAX_DI];
DIO_EXT  DIO_DO       DOTbl[DIO_MAX_DO];
```

DIO.C contains the following declarations:

Listing 1.6   *Example using* **DIO.C**

```
#define  DIO_GLOBALS
#include "includes.h"
```

When the compiler processes DIO.C, it makes the header file (DIO.H) appear as shown below because DIO_EXT is set to "nothing":

Listing 1.7   *Expanding* **DIO.H**

```
       DIO_DI       DITbl[DIO_MAX_DI];
       DIO_DO       DOTbl[DIO_MAX_DO];
```

The compiler is thus told to allocate storage for these variables. When the compiler processes any other .C files, the header file (DIO.H) looks as shown by the following code because DIO_GLOBALS is not defined and thus DIO_EXT is set to extern.

Listing 1.8   *Expanded* **.H** *file other than* **DIO.H**

```
extern DIO_DI       DITbl[DIO_MAX_DI];
extern DIO_DO       DOTbl[DIO_MAX_DO];
```

In this case, no storage is allocated and any .C file can access these variables. The nice thing about this technique is that the declaration for the variables is done in only one file, the .H file.

1.06 OS_ENTER_CRITICAL() and OS_EXIT_CRITICAL()

Throughout the source code provided in this book, you will see calls to the following macros: OS_ENTER_CRITICAL() and OS_EXIT_CRITICAL(). OS_ENTER_CRITICAL() is a macro that disables interrupts and OS_EXIT_CRITICAL() is a macro that enables interrupts. Disabling and enabling interrupts is done to protect critical sections of code. These macros are obviously processor specific and are different for each processor. These macros are found in OS_CPU.H (see Appendix A, Listing A.1) and for the code provided in this book, these macros are defined as follows.

Listing 1.9   Critical section macros

```
#define  OS_ENTER_CRITICAL()   asm {PUSHF; CLI}
#define  OS_EXIT_CRITICAL()    asm  POPF
```

Your application code can make use of these macros as long as you realize that they are used to disable and enable interrupts. Disabling interrupts obviously affects interrupt latency so be careful. You can also protect critical sections using semaphores.

1.07 ESBB Sample Code

The sample code is found in the \SOFTWARE\BLOCKS\SAMPLE\SOURCE of the installation directory. This source directory contains the following files:

- CFG.C (Listing 1.22)
- CFG.H (Listing 1.23)
- INCLUDES.H (Listing 1.24)
- OS_CFG.H (Listing 1.26)
- TEST.C (Listing 1.27)
- TEST.LNK (Listing 1.28)

CFG.C and CFG.H were discussed in section 1.04. INCLUDES.H was discussed in section 1.02. OS_CFG.H is a configuration file needed by µC/OS-II and should not be altered unless you obtain the full source version of µC/OS-II (see Appendix A for details). TEST.LNK is the linker command file and is shown in Listing 1.28.

The sample code is actually found in TEST.C (see Listing 1.27) and will be described in this section.

The sample provided (along with the building blocks used) in this chapter was compiled using the Borland C/C++ V4.51 compiler in a DOS box on a Windows 95 platform. To make the process easy, I created a makefile called TEST.MAK (see Listing 1.29). The makefile is invoked by the batch file MAKETEST.BAT (see Listing 1.25). Both files are found in the \SOFTWARE\BLOCKS\SAMPLE\TEST directory. To build the sample code, you need to change your current directory (using the DOS CD command) to \SOFTWARE\BLOCKS\SAMPLE\TEST and type:

```
C:\SOFTWARE\BLOCKS\SAMPLE\TEST > MAKETEST
```

You should note that my Borland compiler is installed on my `E:` drive, but you can easily change the `makefile` to have it point to the proper directory and drive by changing the following lines in TEST.MAK:

Listing 1.10   Tool declarations in **TEST.MAK**

```
##############################################################################
#                                  TOOLS
##############################################################################


BORLAND=E:\BC45
BORLAND_EXE=E:\BC45\BIN
```

µC/OS-II is a scalable operating system which means that the code size of µC/OS-II can be reduced if you are not using all of its services. However, because µC/OS-II is not provided in source form in this book, you will be limited to the features I needed to run the sample code. You can obtain the full source version of µC/OS-II by obtaining a copy of my other book, *MicroC/OS-II, The Real-Time Kernel*, ISBN 0-87930-543-6.

Once built, you can run the sample code by typing:

```
C:\SOFTWARE\BLOCKS\SAMPLE\TEST > TEST
```

The display on your PC should look as shown in Figure 1.1. You will notice that there is no sample code for Chapter 3 "Keyboards", Chapter 4 "Multiplexed LED Displays", and Chapter 5 "Character LCD Modules" because you would need some special hardware which I didn't want to assume.

Figure 1.1   DOS Window display for Sample code

```
                    EMBEDDED SYSTEMS BUILDING BLOCKS
                   Complete and Ready-to-Use Modules in C
                            Jean J. Labrosse
                             SAMPLE  CODE

  Chapter 3, Keyboards              Chapter  8, Discrete I/Os
  Chapter 4, Multiplexed LED Displays   DO #0:    50% Duty Cycle (Async)
  Chapter 5, Character LCD Modules      DO #1:    50% Duty Cycle (Async)
   -No Sample Code-                     DO #2:    25% Duty Cycle (Sync)

  Chapter 6, Time-Of-Day Clock      Chapter 10, Analog I/Os
    Date: Friday December 31, 1999   AI #0:
    Time: 23:58:00
    TS  : 1999-12-31 23:58:00
    Date:  11 uS  Time:  4 uS
  Chapter 7, Timer Manager          Chapter 11, Async. Serial Comm.
    Tmr0: 01:03.0                      Tx   :
                                       Rx   :
    Tmr1: 02:00.0




  MicroC/OS-II V2.00    #Tasks:  14    #Task switch/sec:   345    CPU Usage: 1 %
                            <-PRESS 'ESC' TO QUIT->
```

The sample code basically consists of 13 tasks as listed in Table 1.1.

Table 1.1   Tasks in sample code

Module/File	Task	Priority
TEST.C	Analog I/O Test Task	10 (Highest)
TEST.C	Clock Test Task	11
TEST.C	Asynchronous Serial Comm. Tx Test Task	12
TEST.C	Asynchronous Serial Comm. Rx Test Task	13
TEST.C	Discrete I/O Test Task	14
TEST.C	Timer Manager Test Task	15
TEST.C	Statistic / PC Keyboard Test Task	16
CLK.C	Time-of-Day Clock Task	51
TMR.C	Timer Manager Task	52
DIO.C	Discrete I/O Manager Task	53
AIO.C	Analog I/O Manager Task	54
µC/OS-II	Statistic Task	62
µC/OS-II	Idle Task	63 (Lowest)

μC/OS-II creates two *internal* tasks: the idle task and a task that determines CPU usage. Four of the building blocks each create a task and TEST.C creates the other 7 tasks.

As can be seen from the screen of Figure 1.1, there is no sample code for Chapters 3, 4, and 5 because they would require hardware not available on a regular PC.

For Chapter 6, the test code sets up the CLK module's current date and time to December 31, 1999 at 11:58 PM to show you that the CLK module is year 2000 (Y2K) compliant by correctly rolling over to Saturday, January 1, 2000 in two minutes. However, by the time you get this book, the Y2K problem should be a thing of the past. You should note that the CLK module doesn't change the actual date and time of your PC. When you run the code, you will also see the timestamp being updated. Also, I used the elapsed time measurement functions in PC.C to determine the execution time of ClkFormatDate() and ClkFormatTime().

The sample code for Chapter 7 sets up 2 timers. The first timer expires after 1 minute and 3 seconds and the second expires after 2 minutes. When the first timer expires, the message, "Timer #0 Timed Out!" will be displayed just below the line showing timer #0. When the second timer expires, the message, "Timer #1 Timed Out!" below its timer. Instead of displaying messages, you could perform any other operation including signaling a task.

For Chapter 8, although the DIO task continuously reads discrete inputs (DI), I don't actually make use of that feature because it would require external hardware. Instead, I only set up 3 discrete outputs (DO) for which I display the state of these outputs on the screen (TRUE or FALSE for DO #0, HIGH or LOW for DO #1 and, ON or OFF for DO #2). The first discrete output is setup to produce a 'blinking' output with a 50% duty-cycle (50% ON, 50% OFF) at a rate of 1 Hz. The second discrete output is also set up to 'blink' but does so at half the rate of the first channel (0.5 Hz). Finally, the third output blinks with a 25% duty cycle but runs in 'synchronous mode' (see Chapter 8).

There is no sample code provided for Chapter 9 because this chapter doesn't actually contain a building block.

For Chapter 10, instead of having you come up with an ADC on a PC, I simply decided to 'simulate' the ramping of an analog input which increases by 10 counts every time an ADC reading is required. When the counts reach 32700 (assuming a simulated 15-bit ADC), the counts are reset back to 0. Note that there aren't too many commercial 15-bit ADCs but, as you will see in Chapter 10, you can fake your software into thinking that all ADCs with less than 16 bits can actually look like they have 15 bits!

For Chapter 11, I created two tasks. One task sends the value of a counter to the other task. However, this message is actually sent through the serial port (COM1 on the PC). To see the operation of the sample code, you'll need to truly run in DOS (i.e., not in a DOS box under Windows 95/98 or NT) and connect the Tx and Rx lines of COM1 on your PC together. In order to accomplish this, I used a 'LapLink' serial cable (you can buy this at any good computer store) that I plugged into my PC. I then shorted pins 2 and 3 of either the DB-9 female or DB25 female connector using a paper clip.

1.07.01 `main()`

A μC/OS-II application looks just like any other DOS application. You compile and link your code just as if you would do a single threaded application running under DOS. The .EXE file that you create is loaded and executed by DOS, and execution of your application starts from main().

The sample code (TEST.EXE) serves two purposes. First, if you invoke the sample code from the DOS prompt and specify either "display" or "DISPLAY" [L1.11(1)] as an argument, your screen will display the corresponding *characters* that corresponds to each byte value from 0x00 to 0xFF. In other words, to see the character mapping simply type:

```
TEST display
```

or,

```
        TEST DISPLAY
```

at the DOS prompt.

If you simply typed TEST at the DOS prompt, then main() clears the screen to ensure we don't have any characters left over from the previous DOS session [L1.11(2)]. Note that I specified to use white letters on a black background. Since the screen will be cleared, I could have simply specified to use a black background and not specify a foreground. If I did this, and you decided to return to DOS then you would not see anything on the screen! It's always better to specify a visible foreground just for this reason.

Listing 1.11 *main()*

```
void  main (int argc, char *argv[])
{
    if (argc > 1) {                                                  (1)
        if (strcmp(argv[1], "display") == 0 ||
            strcmp(argv[1], "DISPLAY") == 0) {
            TestDispMap();
        }
        exit(0);
    }

    PC_DispClrScr(DISP_FGND_WHITE + DISP_BGND_BLACK);                (2)
    OSInit();                                                       (3)
    OSFPInit();                                                     (4)
    PC_DOSSaveReturn();                                             (5)
    PC_VectSet(uCOS, OSCtxSw);                                      (6)
    OSTaskCreateExt(TestStatTask,                                   (7)
                    (void *)0,
                    &TestStatTaskStk[TASK_STK_SIZE],
                    STAT_TASK_PRIO,
                    STAT_TASK_PRIO,
                    &TestStatTaskStk[0],
                    TASK_STK_SIZE,
                    (void *)0,
                    OS_TASK_OPT_SAVE_FP);
    OSStart();                                                      (8)
}
```

A requirement of μC/OS-II is that you call OSInit() [L1.11(3)] before you invoke any of its other services. OSInit() creates two tasks: an idle task which executes when no other task is ready-to-run and a statistic task which computes CPU usage.

Because the code is assumed to run on a 80486 or Pentium class computer, I decided to make use of hardware assisted floating-point and thus, we need to invoke the code that will tell µC/OS-II to initialize the floating-point support [L1.11(4)].

The current DOS environment is then saved by calling PC_DOSSaveReturn() [L1.11(5)]. This allows us to return to DOS as if we had never started µC/OS-II. A lot happens in PC_DOSSaveReturn() and this is all explained in Chapter 12 (section 12.01).

main() then calls PC_VectSet() [L1.11(6)] to install µC/OS-II's context switch handler. Task level context switching is done by issuing an 80x86 INT instruction to this vector location. I decided to use vector 0x80 (i.e., 128) because it's not used by either DOS or the BIOS.

Before starting multitasking, I create one task [L1.11(7)] called TestStatTask(). It is very important that you create at least one task before multitasking begins with OSStart() [L1.11(8)]. Failure to do this will certainly make your application crash. Once OSStart() is called, multitasking begins and µC/OS-II runs the highest priority task that is ready-to-run. This happens to be TestStatTask() which will be described next.

1.07.02 *TestStatTask()*

Initialization of the sample code continues in TestStatTask(). µC/OS-II needs a little more setup which is accomplished by 'installing' the tick handler [L1.12(1)]. Next, I decided to change the tick rate from the default DOS 18.2 Hz to 200 Hz [L1.12(2)]. This allows better granularity when we need to run tasks at regular intervals. You should note that a lot of setup has to be done to move from the DOS environment to the µC/OS-II environment. In an actual embedded system, there would be no need to save the CPU registers to return back to DOS (see PC_DOSSaveReturn()) because we would most likely not return back to DOS to begin with. We would, however, most likely need to install the tick ISR handler and set a hardware timer which would provide a tick source.

Note that main() purposely didn't set the interrupt vector to µC/OS-II's tick handler because you don't want a tick interrupt to occur before the operating system (µC/OS-II) is fully initialized and running. If you run code in an embedded application, you should always enable the ticker (as I have done here) from within the first task.

Before we create any other tasks, we need to determine how fast you particular PC is. This is done by calling the µC/OS-II function OSStatInit() [L1.12(4)]. Calling OSStatInit() allows µC/OS-II to determine the CPU usage (in percent) of your CPU while your application (in this case, the test code) is running.

Once µC/OS-II knows about your CPU, we call TestInitModules() to initialize the building blocks that are used in the sample code. The code for TestInitModules() is shown in Listing 1.13.

Listing 1.12 Beginning of *TestStatTask()*

```
void  TestStatTask (void *pdata)
{
    INT8U   i;
    INT16S  key;
    char    s[81];


    pdata = pdata;
```

Listing 1.12 Beginning of *TestStatTask()*

```
OS_ENTER_CRITICAL();
PC_VectSet(0x08, OSTickISR);                                     (1)
PC_SetTickRate(OS_TICKS_PER_SEC);                                (2)
OS_EXIT_CRITICAL();


PC_DispStr(0, 22, "Determining  CPU's capacity ...",            (3)
           DISP_FGND_WHITE);
OSStatInit();                                                    (4)
PC_DispClrRow(22, DISP_FGND_WHITE + DISP_BGND_BLACK);           (5)


TestInitModules();                                              (6)
```

TestInitModules() starts off by initializing the elapsed time measurement provided in the PC services (see Chapter 12) [L1.13(1)]. Because MODULE_KEY_MN [L1.13(2)], and MODULE_LED [L1.13(3)] and MODULE_LCD [L1.13(4)] are set to 0 in INCLUDES.H, the keyboard, LED, and LCD building blocks are not initialized. All of the other building blocks, however, are initialized because they are enabled in INCLUDES.H [L1.13(5–8)]. The last building block (COMM) uses the RTOS version (see Chapter 11) because it is used in conjunction with μC/OS-II. In this case, I assume that COMM1 on your PC is used for the test and it is setup to communicate at 9600 baud [L1.13(10-13)].

Listing 1.14 is part of TestStatTask() and is responsible for creating the test tasks which will exercise the building blocks used in the sample code. Each task that is to be managed by μC/OS-II must be *created*. This allows μC/OS-II to know about where the task code resides, what stack is to be allocated to the task, what priority is given to the task, and more. You can find out more about OSTaskCreateExt() in Appendix A.

Listing 1.13 *TestInitModules()*

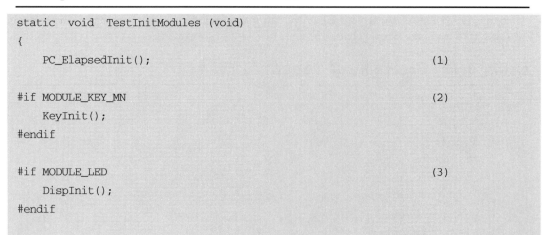

```
static  void  TestInitModules (void)
{
    PC_ElapsedInit();                                           (1)


#if MODULE_KEY_MN                                               (2)
    KeyInit();
#endif


#if MODULE_LED                                                  (3)
    DispInit();
#endif
```

Listing 1.13   TestInitModules()

```
#if MODULE_LCD                                                      (4)
    DispInit(4, 20);
#endif


#if MODULE_CLK
    ClkInit();                                                      (5)
#endif


#if MODULE_TMR
    TmrInit();                                                      (6)
#endif


#if MODULE_DIO
    DIOInit();                                                      (7)
#endif


#if MODULE_AIO
    AIOInit();                                                      (8)
#endif


#if MODULE_COMM_BGND
    CommInit();                                                     (9)
#endif


#if MODULE_COMM_RTOS
    CommInit();                                                     (10)
#endif


#if MODULE_COMM_PC
    CommCfgPort(COMM1, 9600, 8, COMM_PARITY_NONE, 1);              (11)
    CommSetIntVect(COMM1);                                         (12)
    CommRxIntEn(COMM1);                                            (13)
#endif
}
```

Listing 1.14 *Creation of test tasks (*`TestStatTask()`*)*

```
OSTaskCreateExt(TestClkTask,
                (void *)0,
                &TestClkTaskStk[TASK_STK_SIZE],
                TEST_CLK_TASK_PRIO, TEST_CLK_TASK_PRIO,
                &TestClkTaskStk[0],
                TASK_STK_SIZE,
                (void *)0,
                OS_TASK_OPT_SAVE_FP);
OSTaskCreateExt(TestRxTask,
                (void *)0,
                &TestRxTaskStk[TASK_STK_SIZE],
                TEST_RX_TASK_PRIO, TEST_RX_TASK_PRIO,
                &TestRxTaskStk[0],
                TASK_STK_SIZE,
                (void *)0,
                OS_TASK_OPT_SAVE_FP);
OSTaskCreateExt(TestTxTask,
                (void *)0,
                &TestTxTaskStk[TASK_STK_SIZE],
                TEST_TX_TASK_PRIO, TEST_TX_TASK_PRIO,
                &TestTxTaskStk[0],
                TASK_STK_SIZE,
                (void *)0,
                OS_TASK_OPT_SAVE_FP);
OSTaskCreateExt(TestTmrTask,
                (void *)0,
                &TestTmrTaskStk[TASK_STK_SIZE],
                TEST_TMR_TASK_PRIO, TEST_TMR_TASK_PRIO,
                &TestTmrTaskStk[0],
                TASK_STK_SIZE,
                (void *)0,
                OS_TASK_OPT_SAVE_FP);
```

1

Listing 1.14   Creation of test tasks (`TestStatTask()`)

```
    OSTaskCreateExt(TestDIOTask,
                    (void *)0,
                    &TestDIOTaskStk[TASK_STK_SIZE],
                    TEST_DIO_TASK_PRIO, TEST_DIO_TASK_PRIO,
                    &TestDIOTaskStk[0],
                    TASK_STK_SIZE,
                    (void *)0,
                    OS_TASK_OPT_SAVE_FP);
    OSTaskCreateExt(TestAIOTask,
                    (void *)0,
                    &TestAIOTaskStk[TASK_STK_SIZE],
                    TEST_AIO_TASK_PRIO, TEST_AIO_TASK_PRIO,
                    &TestAIOTaskStk[0],
                    TASK_STK_SIZE,
                    (void *)0,
                    OS_TASK_OPT_SAVE_FP);
```

Listing 1.15 is also part of `TestStatTask()`. The *literals* (i.e., text that doesn't change on the screen) are displayed by calling `TestDispLit()` [L1.15(1)]. This is done to avoid wasting CPU time updating the display with information that doesn't change. Next, `TestStatTask()` displays the current version of µC/OS-II at the bottom left hand corner of the screen [L1.15(2)].

`TestStatTask()` then enters an infinite loop. This is the main body of the task code. Every second (you'll see why later) the following information is displayed at the bottom of the screen:

- the number of tasks created (`OSTaskCtr`) [L1.15(3)],
- the number of context switches (i.e., task switches) per second (`OSCtxSwCtr`) [L1.15(4)] and,
- the percentage of the CPU used by the sample code (`OSCPUUsage`) [L1.15(5)].

You may question why I am updating the display of the task counter every second since there are no other tasks created from here on. The reason is to allow you to create other tasks which could be delayed. In other words, you may decide to create a task only after some time has expired.

This task then checks to see if a key has been pressed [L1.15(6)] and if so, determines whether the key pressed was the Esc key [L1.15(7)]. If the Esc key is pressed, the sample code exits back to DOS. Before we can return to DOS, though, we must reinstate the original DOS COMM1 ISR vector [L1.15(8)]. Returning back to DOS is accomplished by calling `PC_DOSReturn()` [L1.15(9)] (see Chapter 12, section 12.01).

In order to display the number of context switches per second, the global variable `OSCtxSwCtr` must be cleared every second [L1.15(10)].

To prevent this task from using all the CPU (remember that we are in an infinite loop), the task calls the µC/OS-II service `OSTimeDlyHMSM()` [L1.15(11)] (see Appendix A). This call *suspends* the current task until some time expires. In our case, the arguments 0, 0, 1, 0 specify a one second delay. When one second expires, µC/OS-II will resume execution of this task immediately after the call to `OSTimeDlyHMSM()` or, at the top of the `for()` loop.

Listing 1.15   Task portion of **TestStatTask()**

```
    TestDispLit();                                              (1)


    sprintf(s, "V%1d.%02d",                                    (2)
             OSVersion() / 100,
             OSVersion() % 100);
    PC_DispStr(13, 23, s, DISP_FGND_YELLOW + DISP_BGND_BLUE);


    for (;;) {
        sprintf(s, "%5d", OSTaskCtr);                          (3)
        PC_DispStr(30, 23, s, DISP_FGND_BLUE + DISP_BGND_CYAN);


        sprintf(s, "%5d", OSCtxSwCtr);                         (4)
        PC_DispStr(56, 23, s, DISP_FGND_BLUE + DISP_BGND_CYAN);


        sprintf(s, "%3d", OSCPUUsage);                         (5)
        PC_DispStr(75, 23, s, DISP_FGND_BLUE + DISP_BGND_CYAN);


        if (PC_GetKey(&key) == TRUE) {                         (6)
            if (key == 0x1B) {                                 (7)
#if MODULE_COMM_PC
                CommRclIntVect(COMM1);                         (8)
#endif
                PC_DOSReturn();                                (9)
            }
        }


        OSCtxSwCtr = 0;                                        (10)


        OSTimeDlyHMSM(0, 0, 1, 0);                             (11)
    }
}
```

1.07.03 *TestClkTask()*

TestClkTask() is shown in Listing 1.16 and this task shows some of the functions of the CLK building block of Chapter 6 which consist of code to maintain a time-of-day clock.

We first set up the current time-of-day and date to December 31, 1999 at 12:58 PM (i.e., 2 minutes before midnight) [L1.16(1)].

The task portion of the code (i.e., the infinite loop) is then entered and the function PC_ ElapsedStart() is invoked [L1.16(2)] to setup the PC's timer #2 so that it can be used to measure the execution time of ClkFormatDate() [L1.16(3)]. ClkFormatDate() formats the current date maintained by the CLK building block into an ASCII string. The format selected (i.e., 2) is "Day Month DD, YYYY" where 'Day' is the day of the week (Monday, Tuesday ...), 'Month' is the month of the year (January, February ...), 'DD' is the calendar day (1, 2, 3 ...), and 'YYYY' is the current year using 4 digits. The execution time of ClkFormatDate() is captured by calling PC_ElapsedStop() [L1.16(4)] which returns the time in microseconds. Both the current date and the execution time are then displayed.

Listing 1.16   *TestClkTask()*

```
void  TestClkTask (void *data)
{
    char    s[81];
    INT16U  time;
    TS      ts;


    data  = data;

    ClkSetDateTime(12, 31, 1999, 23, 58, 0);                          (1)

    for (;;) {
        PC_ElapsedStart();                                           (2)
        ClkFormatDate(2, s);                                         (3)
        time = PC_ElapsedStop();                                     (4)
        PC_DispStr( 8, 11, "                     ", DISP_FGND_WHITE);
        PC_DispStr( 8, 11, s, DISP_FGND_BLUE + DISP_BGND_CYAN);
        sprintf(s, "%3d uS", time);
        PC_DispStr( 8, 14, s, DISP_FGND_RED + DISP_BGND_LIGHT_GRAY);

        PC_ElapsedStart();                                           (5)
        ClkFormatTime(1, s);                                         (6)
        time = PC_ElapsedStop();                                     (7)
        PC_DispStr( 8, 12, s, DISP_FGND_BLUE + DISP_BGND_CYAN);
        sprintf(s, "%3d uS", time);
        PC_DispStr(22, 14, s, DISP_FGND_RED + DISP_BGND_LIGHT_GRAY);

        ts = ClkGetTS();                                             (8)
        ClkFormatTS(2, ts, s);                                       (9)
        PC_DispStr( 8, 13, s, DISP_FGND_BLUE + DISP_BGND_CYAN);

        OSTimeDlyHMSM(0, 0, 0, 100);                                 (10)
    }
}
```

The function PC_ElapsedStart() is called again [L1.16(5)] to setup the PC's timer #2 so that it can be used to measure the execution time of ClkFormatTime() [L1.16(6)]. ClkFormatTime() formats the current time maintained by the CLK building block into an ASCII string. The format selected (i.e., 1) is "HH:MM:SS" which consist of the current time in 24 hour format (i.e., up to 23:59:59). The execution time of ClkFormatTime() is captured by also calling PC_ElapsedStop() [L1.16(7)]. Both the current time and the execution time are then displayed.

The CLK building block also maintains a special format called a *timestamp*. A timestamp basically captures the date and time in a single 32-bit variable as shown in Figure 1.2. This allows your application to mark an event such as the occurrence of an error or the reception of a message and capture when that event occurred. You can thus obtain the current timestamp by calling ClkGetTS() [L1.16(8)]. It is easier to display the timestamp in ASCII which is why ClkFormatTS() is invoked [L1.16(9)]. The format selected "YYYY-MM-DD HH:MM:SS" is new to this second edition. I personally like this format because it displays the year as 4 digits followed by the month and then the day. What's also convenient about this ASCII format is that it can be sorted easily. OSTimeDlyHMSM() is then called to suspend this task for 100 milliseconds. In other words, this task executes 10 times per second [L1.16(10)].

Figure 1.2   *Timestamp format*

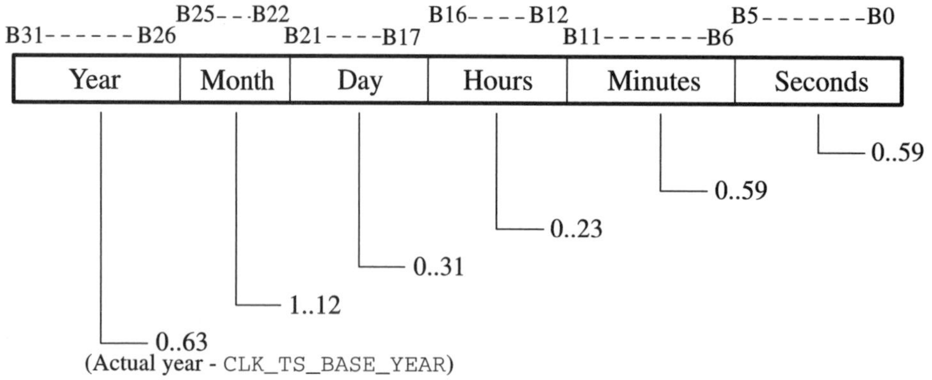

1.07.04 TestTmrTask()

TestTmrTask() is shown in Listing 1.17 and shows some of the functions of the TMR building block of Chapter 7 which consists of code that maintains up to 250 down counters that can be set to any time from 1 tenth of a second to 99 minutes, 59 seconds and 9 tenths of a second or, 99:59.9 (using the nomenclature MM:SS.T). When a timer expires, it can optionally call a user-definable function.

We first start up by configuring timer #0's timeout function [L1.17(1)]. When timer #0 times out, it will call TestTmr0TO() which simply displays "Timer #0 Timed Out!". The timer is then initialized to 1:03.9 [L1.17(2)] and then it's started [L1.17(3)].

We then configure a second timer, timer #1's timeout function [L1.17(4)]. When timer #1 times out, it will call `TestTmr1TO()` which also displays a similar message, "`Timer #1 Timed Out!`". The timer is then initialized to 2:00.0 [L1.17(5)] and then it's started [L1.17(6)].

Listing 1.17   TestTmrTask()

```
void  TestTmrTask (void *data)
{
    char    s[81];
    INT16U  time;


    data = data;

    TmrCfgFnct(0, TestTmr0TO, (void *)0);                          (1)
    TmrSetMST(0, 1, 3, 9);                                         (2)
    TmrStart(0);                                                   (3)

    TmrCfgFnct(1, TestTmr1TO, (void *)0);                          (4)
    TmrSetMST(1, 2, 0, 0);                                         (5)
    TmrStart(1);                                                   (6)

    for (;;) {
        TmrFormat(0, s);                                           (7)
        PC_DispStr(8, 16, s, DISP_FGND_RED+DISP_BGND_LIGHT_GRAY);

        TmrFormat(1, s);                                           (8)
        PC_DispStr(8, 18, s, DISP_FGND_RED+DISP_BGND_LIGHT_GRAY);

        OSTimeDlyHMSM(0, 0, 0, 50);                                (9)
    }
}
```

The time remaining for both timers is displayed [L1.17(7–8)] and the task body continuously loops 20 times per second (it doesn't really need to be this fast though) [L1.17(9)].

1.07.05 TestDIOTask()

`TestDIOTask()` is shown in Listing 1.18 and shows some of the functions of the DIO building block of Chapter 8. The DIO module reads and updates up to 256 discrete inputs and outputs. A discrete input normally represents the state of an external switch (a pushbutton switch, a pressure switch, a temperature switch, etc.). A discrete output generally consists of a single relay output to control a single light, a valve, a motor, etc.

Although the DIO task can read discrete inputs (DI), I don't actually make use of that feature because it would require external hardware. Instead, I only set up 3 discrete outputs (DO) for which I display the state of these outputs on the screen:

- For DO #0 we will display TRUE or FALSE
- For DO #1 we will display HIGH or LOW
- For DO #2 we will display ON or OFF

The DIO task [DIOTask(), see Chapter 8] which is responsible for updating the DIs and DOs will execute 10 times per second (see CFG.H, DIO_TASK_DLY_TICKS). To get a 10 second synchronous count value, we call DOSetSyncCtrMax() [L1.18(1)] which sets DOSyncCtrMax to 100 (100 * 0.1 sec). Note that you wouldn't need to invoke this function if you didn't use the 'synchronous' mode of the DIO module.

I then configure DO #0 to blink at a rate of 1 Hz with a 50% duty cycle [L1.18(2)]. The values specified as arguments to DOCfgBlink() do not correspond to RTOS ticks but instead, they correspond to number of updates of the DIO module. In other words, if the DIO task is updated 10 times per second — then a value of 10 represents 1 second, a value of 20 represents 2 seconds, etc. To finalize the configuration of DO #0, I need to set the mode to *asynchronous* blinking and non-invert the output (see Chapter 8, Figure 8.9) [L1.18(3)]. Configuration of DO #1 is similar to DO #0 except that I set the blink at a rate to 0.5 Hz (i.e., 2 seconds) [L1.18(4)]. DO #1 is also set to *asynchronous* blinking and non-invert the output [L1.18(5)]. Configuration of DO #2 is set to *synchronous* blinking and its output is also non-inverted [L1.18(6-7)].

We then enter the task body which simply obtains the state of each discrete output and displays it on the screen. This happens 10 times per second although this doesn't need to be done this fast considering that none of the outputs change this quickly.

Listing 1.18 *TestDIOTask()*

```
void  TestDIOTask (void *data)
{
    BOOLEAN state;


    data = data;

    DOSetSyncCtrMax(100);                               (1)

    DOCfgBlink(0, DO_BLINK_EN,  5, 10);                 (2)
    DOCfgMode(0,  DO_MODE_BLINK_ASYNC, FALSE);          (3)

    DOCfgBlink(1, DO_BLINK_EN, 10, 20);                 (4)
    DOCfgMode(1,  DO_MODE_BLINK_ASYNC, FALSE);          (5)

    DOCfgBlink(2, DO_BLINK_EN, 25,  0);                 (6)
    DOCfgMode(2,  DO_MODE_BLINK_SYNC, FALSE);           (7)
```

Listing 1.18   *TestDIOTask()*

```
for (;;) {
    state = DOGet(0);
    if (state == TRUE) {
        PC_DispStr(49,   6, "TRUE ",
                    DISP_FGND_YELLOW + DISP_BGND_BLUE);
    } else {
        PC_DispStr(49,   6, "FALSE",
                    DISP_FGND_YELLOW + DISP_BGND_BLUE);
    }
    state = DOGet(1);
    if (state == TRUE) {
        PC_DispStr(49,   7, "HIGH",
                    DISP_FGND_YELLOW + DISP_BGND_BLUE);
    } else {
        PC_DispStr(49,   7, "LOW ",
                    DISP_FGND_YELLOW + DISP_BGND_BLUE);
    }
    state = DOGet(2);
    if (state == TRUE) {
        PC_DispStr(49,   8, "ON ",
                    DISP_FGND_YELLOW + DISP_BGND_BLUE);
    } else {
        PC_DispStr(49,   8, "OFF",
                    DISP_FGND_YELLOW + DISP_BGND_BLUE);
    }

    OSTimeDlyHMSM(0,  0,  0, 100);
}
}
```

1.07.06 *TestAIOTask()*

TestAIOTask() is shown in Listing 1.19 and shows some of the functions of the AIO building block of Chapter 10. The AIO module reads and updates up to 256 analog inputs and outputs. Each analog input can be configured to read just about any type of sensor (temperature, pressure, position, flow, etc.). An analog output can be made to control a large number of devices such as a valve, an actuator, a positioner, etc.

It's difficult to show the operation of this building block without actually having an ADC (Analog to Digital Converter) and a DAC (Digital to Analog Converter) on a PC. What I decided to do is simply simulate a ramping ADC and convert the value to some engineering units. I thought of using the LM-34A (see Chapter 10, Figure 10.7) as my 'simulated' sensor and generate temperatures from –50 to about 300 degrees Farenheit. I assumed that my ADC would be made to look like a 16-bit signed ADC, referenced at 10 volts, the gain would be set to 2.5 and, I have a 1.25 volt offset so that I could read negative temperatures. From Equations 10.9 and 10.10, I obtain a gain of 0.01220740 and an offset of –4095.875 and I configure AI #0 accordingly [L1.19(1)].

The task code simply consists of reading the current engineering value (i.e., the temperature of the simulated LM34A) from the analog channel [L1.19(2)] and displaying it on the screen. You should note that I didn't need to display decimal places and thus, I converted the temperature to an integer.

The task code repeats 100 times per second [L1.19(3)]. Again, this rate is not necessary and has been chosen simply to make the CPU busy.

Listing 1.19   *TestAIOTask()*

```
void  TestAIOTask (void *data)
{
    char     s[81];
    FP32     value;
    INT16S   temp;
    INT8U    err;


    data = data;

    AICfgConv(0, 0.01220740, -4095.875, 10);                         (1)
    AICfgCal(0,  1.00,              0.00);

    for (;;) {
        err  = AIGet(0, &value);                                     (2)
        temp = (INT16S)value;
        sprintf(s, "%5d", temp);
        PC_DispStr(49, 11, s, DISP_FGND_YELLOW + DISP_BGND_BLUE);

        OSTimeDlyHMSM(0, 0, 0, 10);                                  (3)
    }
}
```

1.07.07 *TestTxTask()* and *TestRxTask()*

It is assumed that you would connect a 'LapLink' serial cable on COM1 and short the Tx line (pin #3) to the Rx line (pin #2) on the free end of the DB9F connector.

TestTxTask() is shown in Listing 1.20 and shows some of the functions of the COMM building block of Chapter 11. This task simply increments a 16-bit counter, converts it to ASCII [L1.20(1)] and sends the string on COM1 one character after the other [L1.20(2)]. A delay of 5 ticks is added in case you run this code under Windows 95/98 or NT [L1.20(3)]. This is needed to accommodate overhead imposed by Windows. If you were to run this code either in DOS or on an actual embedded system, you

1

would not need the delay. I actually tested this code on a DOS-based machine all the way to 38400 baud for a few hours without any glitches, however, it crashes with Windows 95/98.

TestRxTask() is shown in Listing 1.21 and is basically the receiving task for the transmitted messages from TestTxTask(). This task waits for characters to be received on COM1 [L1.21(1)]. As each character is received, it is placed in a buffer [L1.21(2)]. When the carriage return character ('\n' or 0x0D) is received, the string is terminated [L1.21(3)] and the string received is displayed [L1.21(4)]. Of course both the transmitted and received messages should match.

Listing 1.20 `TestTxTask()`

```
void  TestTxTask (void *data)
{
    INT16U  ctr;
    char    s[81];
    char    *ps;


    data  = data;
    ctr   = 0;
    for (;;) {
        sprintf(s, "%05d\n", ctr);                                  (1)
        PC_DispStr(49, 16, s, DISP_FGND_YELLOW + DISP_BGND_BLUE);
        ps = s;
        while (*ps != NUL) {
            CommPutChar(COMM1, *ps, OS_TICKS_PER_SEC);              (2)
            OSTimeDly(5);                                           (3)
            ps++;
        }
        ctr++;
    }
}
```

Listing 1.21 *TestRxTask()*

```
void  TestRxTask (void *data)
{
    INT8U  err;
    INT8U  nbytes;
    INT8U  c;
    char   s[81];
    char   *ps;


    data   = data;
    for (;;) {
        ps     = s;
        nbytes = 0;
        do {
            c      = CommGetChar(COMM1, OS_TICKS_PER_SEC, &err);   (1)
            *ps++  = c;                                            (2)
            nbytes++;
        } while (c != '\n' && nbytes < 20);
        *ps = NUL;                                                 (3)
        PC_DispStr(49, 17, s, DISP_FGND_YELLOW + DISP_BGND_BLUE);  (4)
    }
}
```

1.08 Bibliography

JDR Microdevices
1850 South 10th Street
San Jose, CA 95112-4108
(800) 538-5000
(408) 494-1400

PDS-601 link:
http://www.jdr.com/interact/item.asp?itemno=gr-pds

1

Listing 1.22   **CFG.C**

```
/*
********************************************************************************************************
*                              Embedded Systems Building Blocks
*                            Complete and Ready-to-Use Modules in C
*
*                                        Configuration File
*
*                       (c) Copyright 1999, Jean J. Labrosse, Weston, FL
*                                       All Rights Reserved
*
* Filename   : CFG.C
* Programmer : Jean J. Labrosse
********************************************************************************************************
*/

#include "includes.h"

/*$PAGE*/
```

Listing 1.22 (continued) CFG.C

```
/*
*********************************************************************************************
*                                      KEYBOARD
*                                  INITIALIZE I/O PORTS
*********************************************************************************************
*/

#if  MODULE_KEY_MN
void  KeyInitPort (void)
{
    outp(KEY_PORT_CW, 0x82);                      /* Initialize 82C55A: A=OUT, B=IN (COLS), C=OUT (ROWS) */
}

/*
*********************************************************************************************
*                                      KEYBOARD
*                                    SELECT A ROW
*
* Description : This function is called to select a row on the keyboard.
* Arguments   : 'row'  is the row number (0..7) or KEY_ALL_ROWS
* Returns     : none
* Note        : The row is selected by writing a LOW.
*********************************************************************************************
*/

void  KeySelRow (INT8U row)
{
    if (row == KEY_ALL_ROWS) {
        outp(KEY_PORT_ROW, 0x00);                 /* Force all rows LOW                         */
    } else {
        outp(KEY_PORT_ROW, ~(1 << row));          /* Force desired row LOW                      */
    }
}

/*
*********************************************************************************************
*                                      KEYBOARD
*                                     READ COLUMNS
*
* Description : This function is called to read the column port.
* Arguments   : none
* Returns     : the complement of the column port thus, ones are keys pressed
*********************************************************************************************
*/

INT8U  KeyGetCol (void)
{
    return (~inp(KEY_PORT_COL));                  /* Complement columns (ones indicate key is pressed)  */
}
#endif

/*$PAGE*/
```

Listing 1.22 (continued) CFG.C

```c
/*
*********************************************************************************************
*                               MULTIPLEXED LED DISPLAY
*                               I/O PORTS INITIALIZATION
*
* Description: This is called by DispInit() to initialize the output ports used in the LED multiplexing.
* Arguments  : none
* Returns    : none
* Notes      : 74HC573  8 bit latches are used for both the segments and digits outputs.
*********************************************************************************************
*/

#if  MODULE_LED
void  DispInitPort (void)
{
    outp(DISP_PORT_SEG, 0x00);          /* Turn OFF segments                              */
    outp(DISP_PORT_DIG, 0x00);          /* Turn OFF digits                                */
}


/*
*********************************************************************************************
*                               MULTIPLEXED LED DISPLAY
*                                     SEGMENTS output
*
* Description: This function outputs seven-segment patterns.
* Arguments  : seg    is the seven-segment patterns to output
* Returns    : none
*********************************************************************************************
*/

void  DispOutSeg (INT8U seg)
{
    outp(DISP_PORT_SEG, seg);
}


/*
*********************************************************************************************
*                               MULTIPLEXED LED DISPLAY
*                                     DIGIT output
*
* Description: This function outputs the digit selector.
* Arguments  : msk    is the mask used to select the current digit.
* Returns    : none
*********************************************************************************************
*/

void  DispOutDig (INT8U msk)
{
    outp(DISP_PORT_DIG, msk);
}
#endif

/*$PAGE*/
```

Listing 1.22 (continued)   CFG.C

```c
/*
*********************************************************************************
*                                  LCD DISPLAY MODULE
*                            INITIALIZE DISPLAY DRIVER I/O PORTS
*
* Description : This initializes the I/O ports used by the display driver.
* Arguments   : none
* Returns     : none
*********************************************************************************
*/

#if  MODULE_LCD
void  DispInitPort (void)
{
    outp(DISP_PORT_CMD, 0x82);          /* Set to Mode 0: A are output, B are inputs, C are outputs    */
}


/*
*********************************************************************************
*                                  LCD DISPLAY MODULE
*                             WRITE DATA TO DISPLAY DEVICE
*
* Description : This function sends a single BYTE to the display device.
* Arguments   : 'data'  is the BYTE to send to the display device
* Returns     : none
* Notes       : You will need to adjust the value of DISP_DLY_CNTS (LCD.H) to produce a delay between
*               writes of at least 40 uS.  The display I used for the test actually required a delay of
*               80 uS!  If characters seem to appear randomly on the screen, you might want to increase
*               the value of DISP_DLY_CNTS.
*********************************************************************************
*/
void  DispDataWr (INT8U data)
{
    INT8U  dly;


    outp(DISP_PORT_DATA, data);              /* Write data to display module        */
    outp(DISP_PORT_CMD,  0x01);              /* Set E   line HIGH                   */
    DispDummy();                             /* Delay about 1 uS                    */
    outp(DISP_PORT_CMD,  0x00);              /* Set E   line LOW                    */
    for (dly = DISP_DLY_CNTS; dly > 0; dly--) {  /* Delay for at least 40 uS        */
        DispDummy();
    }
}
```

Listing 1.22 (continued) *CFG.C*

```c
/*
*********************************************************************************************
*                                 LCD DISPLAY MODULE
*                            SELECT COMMAND OR DATA REGISTER
*
* Description : This function read a BYTE from the display device.
* Arguments   : none
*********************************************************************************************
*/
void  DispSel (INT8U sel)
{
    if (sel == DISP_SEL_CMD_REG) {
        outp(DISP_PORT_CMD, 0x02);     /* Select the command register (RS low)              */
    } else {
        outp(DISP_PORT_CMD, 0x03);     /* Select the data    register (RS high)             */
    }
}
#endif


/*
*********************************************************************************************
*                                 CLOCK/CALENDAR MODULE
*********************************************************************************************
*/

#if MODULE_CLK
#endif

/*
*********************************************************************************************
*                                    TIMER MANAGER
*********************************************************************************************
*/

#if MODULE_TMR
#endif

/*$PAGE*/
```

Listing 1.22 (continued) CFG.C

```
/*
*********************************************************************************************
*                               DISCRETE I/O MODULE
*                               INITIALIZE PHYSICAL I/Os
*
* Description : This function is by DIOInit() to initialze the physical I/O used by the DIO driver.
* Arguments   : None.
* Returns     : None.
* Notes       : The physical I/O is assumed to be an 82C55A chip initialized as follows:
*                       Port A = OUT  (Discrete outputs)   (Address 0x0300)
*                       Port B = IN   (Discrete inputs)    (Address 0x0301)
*                       Port C = OUT  (not used)           (Address 0x0302)
*                       Control Word                       (Address 0x0303)
*                   Refer to the Intel 82C55A data sheet.
*********************************************************************************************
*/

#if  MODULE_DIO
void  DIOInitIO (void)
{
    outp(0x0303, 0x82);                        /* Port A = OUT, Port B = IN, Port C = OUT       */
}


/*
*********************************************************************************************
*                               DISCRETE I/O MODULE
*                               READ PHYSICAL INPUTS
*
* Description : This function is called to read and map all of the physical inputs used for discrete
*               inputs and map these inputs to their appropriate discrete input data structure.
* Arguments   : None.
* Returns     : None.
*********************************************************************************************
*/

void  DIRd (void)
{
    DIO_DI *pdi;
    INT8U  i;
    INT8U  in;
    INT8U  msk;


    pdi = &DITbl[0];                            /* Point at beginning of discrete inputs    */
    msk = 0x01;                                 /* Set mask to extract bit 0                */
    in  = inp(0x0301);                          /* Read the physical port (8 bits)          */
    for (i = 0; i < 8; i++) {                   /* Map all 8 bits to first 8 DI channels    */
        pdi->DIIn   = (BOOLEAN)(in & msk) ? 1 : 0;
        msk        <<= 1;
        pdi++;
    }
}
```

Listing 1.22 (continued) CFG.C

```
/*
*********************************************************************************************
*                                   DISCRETE I/O MODULE
*                                 UPDATE PHYSICAL OUTPUTS
*
* Description : This function is called to map all of the discrete output channels to their appropriate
*               physical destinations.
* Arguments   : None.
* Returns     : None.
*********************************************************************************************
*/

void  DOWr (void)
{
    DIO_DO *pdo;
    INT8U   i;
    INT8U   out;
    INT8U   msk;


    pdo = &DOTbl[0];                        /* Point at first discrete output channel              */
    msk = 0x01;                             /* First DO will be mapped to bit 0                    */
    out = 0x00;                             /* Local 8 bit port image                              */
    for (i = 0; i < 8; i++) {               /* Map first 8 DOs to 8 bit port image                 */
        if (pdo->DOOut == TRUE) {
            out |= msk;
        }
        msk <<= 1;
        pdo++;
    }
    outp(0x0300, out);                      /* Output port image to physical port                  */
}
#endif

/*$PAGE*/
```

Listing 1.22 (continued)   CFG.C

```
/*
*********************************************************************************************
*                                ANALOG I/O MODULE
*                               INITIALIZE PHYSICAL I/Os
*
* Description : This function is called by AIOInit() to initialize the physical I/O used by the AIO
*               driver.
* Arguments   : None.
* Returns     : None.
*********************************************************************************************
*/

#if  MODULE_AIO
void  AIOInitIO (void)
{
    /* This is where you will need to put you initialization code for the ADCs and DACs        */
    /* You should also consider initializing the contents of your DAC(s) to a known value.     */
}


/*
*********************************************************************************************
*                                ANALOG I/O MODULE
*                               READ PHYSICAL INPUTS
*
* Description : This function is called to read a physical ADC channel.  The function is assumed to
*               also control a multiplexer if more than one analog input is connected to the ADC.
* Arguments   : ch    is the ADC logical channel number (0..AIO_MAX_AI-1).
* Returns     : The raw ADC counts from the physical device.
*********************************************************************************************
*/

INT16S  AIRd (INT8U ch)
{
    /* This is where you will need to provide the code to read your ADC(s).                     */
    /* AIRd() is passed a 'LOGICAL' channel number.  You will have to convert this logical channel */
    /* number into actual physical port locations (or addresses) where your MUX. and ADCs are located. */
    /* AIRd() is responsible for:                                                               */
    /*      1) Selecting the proper MUX. channel,                                               */
    /*      2) Waiting for the MUX. to stabilize,                                               */
    /*      3) Starting the ADC,                                                                */
    /*      4) Waiting for the ADC to complete its conversion,                                  */
    /*      5) Reading the counts from the ADC and,                                             */
    /*      6) Returning the counts to the calling function.                                    */

    return (ch);
}

/*$PAGE*/
```

Listing 1.22 (continued) *CFG.C*

```
/*
*********************************************************************************************
*                                  ANALOG I/O MODULE
*                                UPDATE PHYSICAL OUTPUTS
*
* Description : This function is called to write the 'raw' counts to the proper analog output device
*               (i.e. DAC).  It is up to this function to direct the DAC counts to the proper DAC if more
*               than one DAC is used.
* Arguments   : ch     is  the DAC logical channel number (0..AIO_MAX_AO-1).
*               cnts   are the DAC counts to write to the DAC
* Returns     : None.
*********************************************************************************************
*/

void  AOWr (INT8U ch, INT16S cnts)
{
    ch   = ch;
    cnts = cnts;

    /* This is where you will need to provide the code to update your DAC(s).              */
    /* AOWr() is passed a 'LOGICAL' channel number.  You will have to convert this logical channel */
    /* number into actual physical port locations (or addresses) where your DACs are located.      */
    /* AOWr() is responsible for writing the counts to the selected DAC based on a logical number.  */
}
#endif
```

Listing 1.23   CFG.H

```
/*
********************************************************************************************
*                           Embedded Systems Building Blocks
*                          Complete and Ready-to-Use Modules in C
*
*                               Configuration Header File
*
*                       (c) Copyright 1999, Jean J. Labrosse, Weston, FL
*                                   All Rights Reserved
*
* Filename   : CFG.H
* Programmer : Jean J. Labrosse
********************************************************************************************
*/


/*
********************************************************************************************
*                          KEYBOARD CONFIGURATION CONSTANTS
*                                     (Chapter 3)
*
* Note: These #defines would normally reside in your application specific code.
********************************************************************************************
*/

#if      MODULE_KEY_MN

#define  KEY_BUF_SIZE              10       /* Size of the KEYBOARD buffer                         */

#define  KEY_MAX_ROWS              4        /* The maximum number of rows     on the keyboard      */
#define  KEY_MAX_COLS              6        /* The maximum number of columns on the keyboard       */

#define  KEY_PORT_ROW          0x0312       /* The port address of the keyboard matrix ROWs        */
#define  KEY_PORT_COL          0x0311       /* The port address of the keyboard matrix COLUMNs     */
#define  KEY_PORT_CW           0x0313       /* The port address of the I/O ports control word      */

#define  KEY_RPT_DLY              20        /* Number of scan times before auto repeat executes again  */
#define  KEY_RPT_START_DLY       100        /* Number of scan times before auto repeat function engages*/

#define  KEY_SCAN_TASK_DLY        50        /* Number of milliseconds between keyboard scans       */
#define  KEY_SCAN_TASK_PRIO       50        /* Set priority of keyboard scan task                  */
#define  KEY_SCAN_TASK_STK_SIZE 1024        /* Size of keyboard scan task stack                    */

#define  KEY_SHIFT1_MSK         0x80        /* The SHIFT1 key is on bit B7 of the column input port  */
                                            /*    (A 0x00 indicates that a SHIFT1 key is not present) */
#define  KEY_SHIFT1_OFFSET        24        /* The scan code offset to add when SHIFT1 is pressed  */

#define  KEY_SHIFT2_MSK         0x40        /* The SHIFT2 key is on bit B6 of the column input port  */
                                            /*    (A 0x00 indicates that an SHIFT2 key is not present)*/
#define  KEY_SHIFT2_OFFSET        48        /* The scan code offset to add when SHIFT2 is pressed  */

#define  KEY_SHIFT3_MSK         0x00        /* The SHIFT3 key is on bit B5 of the column input port  */
                                            /*    (A 0x00 indicates that a SHIFT3 key is not present) */
#define  KEY_SHIFT3_OFFSET         0        /* The scan code offset to add when SHIFT3 is pressed  */

#endif

/*$PAGE*/
```

Listing 1.23 (continued)   CFG.H

```
/*
********************************************************************************************************
*                         MULTIPLEXED LED DISPLAY DRIVER CONFIGURATION CONSTANTS
*                                              (Chapter 4)
********************************************************************************************************
*/

#if      MODULE_LED

#define  DISP_PORT_SEG          0x0300      /* Port address of SEGMENTS output                      */
#define  DISP_PORT_DIG          0x0301      /* Port address of DIGITS    output                     */

#define  DISP_N_DIG                  8      /* Total number of digits (including status indicators)  */
#define  DISP_N_SS                   7      /* Total number of seven-segment digits                  */

#endif

/*
********************************************************************************************************
*                          LCD DISPLAY MODULE DRIVER CONFIGURATION CONSTANTS
*                                              (Chapter 5)
********************************************************************************************************
*/

#if      MODULE_LCD

#define  DISP_DLY_CNTS             100      /* Number of iterations to delay for 40 uS (software loop) */

#define  DISP_PORT_DATA         0x0300      /* Port address of the DATA port of the LCD module       */
#define  DISP_PORT_CMD          0x0303      /* Address of the Control Word (82C55) to control RS & E  */

#endif

/*$PAGE*/
```

Listing 1.23 (continued) CFG.H

```
/*
*********************************************************************************************************
*                              CLOCK/CALENDAR MODULE CONFIGURATION CONSTANTS
*                                          (Chapter 6)
*********************************************************************************************************
*/

#if        MODULE_CLK

#define   CLK_TASK_PRIO             45      /* This defines the priority of ClkTask()                */
#define   CLK_DLY_TICKS    OS_TICKS_PER_SEC /* # of clock ticks to obtain 1 second                   */
#define   CLK_TASK_STK_SIZE        512      /* Stack size in BYTEs for ClkTask()                     */

#define   CLK_DATE_EN                1      /* Enable DATE (when 1)                                  */
#define   CLK_TS_EN                  1      /* Enable TIME-STAMPS (when 1)                           */
#define   CLK_USE_DLY                1      /* Task will use OSTimeDly() instead of pend on sem.     */

#endif

/*
*********************************************************************************************************
*                                         TIMER MANAGER
*                                          (Chapter 7)
*********************************************************************************************************
*/

#if        MODULE_TMR

#define   TMR_TASK_PRIO             40
#define   TMR_DLY_TICKS            (OS_TICKS_PER_SEC / 10)
#define   TMR_TASK_STK_SIZE        512

#define   TMR_MAX_TMR               20

#define   TMR_USE_SEM                0

#endif

/*$PAGE*/
```

Listing 1.23 (continued) `CFG.H`

```
/*
********************************************************************************************************
*                          DISCRETE I/O MODULE CONFIGURATION CONSTANTS
*                                        (Chapter 8)
********************************************************************************************************
*/

#if      MODULE_DIO

#define  DIO_TASK_PRIO             35
#define  DIO_TASK_DLY_TICKS        (OS_TICKS_PER_SEC / 10)
#define  DIO_TASK_STK_SIZE         512

#define  DIO_MAX_DI                8       /* Maximum number of Discrete Input  Channels (1..255)   */
#define  DIO_MAX_DO                8       /* Maximum number of Discrete Output Channels (1..255)   */

#define  DI_EDGE_EN                1       /* Enable code generation to support edge trig. (when 1) */

#define  DO_BLINK_MODE_EN          1       /* Enable code generation to support blink mode (when 1) */

#endif

/*
********************************************************************************************************
*                          ANALOG I/O MODULE CONFIGURATION CONSTANTS
*                                        (Chapter 10)
********************************************************************************************************
*/

#if      MODULE_AIO

#define  AIO_TASK_PRIO             30
#define  AIO_TASK_DLY              100     /* Execute every 100 mS                                  */
#define  AIO_TASK_STK_SIZE         512

#define  AIO_MAX_AI                8       /* Maximum number of Analog Input  Channels (1..250)     */
#define  AIO_MAX_AO                8       /* Maximum number of Analog Output Channels (1..250)     */

#endif

/*$PAGE*/
```

Listing 1.23 (continued)   CFG.H

```
/*
********************************************************************************************
*              ASYNCHRONOUS SERIAL COMMUNICATIONS MODULE CONFIGURATION CONSTANTS
*                                        (Chapter 11)
********************************************************************************************
*/

#if        MODULE_COMM_PC

#define  COMM1_BASE          0x03F8          /* Base address of PC's COM1                */
#define  COMM2_BASE          0x02F8          /* Base address of PC's COM2                */

#define  COMM_MAX_RX         2               /* Maximum number of characters in Rx buffer of ...  */
                                             /* ... NS16450 UART.  2 for 16450, 16 for 16550.     */
#endif


#if        MODULE_COMM_BGND

#define  COMM1               1
#define  COMM2               2

#define  COMM_RX_BUF_SIZE    64              /* Number of characters in Rx ring buffer   */
#define  COMM_TX_BUF_SIZE    64              /* Number of characters in Tx ring buffer   */

#endif


#if        MODULE_COMM_RTOS

#define  COMM1               1
#define  COMM2               2

#define  COMM_RX_BUF_SIZE    64              /* Number of characters in Rx ring buffer   */
#define  COMM_TX_BUF_SIZE    64              /* Number of characters in Tx ring buffer   */

#endif
```

Listing 1.24   INCLUDES.H

```
/*
*********************************************************************************
*                          Embedded Systems Building Blocks
*                          Complete and Ready-to-Use Modules in C
*
*                          Master Include File
*
*                          (c) Copyright 1999, Jean J. Labrosse, Weston, FL
*                          All Rights Reserved
*
* Filename   : INCLUDES.H
* Programmer : Jean J. Labrosse
*********************************************************************************
*/


/*
*********************************************************************************
*                                    CONSTANTS
*********************************************************************************
*/

                                    /* MODULE ENABLED (1) or DISABLED (0)                    */
#define   MODULE_KEY_MN     1       /* Keyboard module                                       */
#define   MODULE_LED        0       /* Multiplexed LED module                                */
#define   MODULE_LCD        1       /* LCD Character module                                  */
#define   MODULE_CLK        1       /* Clock/Calendar module                                 */
#define   MODULE_TMR        1       /* Timer Manager module                                  */
#define   MODULE_DIO        1       /* Discrete I/O module                                   */
#define   MODULE_AIO        1       /* Analog   I/O module                                   */
#define   MODULE_COMM_PC    1       /* Asynchronous Serial Communications module             */
#define   MODULE_COMM_BGND  0       /*     Foreground/Background buffered serial I/O         */
#define   MODULE_COMM_RTOS  1       /*     Real-Time Kernel       buffered serial I/O        */

#define   MODULE_ELAPSED    1       /* Elapsed time measurement module                       */

#define   CFG_C                     /* Indicate that application specific code is found in CFG.C  */
#define   CFG_H                     /* Indicate that configuration #defines is found in CFG.H     */

/*
*********************************************************************************
*                                    CONSTANTS
*********************************************************************************
*/

#define  FALSE    0
#define  TRUE     1

/*$PAGE*/
```

Listing 1.24 (continued) INCLUDES.H

```
/*
*********************************************************************************************
*                                Standard Libraries (DOS)
*********************************************************************************************
*/

#include    <stdio.h>
#include    <string.h>
#include    <ctype.h>
#include    <stdlib.h>
#include    <conio.h>
#include    <dos.h>
#include    <setjmp.h>


/*
*********************************************************************************************
*                                uC/OS Header Files
*********************************************************************************************
*/

#include    "\software\ucos-ii\ix861-fp\bc45\os_cpu.h"
#include    "\software\blocks\sample\source\os_cfg.h"
#include    "\software\ucos-ii\source\ucos_ii.h"
#include    "\software\blocks\pc\bc45\pc.h"

/*$PAGE*/
```

1

Listing 1.24 (continued) `INCLUDES.H`

```
/*
*********************************************************************************************************
*                                       Building Blocks Header Files
*********************************************************************************************************
*/
#ifdef      CFG_H
#include    "\software\blocks\sample\source\cfg.h"
#endif

#if         MODULE_KEY_MN
#include    "\software\blocks\key_mn\source\key.h"
#endif

#if         MODULE_LCD
#include    "\software\blocks\lcd\source\lcd.h"
#endif

#if         MODULE_LED
#include    "\software\blocks\led\source\led.h"
#endif

#if         MODULE_CLK
#include    "\software\blocks\clk\source\clk.h"
#endif

#if         MODULE_TMR
#include    "\software\blocks\tmr\source\tmr.h"
#endif

#if         MODULE_DIO
#include    "\software\blocks\dio\source\dio.h"
#endif

#if         MODULE_AIO
#include    "\software\blocks\aio\source\aio.h"
#endif

#if         MODULE_COMM_PC
#include    "\software\blocks\comm\source\comm_pc.h"
#endif

#if         MODULE_COMM_BGND
#include    "\software\blocks\comm\source\commbgnd.h"
#endif

#if         MODULE_COMM_RTOS
#include    "\software\blocks\comm\source\commrtos.h"
#endif
```

Listing 1.25 MAKETEST.BAT

```
ECHO OFF
CLS
ECHO ***************************************************************************
ECHO *                       Embedded Systems Building Blocks
ECHO *
ECHO *               (c) Copyright 1999, Jean J. Labrosse, Weston, FL
ECHO *                       All Rights Reserved
ECHO *
ECHO *
ECHO * Filename    : MAKETEST.BAT
ECHO * Description : Batch file to create the application.
ECHO * Output      : TEST.EXE will contain the DOS executable
ECHO * Usage       : MAKETEST
ECHO * Note(s)     : 1) This file assume that we use a MAKE utility.
ECHO ***************************************************************************
ECHO *
ECHO ON
MD      ..\WORK
MD      ..\OBJ
MD      ..\LST
CD      ..\WORK
COPY  ..\TEST\TEST.MAK    TEST.MAK
E:\UTILS\MAKE -R -C TEST.BAT -#4 -F TEST.MAK
IF NOT EXIST TEST.BAT GOTO END
COPY  TEST.BAT  ..\TEST /y
CALL  TEST.BAT
:END
CD      ..\TEST
```

1

Listing 1.26 OS_CFG.H

```
/*
********************************************************************************************
*                                       uC/OS-II
*                                 The Real-Time Kernel
*
*                  (c) Copyright 1992-1998, Jean J. Labrosse, Plantation, FL
*                                   All Rights Reserved
*
*                          Configuration for Intel 80x86 (Large)
*
* File : OS_CFG.H
* By   : Jean J. Labrosse
********************************************************************************************
*/


/*
********************************************************************************************
*                                  uC/OS-II CONFIGURATION
********************************************************************************************
*/

#define OS_MAX_EVENTS             5    /* Max. number of event control blocks in your application ...  */
                                       /* ... MUST be >= 2                                             */
#define OS_MAX_MEM_PART           5    /* Max. number of memory partitions ...                         */
                                       /* ... MUST be >= 2                                             */
#define OS_MAX_QS                 5    /* Max. number of queue control blocks in your application ...  */
                                       /* ... MUST be >= 2                                             */
#define OS_MAX_TASKS             20    /* Max. number of tasks in your application ...                 */
                                       /* ... MUST be >= 2                                             */

#define OS_LOWEST_PRIO           63    /* Defines the lowest priority that can be assigned ...         */
                                       /* ... MUST NEVER be higher than 63!                            */

#define OS_TASK_IDLE_STK_SIZE   512    /* Idle task stack size (# of 16-bit wide entries)              */

#define OS_TASK_STAT_EN           1    /* Enable (1) or Disable(0) the statistics task                 */
#define OS_TASK_STAT_STK_SIZE   512    /* Statistics task stack size (# of 16-bit wide entries)        */

#define OS_CPU_HOOKS_EN           1    /* uC/OS-II hooks are found in the processor port files         */
#define OS_MBOX_EN                0    /* Include code for MAILBOXES                                   */
#define OS_MEM_EN                 1    /* Include code for MEMORY MANAGER (fixed sized memory blocks)  */
#define OS_Q_EN                   1    /* Include code for QUEUES                                      */
#define OS_SEM_EN                 1    /* Include code for SEMAPHORES                                  */
#define OS_TASK_CHANGE_PRIO_EN    0    /* Include code for OSTaskChangePrio()                          */
#define OS_TASK_CREATE_EN         1    /* Include code for OSTaskCreate()                              */
#define OS_TASK_CREATE_EXT_EN     1    /* Include code for OSTaskCreateExt()                           */
#define OS_TASK_DEL_EN            0    /* Include code for OSTaskDel()                                 */
#define OS_TASK_SUSPEND_EN        0    /* Include code for OSTaskSuspend() and OSTaskResume()          */

#define OS_TICKS_PER_SEC        200    /* Set the number of ticks in one second                       */
```

Listing 1.27 TEST.C

```
/*
*********************************************************************************************
*                              Embedded Systems Building Blocks
*                              Complete and Ready-to-Use Modules in C
*
*                              (c) Copyright 1999, Jean J. Labrosse, Weston, FL
*                                       All Rights Reserved
*
* Filename   : TEST.C
* Programmer : Jean J. Labrosse
*********************************************************************************************
*/

#include "includes.h"

/*
*********************************************************************************************
*                                       CONSTANTS
*********************************************************************************************
*/

#define        TASK_STK_SIZE    512          /* Size of each task's stacks (# of 16-bit words)    */

#define        TEST_TASK_PRIO   10
#define        STAT_TASK_PRIO   20
#define        RND_TASK_PRIO    30

/*
*********************************************************************************************
*                                       VARIABLES
*********************************************************************************************
*/

OS_STK         TestStatTaskStk[TASK_STK_SIZE];
OS_STK         TestTaskStk[TASK_STK_SIZE];
OS_STK         TestRndTaskStk[10][TASK_STK_SIZE];

/*
*********************************************************************************************
*                                       FUNCTION PROTOTYPES
*********************************************************************************************
*/

       void      TestStatTask(void *data);
       void      TestTask(void *data);
       void      TestRndTask(void *data);
static void      TestInitModules(void);
static void      TestTmr0TO(void *arg);
static void      TestTmr1TO(void *arg);

/*$PAGE*/
```

Listing 1.27 (continued) *TEST.C*

```
/*
*********************************************************************************************
*                                          MAIN
*********************************************************************************************
*/

void  main (void)
{
    PC_DispClrScr(DISP_FGND_WHITE + DISP_BGND_BLACK);     /* Clear the screen                      */
    OSInit();                                             /* Initialize uC/OS-II                   */
    OSFPInit();                                           /* Initialize floating-point support     */
    PC_DOSSaveReturn();                                  /* Save environment to return to DOS     */
    PC_VectSet(uCOS, OSCtxSw);                            /* Install uC/OS-II's context switch vector */
    OSTaskCreateExt(TestStatTask, (void *)0, &TestStatTaskStk[TASK_STK_SIZE], STAT_TASK_PRIO,
                STAT_TASK_PRIO, &TestStatTaskStk[0], TASK_STK_SIZE, (void *)0, OS_TASK_OPT_SAVE_FP);
    OSStart();                                            /* Start multitasking                    */
}

/*$PAGE*/
```

Listing 1.27 (continued) `TEST.C`

```c
/*
*********************************************************************************************************
*                                         STATISTICS TASK
*********************************************************************************************************
*/

void  TestStatTask (void *pdata)
{
    INT8U   i;
    INT16S  key;
    char    s[100];


    pdata = pdata;                                           /* Prevent compiler warning              */

    PC_DispStr(21,  0, "   EMBEDDED SYSTEMS BUILDING BLOCKS   ",
                       DISP_FGND_WHITE + DISP_BGND_RED + DISP_BLINK);
    PC_DispStr(21,  1, "Complete and Ready-to-Use Modules in C", DISP_FGND_WHITE);
    PC_DispStr(21,  2, "          Jean J. Labrosse",           DISP_FGND_WHITE);
    PC_DispStr(21,  3, "             SAMPLE  CODE",            DISP_FGND_WHITE);


    OS_ENTER_CRITICAL();
    PC_VectSet(0x08, OSTickISR);                             /* Install uC/OS-II's clock tick ISR     */
    PC_SetTickRate(OS_TICKS_PER_SEC);                        /* Reprogram tick rate                   */
    OS_EXIT_CRITICAL();

    PC_DispStr(0, 22, "Determining  CPU's capacity ...", DISP_FGND_WHITE);
    OSStatInit();                                           /* Initialize uC/OS-II's statistics       */
    PC_DispClrLine(22, DISP_FGND_WHITE + DISP_BGND_BLACK);

    PC_DispStr( 0, 22, "#Tasks        : xxxxx  CPU Usage: xxx %", DISP_FGND_WHITE);
    PC_DispStr( 0, 23, "#Task switch/sec: xxxxx", DISP_FGND_WHITE);
    PC_DispStr(28, 24, "<-PRESS 'ESC' TO QUIT->", DISP_FGND_WHITE + DISP_BLINK);

    OSTaskCreateExt(TestTask,  (void *)0, &TestTaskStk[TASK_STK_SIZE],  TEST_TASK_PRIO,
              TEST_TASK_PRIO, &TestTaskStk[0],  TASK_STK_SIZE,  (void *)0,  OS_TASK_OPT_SAVE_FP);
    for (i = 0; i < 10; i++) {
       OSTaskCreateExt(TestRndTask,  (void *)0, &TestRndTaskStk[i][TASK_STK_SIZE],  RND_TASK_PRIO + i,
              RND_TASK_PRIO + i, &TestRndTaskStk[i][0], TASK_STK_SIZE, (void *)0, OS_TASK_OPT_SAVE_FP);
    }

    for (;;) {
       sprintf(s, "%5d", OSTaskCtr);                        /* Display #tasks running                 */
       PC_DispStr(18, 22, s, DISP_FGND_BLUE + DISP_BGND_CYAN);

       sprintf(s, "%3d", OSCPUUsage);                       /* Display CPU usage in %                 */
       PC_DispStr(36, 22, s, DISP_FGND_BLUE + DISP_BGND_CYAN);

       sprintf(s, "%5d", OSCtxSwCtr);                       /* Display #context switches per second   */
       PC_DispStr(18, 23, s, DISP_FGND_BLUE + DISP_BGND_CYAN);

       OSCtxSwCtr = 0;
       sprintf(s, "V%1d.%02d", OSVersion() / 100, OSVersion() % 100);
       PC_DispStr(75, 24, s, DISP_FGND_YELLOW + DISP_BGND_BLUE);
```

Listing 1.27 (continued) **TEST.C**

```
        PC_GetDateTime(s);                              /* Get and display date and time        */
        PC_DispStr(0, 24, s, DISP_FGND_BLUE + DISP_BGND_CYAN);

        if (PC_GetKey(&key) == TRUE) {                  /* See if key has been pressed          */
            if (key == 0x1B) {                          /* Yes, see if it's the ESCAPE key      */
                PC_DOSReturn();                         /* Return to DOS                        */
            }
        }

        OSTimeDlyHMSM(0, 0, 1, 0);                      /* Wait one second                      */
    }
}
/*$PAGE*/
```

Listing 1.27 (continued) `TEST.C`

```
/*
*********************************************************************************************************
*                                            TEST TASK
*********************************************************************************************************
*/

void  TestTask (void *data)
{
    char    s[81];
    INT16U  time;


    data = data;                                        /* Prevent compiler warning               */
    PC_DispStr( 0,  6, "Date :", DISP_FGND_WHITE);
    PC_DispStr( 0,  7, "Time :", DISP_FGND_WHITE);
    PC_DispStr( 0,  8, "Tmr#0:                          Task that displays numbers randomly!",
                DISP_FGND_WHITE);
    PC_DispStr( 0,  9, "Tmr#1:                          ------------------------------------",
                DISP_FGND_WHITE);
    PC_DispStr( 0, 10, "DO #0:", DISP_FGND_WHITE);
    PC_DispStr( 0, 11, "DO #1:", DISP_FGND_WHITE);

    TestInitModules();                                  /* Initialize all building blocks used    */

    ClkSetDateTime(12, 31, 1999, 23, 57, 55);           /* Set the clock/calendar                 */
    TmrCfgFnct(0, TestTmr0TO, (void *)0);               /* Execute when Timer #0 times out        */
    TmrCfgFnct(1, TestTmr1TO, (void *)0);               /* Execute when Timer #1 times out        */
    TmrSetMST(0, 1, 3, 9);                              /* Set timer #0 to 1 min., 3 sec. 9/10 sec. */
    TmrStart(0);
    TmrSetMST(1, 2, 0, 0);                              /* Set timer #1 to 2 minutes              */
    TmrStart(1);
    DOCfgBlink(0, DO_BLINK_EN,  9, 18);                 /* Initialize Discrete Outputs #0 and #1  */
    DOCfgBlink(1, DO_BLINK_EN, 45, 90);
    DOCfgMode(0,  DO_MODE_BLINK_ASYNC, FALSE);
    DOCfgMode(1,  DO_MODE_BLINK_ASYNC, FALSE);
```

Listing 1.27 (continued) **TEST.C**

```
    for (;;) {
        PC_ElapsedStart();
        ClkFormatDate(2, s);                              /* Get formatted date from clock/calendar   */
        time = PC_ElapsedStop();
        PC_DispStr(10,  6, "                         ", DISP_FGND_WHITE);
        PC_DispStr(10,  6, s, DISP_FGND_WHITE);

        sprintf(s, "ClkFormatDate() takes %3d uS", time);
        PC_DispStr( 0, 15, s, DISP_FGND_WHITE);

        PC_ElapsedStart();
        ClkFormatTime(1, s);                              /* Get formatted time from clock/calendar   */
        time = PC_ElapsedStop();
        PC_DispStr(10,  7, s, DISP_FGND_WHITE);

        sprintf(s, "ClkFormatTime() takes %3d uS", time);
        PC_DispStr( 0, 16, s, DISP_FGND_WHITE);
        TmrFormat(0, s);                                  /* Get formatted remaining time for Tmr#0   */
        PC_DispStr(10,  8, s, DISP_FGND_WHITE);
        TmrFormat(1, s);                                  /* Get formatted remaining time for Tmr#1   */
        PC_DispStr(10,  9, s, DISP_FGND_WHITE);

        PC_DispChar(10, 10, DOGet(0) + '0', DISP_FGND_WHITE);   /* Display state of discrete outputs  */
        PC_DispChar(10, 11, DOGet(1) + '0', DISP_FGND_WHITE);   /* Display state of discrete outputs  */

        OSTimeDlyHMSM(0, 0, 0, 100);
    }
}

/*$PAGE*/
```

Listing 1.27 (continued) TEST.C

```
/*
*********************************************************************************************
*                                    RANDOM NUMBER TASK
*********************************************************************************************
*/

void  TestRndTask (void *data)
{
    INT8U  x;
    INT8U  y;
    INT8U  z;


    data = data;
    for (;;) {
        OSTimeDly(1);
        x = random(36);                      /* Find X position where task number will appear   */
        y = random(10);                      /* Find Y position where task number will appear   */
        z = random(10);                      /* Find random number from 0 to 9                  */
        PC_DispChar(x + 43, y + 10, z + '0', DISP_FGND_WHITE);  /* Display number at random locations   */
    }
}
/*$PAGE*/
```

Listing 1.27 (continued)   TEST.C

```c
/*
*********************************************************************************************
*                              EMBEDDED SYSTEMS BUILDING BLOCKS
*                                  Modules Initialization
*********************************************************************************************
*/

static  void  TestInitModules (void)
{
#if MODULE_ELAPSED
    PC_ElapsedInit();                                    /* Initialize the elapsed time module       */
#endif

#if MODULE_KEY_MN
    KeyInit();                                           /* Initialize the keyboard scanning module  */
#endif

#if MODULE_LCD
    DispInit(4, 20);                                     /* Initialize the LCD module (4 x 20 disp.) */
#endif

#if MODULE_CLK
    ClkInit();                                           /* Initialize the clock/calendar module     */
#endif

#if MODULE_TMR
    TmrInit();                                           /* Initialize the timer manager module      */
#endif

#if MODULE_DIO
    DIOInit();                                           /* Initialize the discrete I/O module       */
#endif

#if MODULE_AIO
    AIOInit();                                           /* Initialize the analog I/O module         */
#endif

#if MODULE_COMM_PC
    CommCfgPort(COMM1, 9600, 8, COMM_PARITY_NONE, 1);    /* Initialize COM1 on the PC                */
#endif

#if MODULE_COMM_BGND
    CommInit();                                          /* Initialize the buffered serial I/O module*/
#endif

#if MODULE_COMM_RTOS
    CommInit();                                          /* Initialize the buffered serial I/O module*/
#endif
}
/*$PAGE*/
```

Listing 1.27 (continued)   *TEST.C*

```
/*
*********************************************************************************************
*                           Function executed when Timers Time Out
*********************************************************************************************
*/

static  void  TestTmr0TO (void *arg)
{
    arg = arg;
    PC_DispStr(22, 8, "Timer #0 Timed Out!", DISP_FGND_WHITE);
}




static  void  TestTmr1TO (void *arg)
{
    arg = arg;
    PC_DispStr(22, 9, "Timer #1 Timed Out!", DISP_FGND_WHITE);
}
```

Listing 1.28 **TEST.LNK**

```
/v /s /c /P- /LE:\BC45\LIB +
COL.OBJ +
..\OBJ\CFG.OBJ          +
..\OBJ\CLK.OBJ          +
..\OBJ\COMM_PC.OBJ      +
..\OBJ\COMM_PCA.OBJ     +
..\OBJ\COMMRTOS.OBJ     +
..\OBJ\AIO.OBJ          +
..\OBJ\DIO.OBJ          +
..\OBJ\KEY.OBJ          +
..\OBJ\LCD.OBJ          +
..\OBJ\OS_CPU_A.OBJ     +
..\OBJ\OS_CPU_C.OBJ     +
..\OBJ\PC.OBJ           +
..\OBJ\TEST.OBJ         +
..\OBJ\TMR.OBJ          +
..\OBJ\uCOS_II.OBJ,..\OBJ\TEST,..\OBJ\TEST,CL.LIB +
FP87.LIB                +
MATHL.LIB
```

Listing 1.29   TEST.MAK

```
###############################################################################
#                      Embedded Systems Building Blocks
#
#           (c) Copyright 1999, Jean J. Labrosse, Weston, FL
#                           All Rights Reserved
#
#
# Filename    : TEST.MAK
###############################################################################
#
#/*$PAGE*/
###############################################################################
#                                 TOOLS
###############################################################################
#

CC=E:\BC45\BIN\BCC
ASM=E:\BC45\BIN\TASM
LINK=E:\BC45\BIN\TLINK


###############################################################################
#                               DIRECTORIES
###############################################################################
#

TARGET=..\TEST
SOURCE=..\SOURCE
TEST=..\TEST
WORK=..\WORK
OBJ=..\OBJ
LST=..\LST

#

AIO=\SOFTWARE\BLOCKS\AIO\SOURCE
CLK=\SOFTWARE\BLOCKS\CLK\SOURCE
COMM=\SOFTWARE\BLOCKS\COMM\SOURCE
DIO=\SOFTWARE\BLOCKS\DIO\SOURCE
KEY=\SOFTWARE\BLOCKS\KEY_MN\SOURCE
LCD=\SOFTWARE\BLOCKS\LCD\SOURCE
LED=\SOFTWARE\BLOCKS\LED\SOURCE
OS=\SOFTWARE\uCOS-II\SOURCE
PC=\SOFTWARE\BLOCKS\PC\BC45
PORT=\SOFTWARE\uCOS-II\Ix86L-FP\BC45
TMR=\SOFTWARE\BLOCKS\TMR\SOURCE

#

LIB_PATH = E:\BC45\LIB
INCLUDE_PATH = E:\BC45\INCLUDE

#
#/*$PAGE*/
```

Listing 1.29 (continued) TEST.MAK

```
###########################################################################
#                             ASSEMBLER FLAGS
#
# /ml                   Large model
# /zi                   Full debug info
###########################################################################
#

ASM_FLAGS=/ml /zi


###########################################################################
#                             COMPILER FLAGS
#
# -1                    Generate 80186 code
# -B                    Compile and call assembler
# -c                    Compiler to .OBJ
# -d                    Duplicate strings merged
# -dc                   Put strings in code segment
# -G                    Select code for speed
# -I                    Path to include directory
# -k-                   Don't use standard stack frame
# -ml                   Large memory model
# -N-                   Do not check for stack overflow
# -n                    Path to object directory
# -O                    Optimize jumps
# -S                    Generate assembler source
# -v                    Source debugging ON
# -vi                   Turn inline expansion ON
# -wpro                 Error reporting: call to functions with no prototype
# -Z                    Suppress redundant loads
###########################################################################
#

C_FLAGS=-f287 -c -ml -1 -G -O -Ogemvlbpi -Z -d -n..\obj -k- -v -vi- -wpro -I$(INCLUDE_PATH)


###########################################################################
#                             LINKER FLAGS
###########################################################################
#
LINK_FLAGS=

#/*$PAGE*/
```

Listing 1.29 (continued)   TEST.MAK

```
#############################################################################
#                        CREATION OF .HEX FILES
#############################################################################

$(TARGET)\TEST.EXE:      $(OBJ)\AIO.OBJ          \
                         $(OBJ)\CFG.OBJ          \
                         $(OBJ)\CLK.OBJ          \
                         $(OBJ)\COMM_PC.OBJ      \
                         $(OBJ)\COMM_PCA.OBJ     \
                         $(OBJ)\COMMRTOS.OBJ     \
                         $(OBJ)\DIO.OBJ          \
                         $(OBJ)\KEY.OBJ          \
                         $(OBJ)\LCD.OBJ          \
                         $(OBJ)\LED.OBJ          \
                         $(OBJ)\LED_IA.OBJ       \
                         $(OBJ)\OS_CPU_A.OBJ     \
                         $(OBJ)\OS_CPU_C.OBJ     \
                         $(OBJ)\PC.OBJ           \
                         $(OBJ)\TEST.OBJ         \
                         $(OBJ)\TMR.OBJ          \
                         $(OBJ)\uCOS_II.OBJ      \
                         $(SOURCE)\TEST.LNK
                         COPY    $(SOURCE)\TEST.LNK
                         DEL     $(TARGET)\TEST.MAP
                         DEL     $(TARGET)\TEST.EXE
                         $(LINK) $(LINK_FLAGS) @TEST.LNK
                         COPY    $(OBJ)\TEST.EXE  $(WORK)\TEST.EXE    /y
                         E:\PD\PDCONVRT TEST
                         COPY    $(OBJ)\TEST.MAP  $(TARGET)\TEST.MAP  /y
                         COPY    $(OBJ)\TEST.EXE  $(TARGET)\TEST.EXE  /y
                         DEL     TEST.MAK


#############################################################################
#                        CREATION OF .O (Object) FILES
#############################################################################

$(OBJ)\AIO.OBJ:          $(AIO)\AIO.C     \
                         INCLUDES.H
                         COPY  $(AIO)\AIO.C       AIO.C
                         DEL   $(OBJ)\AIO.OBJ
                         $(CC) $(C_FLAGS)         AIO.C


$(OBJ)\CFG.OBJ:          $(SOURCE)\CFG.C    \
                         INCLUDES.H
                         COPY  $(SOURCE)\CFG.C    CFG.C
                         DEL   $(OBJ)\CFG.OBJ
                         $(CC) $(C_FLAGS)         CFG.C


$(OBJ)\CLK.OBJ:          $(CLK)\CLK.C     \
                         INCLUDES.H
                         COPY  $(CLK)\CLK.C       CLK.C
                         DEL   $(OBJ)\CLK.OBJ
                         $(CC) $(C_FLAGS)         CLK.C
```

Listing 1.29 (continued) ***TEST.MAK***

```
$(OBJ)\COMM_PC.OBJ:      $(COMM)\COMM_PC.C    \
                         INCLUDES.H
                         COPY   $(COMM)\COMM_PC.C      COMM_PC.C
                         DEL    $(OBJ)\COMM_PC.OBJ
                         $(CC)  $(C_FLAGS)             COMM_PC.C


$(OBJ)\COMM_PCA.OBJ:     $(COMM)\COMM_PCA.ASM
                         COPY   $(COMM)\COMM_PCA.ASM  COMM_PCA.ASM
                         DEL    $(OBJ)\COMM_PCA.OBJ
                         $(ASM) $(ASM_FLAGS)          $(COMM)\COMM_PCA.ASM, $(OBJ)\COMM_PCA.OBJ


$(OBJ)\COMMRTOS.OBJ:     $(COMM)\COMMRTOS.C    \
                         INCLUDES.H
                         COPY   $(COMM)\COMMRTOS.C    COMMRTOS.C
                         DEL    $(OBJ)\COMMRTOS.OBJ
                         $(CC)  $(C_FLAGS)            COMMRTOS.C


$(OBJ)\DIO.OBJ:          $(DIO)\DIO.C    \
                         INCLUDES.H
                         COPY   $(DIO)\DIO.C          DIO.C
                         DEL    $(OBJ)\DIO.OBJ
                         $(CC)  $(C_FLAGS)            DIO.C


$(OBJ)\KEY.OBJ:          $(KEY)\KEY.C    \
                         INCLUDES.H
                         COPY   $(KEY)\KEY.C          KEY.C
                         DEL    $(OBJ)\KEY.OBJ
                         $(CC)  $(C_FLAGS)            KEY.C


$(OBJ)\LCD.OBJ:          $(LCD)\LCD.C    \
                         INCLUDES.H
                         COPY   $(LCD)\LCD.C          LCD.C
                         DEL    $(OBJ)\LCD.OBJ
                         $(CC)  $(C_FLAGS)            LCD.C


$(OBJ)\LED.OBJ:          $(LED)\LED.C    \
                         INCLUDES.H
                         COPY   $(LED)\LED.C          LED.C
                         DEL    $(OBJ)\LED.OBJ
                         $(CC)  $(C_FLAGS)            LED.C


$(OBJ)\LED_IA.OBJ:       $(LED)\LED_IA.ASM
                         COPY   $(LED)\LED_IA.ASM     LED_IA.ASM
                         DEL    $(OBJ)\LED_IA.OBJ
                         $(ASM) $(ASM_FLAGS)          $(LED)\LED_IA.ASM, $(OBJ)\LED_IA.OBJ
```

Listing 1.29 (continued) **TEST.MAK**

```
$(OBJ)\OS_CPU_A.OBJ:    $(PORT)\OS_CPU_A.ASM   \
                        INCLUDES.H
                        COPY    $(PORT)\OS_CPU_A.ASM  OS_CPU_A.ASM
                        DEL     $(OBJ)\OS_CPU_A.OBJ
                         $(ASM) $(ASM_FLAGS)  $(PORT)\OS_CPU_A.ASM,$(OBJ)\OS_CPU_A.OBJ


$(OBJ)\OS_CPU_C.OBJ:    $(PORT)\OS_CPU_C.C   \
                        INCLUDES.H
                        COPY    $(PORT)\OS_CPU_C.C   OS_CPU_C.C
                        DEL     $(OBJ)\OS_CPU_C.OBJ
                        $(CC)   $(C_FLAGS)           OS_CPU_C.C


$(OBJ)\PC.OBJ:          $(PC)\PC.C    \
                        INCLUDES.H
                        COPY    $(PC)\PC.C           PC.C
                        DEL     $(OBJ)\PC.OBJ
                        $(CC)   $(C_FLAGS)           PC.C


$(OBJ)\TEST.OBJ:        $(SOURCE)\TEST.C   \
                        INCLUDES.H
                        COPY    $(SOURCE)\TEST.C     TEST.C
                        DEL     $(OBJ)\TEST.OBJ
                        $(CC)   $(C_FLAGS)           TEST.C


$(OBJ)\TMR.OBJ:         $(TMR)\TMR.C    \
                        INCLUDES.H
                        COPY    $(TMR)\TMR.C         TMR.C
                        DEL     $(OBJ)\TMR.OBJ
                        $(CC)   $(C_FLAGS)           TMR.C


$(OBJ)\uCOS_II.OBJ:     $(OS)\uCOS_II.C   \
                        INCLUDES.H
                        COPY    $(OS)\uCOS_II.C      uCOS_II.C
                        DEL     $(OBJ)\uCOS_II.OBJ
                        $(CC)   $(C_FLAGS)           uCOS_II.C


#/*$PAGE*/
```

Listing 1.29 (continued)   TEST.MAK

```
##########################################################################
#                               HEADER FILES
##########################################################################

INCLUDES.H:            $(SOURCE)\INCLUDES.H \
                       AIO.H                \
                       CLK.H                \
                       COMM_PC.H            \
                       COMMRTOS.H           \
                       DIO.H                \
                       KEY.H                \
                       LCD.H                \
                       LED.H                \
                       OS_CFG.H             \
                       OS_CPU.H             \
                       PC.H                 \
                       TMR.H                \
                       uCOS_II.H
                       C:\POLYTRON\POLYMAKE\TOUCH -V $(SOURCE)\INCLUDES.H
                       COPY $(SOURCE)\INCLUDES.H    INCLUDES.H

AIO.H:                 $(AIO)\AIO.H
                       COPY $(AIO)\AIO.H            AIO.H

CLK.H:                 $(CLK)\CLK.H
                       COPY $(CLK)\CLK.H            CLK.H

COMM_PC.H:             $(COMM)\COMM_PC.H
                       COPY $(COMM)\COMM_PC.H       COMM_PC.H

COMMRTOS.H:            $(COMM)\COMMRTOS.H
                       COPY $(COMM)\COMMRTOS.H      COMMRTOS.H

DIO.H:                 $(DIO)\DIO.H
                       COPY $(DIO)\DIO.H            DIO.H

KEY.H:                 $(KEY)\KEY.H
                       COPY $(KEY)\KEY.H            KEY.H

LCD.H:                 $(LCD)\LCD.H
                       COPY $(LCD)\LCD.H            LCD.H

LED.H:                 $(LED)\LED.H
                       COPY $(LED)\LED.H            LED.H

OS_CFG.H:              $(SOURCE)\OS_CFG.H
                       COPY $(SOURCE)\OS_CFG.H      OS_CFG.H

OS_CPU.H:              $(PORT)\OS_CPU.H
                       COPY $(PORT)\OS_CPU.H        OS_CPU.H

PC.H:                  $(PC)\PC.H
                       COPY $(PC)\PC.H              PC.H

TMR.H:                 $(TMR)\TMR.H
                       COPY $(TMR)\TMR.H            TMR.H

uCOS_II.H:             $(OS)\uCOS_II.H
                       COPY $(OS)\uCOS_II.H         uCOS_II.H
```

## Chapter 2

Real-Time Systems Concepts

Real-time systems are characterized by the severe consequences that result if logical as well as timing correctness properties of the system are not met. There are two types of real-time systems: SOFT and HARD. In a SOFT real-time system, tasks are performed by the system as fast as possible, but the tasks don't have to finish by specific times. In HARD real-time systems, tasks have to be performed not only correctly but on time. Most real-time systems have a combination of SOFT and HARD requirements. Real-time applications cover a wide range, but most real-time systems are *embedded*. This means that the computer is built into a system and is not seen by the user as being a computer. The following list shows a few examples of embedded systems.

Process control
    Food processing
    Chemical plants
Automotive
    Engine controls
    Antilock braking systems
Office automation
    FAX machines
    Copiers
Computer peripherals
    Printers
    Terminals
    Scanners
    Modems

Communication
    Switches
    Routers
Robots
Aerospace
    Flight management systems
    Weapons systems
    Jet engine controls
Domestic
    Microwave ovens
    Dishwashers
    Washing machines
    Thermostats

Real-time software applications are typically more difficult to design than non-real-time applications. This chapter describes real-time concepts.

2.00 Foreground/Background Systems

Small systems of low complexity are generally designed as shown in Figure 2.1. These systems are called *foreground/background* or *super-loops*. An application consists of an infinite loop that calls modules (i.e., functions) to perform the desired operations (background). Interrupt Service Routines (ISRs) handle asynchronous events (foreground). Foreground is also called *interrupt level*; background is called *task level*. Critical operations must be performed by the ISRs to ensure that they are dealt with in a timely fashion. Because of this, ISRs have a tendency to take longer than they should. Also, information for a background module made available by an ISR is not processed until the background routine gets its turn to execute. This is called the *task level response*. The worst case task-level response time depends on how long the background loop takes to execute. Because the execution time of typical code is not constant, the time for successive passes through a portion of the loop is nondeterministic. Furthermore, if a code change is made, the timing of the loop is affected.

Figure 2.1   Foreground/background systems.

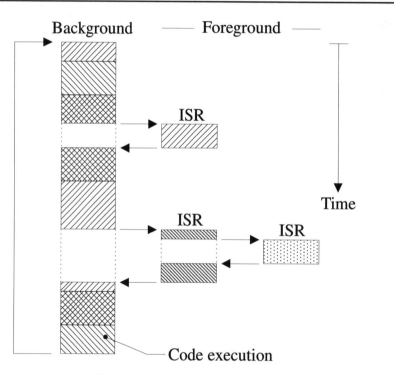

Most high-volume microcontroller-based applications (e.g., microwave ovens, telephones, toys, and so on) are designed as foreground/background systems. Also, in microcontroller-based applications, it may be better (from a power consumption point of view) to halt the processor and perform all of the processing in ISRs.

2.01 *Critical Section of Code*

A critical section of code, also called a *critical region*, is code that needs to be treated indivisibly. Once the section of code starts executing, it must not be interrupted. To ensure this, interrupts are typically disabled before the critical code is executed and enabled when the critical code is finished (see also section 2.03, Shared Resource).

2.02 *Resource*

A resource is any entity used by a task. A resource can thus be an I/O device, such as a printer, a keyboard, or a display, or a variable, a structure, or an array.

2.03 *Shared Resource*

A shared resource is a resource that can be used by more than one task. Each task should gain exclusive access to the shared resource to prevent data corruption. This is called *mutual exclusion*, and techniques to ensure mutual exclusion are discussed in section 2.18, Mutual Exclusion.

2.04 *Multitasking*

Multitasking is the process of scheduling and switching the CPU (Central Processing Unit) between several tasks; a single CPU switches its attention between several sequential tasks. Multitasking is like foreground/background with multiple backgrounds. Multitasking maximizes the utilization of the CPU and also provides for modular construction of applications. One of the most important aspects of multitasking is that it allows the application programmer to manage complexity inherent in real-time applications. Application programs are typically easier to design and maintain if multitasking is used.

2.05 *Task*

A task, also called a *thread*, is a simple program that thinks it has the CPU all to itself. The design process for a real-time application involves splitting the work to be done into tasks responsible for a portion of the problem. Each task is assigned a priority, its own set of CPU registers, and its own stack area (as shown in Figure 2.2).

Each task typically is an infinite loop that can be in any one of five states: *DORMANT, READY, RUNNING, WAITING* (for an event), or *ISR* (interrupted) (Figure 2.3). The DORMANT state corresponds to a task that resides in memory but has not been made available to the multitasking kernel. A task is READY when it can execute but its priority is less than the currently running task. A task is RUNNING when it has control of the CPU. A task is WAITING when it requires the occurrence of an event (waiting for an I/O operation to complete, a shared resource to be available, a timing pulse to occur, time to expire, etc.). Finally, a task is in the ISR state when an interrupt has occurred and the CPU is in the process of servicing the interrupt. Figure 2.3 also shows the functions provided by µC/OS-II to make a task move from one state to another.

Figure 2.2   Multiple tasks.

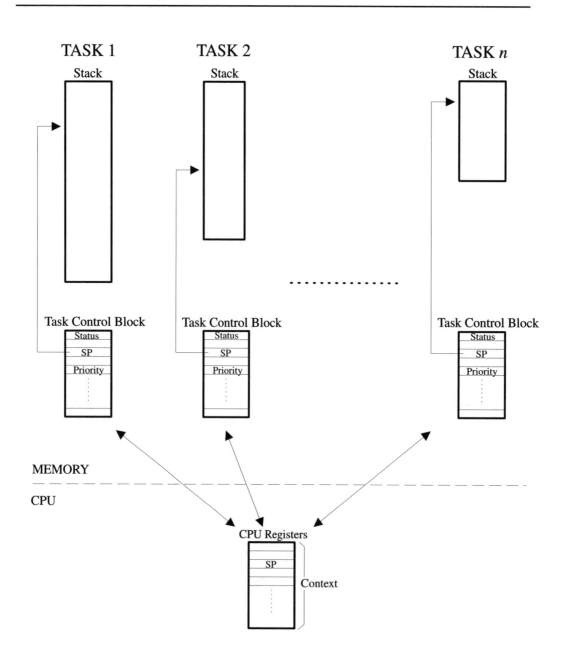

Figure 2.3   Task states.

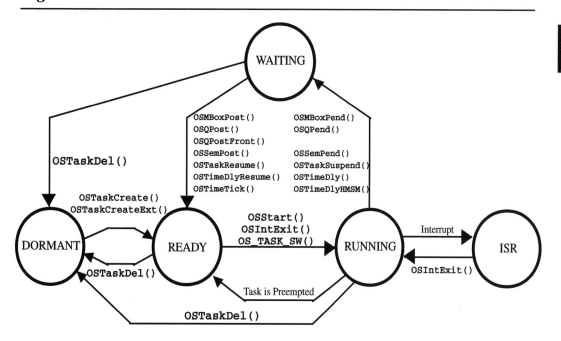

2.06 Context Switch (or Task Switch)

When a multitasking kernel decides to run a different task, it simply saves the current task's *context* (CPU registers) in the current task's context storage area — its stack (Figure 2.2). Once this operation is performed, the new task's context is restored from its storage area then resumes execution of the new task's code. This process is called a *context switch* or a *task switch*. Context switching adds overhead to the application. The more registers a CPU has, the higher the overhead. The time required to perform a context switch is determined by how many registers have to be saved and restored by the CPU. Performance of a real-time kernel should not be judged by how many context switches the kernel is capable of doing per second.

2.07 Kernel

The kernel is the part of a multitasking system responsible for the management of tasks (i.e., for managing the CPU's time) and communication between tasks. The fundamental service provided by the kernel is context switching. The use of a real-time kernel generally simplifies the design of systems by allowing the application to be divided into multiple tasks managed by the kernel. A kernel adds overhead to your system because it requires extra ROM (code space) and additional RAM for the kernel data structures. But most importantly, each task requires its own stack space, which has a tendency to eat up RAM quite quickly. A kernel will also consume CPU time (typically between 2 and 5 percent).

Single-chip microcontrollers are generally not able to run a real-time kernel because they have very little RAM. A kernel allows you to make better use of your CPU by providing you with indispensable services such as semaphore management, mailboxes, queues, time delays, etc. Once you design a system using a real-time kernel, you will not want to go back to a foreground/background system.

2.08 Scheduler

The scheduler, also called the *dispatcher*, is the part of the kernel responsible for determining which task will run next. Most real-time kernels are priority based. Each task is assigned a priority based on its importance. The priority for each task is application specific. In a priority-based kernel, control of the CPU is always given to the highest priority task ready to run. *When* the highest priority task gets the CPU, however, is determined by the type of kernel used. There are two types of priority-based kernels: *non-preemptive* and *preemptive*.

2.09 Non-Preemptive Kernel

Non-preemptive kernels require that each task does something to explicitly give up control of the CPU. To maintain the illusion of concurrency, this process must be done frequently. Non-preemptive scheduling is also called *cooperative multitasking*; tasks cooperate with each other to share the CPU. Asynchronous events are still handled by ISRs. An ISR can make a higher priority task ready to run, but the ISR always returns to the interrupted task. The new higher priority task will gain control of the CPU only when the current task gives up the CPU.

One of the advantages of a non-preemptive kernel is that interrupt latency is typically low (see the later discussion on interrupts). At the task level, non-preemptive kernels can also use non-reentrant functions (discussed later). Non-reentrant functions can be used by each task without fear of corruption by another task. This is because each task can run to completion before it relinquishes the CPU. However, non-reentrant functions should not be allowed to give up control of the CPU.

Task-level response using a non-preemptive kernel can be much lower than with foreground/background systems because task-level response is now given by the time of the longest task.

Another advantage of non-preemptive kernels is the lesser need to guard shared data through the use of semaphores. Each task owns the CPU, and you don't have to fear that a task will be preempted. This is not an absolute rule, and in some instances, semaphores should still be used. Shared I/O devices may still require the use of mutual exclusion semaphores; for example, a task might still need exclusive access to a printer.

The execution profile of a non-preemptive kernel is shown in Figure 2.4. A task is executing [F2.4(1)] but gets interrupted. If interrupts are enabled, the CPU vectors (jumps) to the ISR [L2.4(2)]. The ISR handles the event [F2.4(3)] and makes a higher priority task ready to run. Upon completion of the ISR, a *Return From Interrupt* instruction is executed, and the CPU returns to the interrupted task [F2.4(4)]. The task code resumes at the instruction following the interrupted instruction [F2.4(5)]. When the task code completes, it calls a service provided by the kernel to relinquish the CPU to another task [F2.4(6)]. The new higher priority task then executes to handle the event signaled by the ISR [F2.4(7)].

Figure 2.4   Non-preemptive kernel.

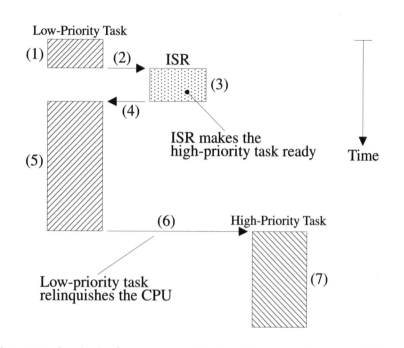

The most important drawback of a non-preemptive kernel is responsiveness. A higher priority task that has been made ready to run may have to wait a long time to run because the current task must give up the CPU when it is ready to do so. As with background execution in foreground/background systems, task-level response time in a non-preemptive kernel is nondeterministic; you never really know when the highest priority task will get control of the CPU. It is up to your application to relinquish control of the CPU.

To summarize, a non-preemptive kernel allows each task to run until it voluntarily gives up control of the CPU. An interrupt preempts a task. Upon completion of the ISR, the ISR returns to the interrupted task. Task-level response is much better than with a foreground/background system but is still nondeterministic. Very few commercial kernels are non-preemptive.

2.10 Preemptive Kernel

A preemptive kernel is used when system responsiveness is important. Because of this, µC/OS-II and most commercial real-time kernels are preemptive. The highest priority task ready to run is always given control of the CPU. When a task makes a higher priority task ready to run, the current task is pre-empted (suspended) and the higher priority task is *immediately* given control of the CPU. If an ISR makes a higher priority task ready, when the ISR completes, the interrupted task is suspended and the new higher priority task is resumed. This is illustrated in Figure 2.5.

Figure 2.5   Preemptive kernel.

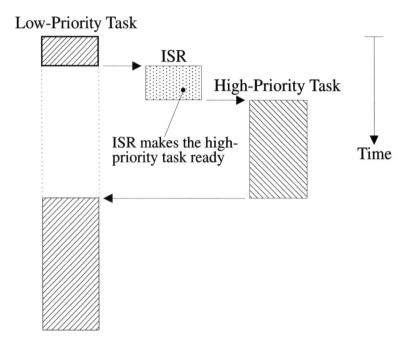

With a preemptive kernel, execution of the highest priority task is deterministic; you can determine when it will get control of the CPU. Task-level response time is thus minimized by using a preemptive kernel.

Application code using a preemptive kernel should not use non-reentrant functions, unless exclusive access to these functions is ensured through the use of mutual exclusion semaphores, because both a low- and a high-priority task can use a common function. Corruption of data may occur if the higher priority task preempts a lower priority task that is using the function.

To summarize, a preemptive kernel always executes the highest priority task that is ready to run. An interrupt preempts a task. Upon completion of an ISR, the kernel resumes execution to the highest priority task ready to run (not the interrupted task). Task-level response is optimum and deterministic. µC/OS-II is a preemptive kernel.

2.11 Reentrancy

A *reentrant function* can be used by more than one task without fear of data corruption. A reentrant function can be interrupted at any time and resumed at a later time without loss of data. Reentrant functions either use local variables (i.e., CPU registers or variables on the stack) or protect data when global variables are used. An example of a reentrant function is shown in Listing 2.1.

Listing 2.1   Reentrant function.

```
void strcpy(char *dest, char *src)
{
    while (*dest++ = *src++) {
        ;
    }
    *dest = NUL;
}
```

Because copies of the arguments to strcpy() are placed on the task's stack, strcpy() can be invoked by multiple tasks without fear that the tasks will corrupt each other's pointers.

An example of a non-reentrant function is shown in Listing 2.2. swap() is a simple function that swaps the contents of its two arguments. For the sake of discussion, I assume that you are using a preemptive kernel, that interrupts are enabled, and that Temp is declared as a global integer:

Listing 2.2   Non-reentrant function.

```
int Temp;

void swap(int *x, int *y)
{
    Temp = *x;
    *x   = *y;
    *y   = Temp;
}
```

The programmer intended to make swap() usable by any task. Figure 2.6 shows what could happen if a low-priority task is interrupted while swap() [F2.6(1)] is executing. Note that at this point Temp contains 1. The ISR makes the higher priority task ready to run, so at the completion of the ISR [F2.6(2)], the kernel (assuming μC/OS-II) is invoked to switch to this task [F2.6(3)]. The high-priority task sets Temp to 3 and swaps the contents of its variables correctly (i.e., z is 4 and t is 3). The high-priority task eventually relinquishes control to the low-priority task [F2.6(4)] by calling a kernel service to delay itself for one clock tick (described later). The lower priority task is thus resumed [F2.6(5)]. Note that at this point, Temp is still set to 3! When the low-priority task resumes execution, it sets y to 3 instead of 1.

Note that this a simple example, so it is obvious how to make the code reentrant. However, other situations are not as easy to solve. An error caused by a non-reentrant function may not show up in your application during the testing phase; it will most likely occur once the product has been delivered! If you are new to multitasking, you will need to be careful when using non-reentrant functions.

You can make swap() reentrant with one of the following techniques:

- Declare Temp local to swap().

- Disable interrupts before the operation and enable them afterwards.

- Use a semaphore (described later).

Figure 2.6   Non-reentrant function.

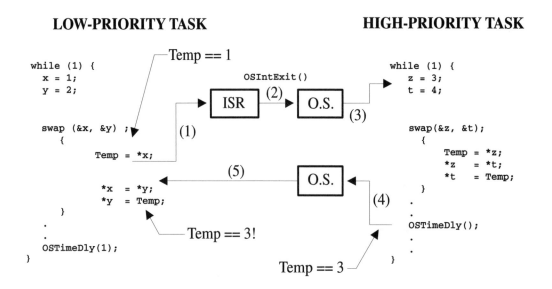

If the interrupt occurs either before or after swap(), the x and y values for both tasks will be correct.

2.12   Round-Robin Scheduling

When two or more tasks have the same priority, the kernel allows one task to run for a predetermined amount of time, called a *quantum*, then selects another task. This is also called *time slicing*. The kernel gives control to the next task in line if

- the current task has no work to do during its time slice or
- the current task completes before the end of its time slice.

µC/OS-II does not currently support round-robin scheduling. Each task must have a unique priority in your application.

2.13   Task Priority

A priority is assigned to each task. The more important the task, the higher the priority given to it.

2.14   Static Priorities

Task priorities are said to be *static* when the priority of each task does not change during the application's execution. Each task is thus given a fixed priority at compile time. All the tasks and their timing constraints are known at compile time in a system where priorities are static.

2.15 Dynamic Priorities

Task priorities are said to be dynamic if the priority of tasks can be changed during the application's execution; each task can change its priority at run time. This is a desirable feature to have in a real-time kernel to avoid priority inversions.

2.16 Priority Inversions

Priority inversion is a problem in real-time systems and occurs mostly when you use a real-time kernel. Figure 2.7 illustrates a priority inversion scenario. Task 1 has a higher priority than Task 2, which in turn has a higher priority than Task 3. Task 1 and Task 2 are both waiting for an event to occur and Task 3 is executing [F2.7(1)]. At some point, Task 3 acquires a semaphore (see section 2.18.04, Semaphores), which it needs before it can access a shared resource [F2.7(2)]. Task 3 performs some operations on the acquired resource [F2.7(4)] until it is preempted by the high-priority task, Task 1 [F2.7(3)]. Task 1 executes for a while until it also wants to access the resource [F2.7(5)]. Because Task 3 owns the resource, Task 1 has to wait until Task 3 releases the semaphore. As Task 1 tries to get the semaphore, the kernel notices that the semaphore is already owned; thus, Task 1 is suspended and Task 3 is resumed [F2.7(6)]. Task 3 continues execution until it is preempted by Task 2 because the event that Task2 was waiting for occurred [F2.7(7)]. Task 2 handles the event [F2.7(8)] and when it's done, Task 2 relinquishes the CPU back to Task 3 [F2.7(9)]. Task 3 finishes working with the resource [F2.7(10)] and releases the semaphore [F2.7(11)]. At this point, the kernel knows that a higher priority task is waiting for the semaphore, and a context switch is done to resume Task 1. At this point, Task 1 has the semaphore and can access the shared resource [F2.7(12)].

The priority of Task 1 has been virtually reduced to that of Task 3 because it was waiting for the resource that Task 3 owned. The situation was aggravated when Task 2 preempted Task 3, which further delayed the execution of Task 1.

You can correct this situation by raising the priority of Task 3, just for the time it takes to access the resource, then restoring the original priority level when the task is finished. The priority of Task 3 must be raised up to or above the highest priority of the other tasks competing for the resource. A multitasking kernel should allow task priorities to change dynamically to help prevent priority inversions. However, it takes some time to change a task's priority. What if Task 3 had completed access of the resource before it was preempted by Task 1 and then by Task 2? Had you raised the priority of Task 3 before accessing the resource and then lowered it back when done, you would have wasted valuable CPU time. What is really needed to avoid priority inversion is a kernel that changes the priority of a task automatically. This is called *priority inheritance*, which μC/OS-II unfortunately does not support. There are, however, some commercial kernels that do.

Figure 2.7   *Priority inversion problem.*

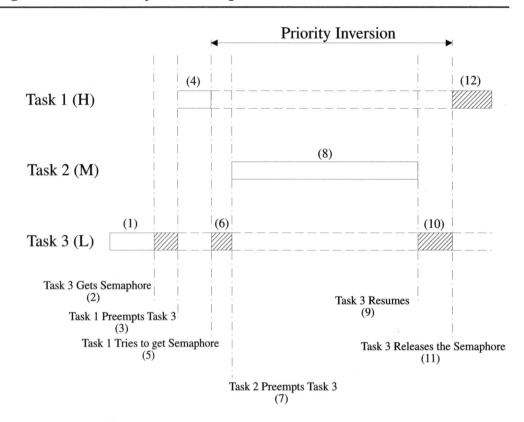

Figure 2.8 illustrates what happens when a kernel supports priority inheritance. As with the previous example, Task 3 is running [F2.8(1)] and acquires a semaphore to access a shared resource [F2.8(2)]. Task 3 accesses the resource [F2.8(3)] and then is preempted by Task 1 [F2.8(4)]. Task 1 executes [F2.8(5)] and tries to obtain the semaphore [F2.8(6)]. The kernel sees that Task 3 has the semaphore but has a lower priority than Task 1. In this case, the kernel raises the priority of Task 3 to the same level as Task 1. The kernel then switches back to Task 3 so that this task can continue with the resource [F2.8(7)]. When Task 3 is done with the resource, it releases the semaphore [F2.8(8)]. At this point, the kernel reduces the priority of Task 3 to its original value and gives the semaphore to Task 1 which is now free to continue [F2.8(9)]. When Task 1 is done executing [F2.8(10)], the medium-priority task (i.e., Task 2) gets the CPU [F2.8(11)]. Note that Task 2 could have been ready to run any time between F2.8(3) and (10) without affecting the outcome. There is still some level of priority inversion that cannot be avoided.

Figure 2.8   Kernel that supports priority inheritance.

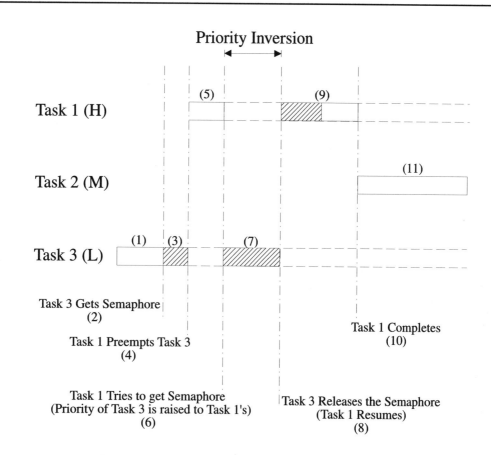

2.17 Assigning Task Priorities

Assigning task priorities is not a trivial undertaking because of the complex nature of real-time systems. In most systems, not all tasks are considered critical. Noncritical tasks should obviously be given low priorities. Most real-time systems have a combination of SOFT and HARD requirements. In a SOFT real-time system, tasks are performed as quickly as possible, but they don't have to finish by specific times. In HARD real-time systems, tasks have to be performed not only correctly, but on time.

An interesting technique called *Rate Monotonic Scheduling* (RMS) has been established to assign task priorities based on how often tasks execute. Simply put, tasks with the highest rate of execution are given the highest priority (Figure 2.9).

Figure 2.9   *Assigning task priorities based on task execution rate.*

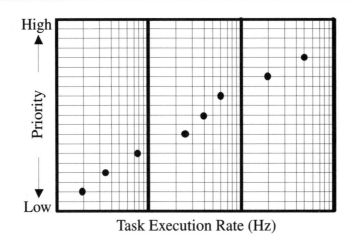

RMS makes a number of assumptions:

- All tasks are periodic (they occur at regular intervals).
- Tasks do not synchronize with one another, share resources, or exchange data.
- The CPU must always execute the highest priority task that is ready to run. In other words, preemptive scheduling must be used.

Given a set of *n* tasks that are assigned RMS priorities, the basic RMS theorem states that all task HARD real-time deadlines will always be met if the inequality in Equation [2.1] is verified.

[2.1] $$\sum_i \frac{E_i}{T_i} \le n(2^{1/n} - 1)$$

where, E_i corresponds to the maximum execution time of task *i* and T_i corresponds to the execution period of task *i*. In other words, E_i/T_i corresponds to the fraction of CPU time required to execute task *i*. Table 2.1 shows the value for size $n(2^{1/n} - 1)$ based on the number of tasks. The upper bound for an infinite number of tasks is given by ln(2), or 0.693. This means that to meet all HARD real-time deadlines based on RMS, CPU utilization of all time-critical tasks should be less than 70 percent! Note that you can still have non-time-critical tasks in a system and thus use 100 percent of the CPU's time. Using 100 percent of your CPU's time is not a desirable goal because it does not allow for code changes and added features. As a rule of thumb, you should always design a system to use less than 60 to 70 percent of your CPU.

RMS says that the highest rate task has the highest priority. In some cases, the highest rate task may not be the most important task. Your application will thus dictate how you need to assign priorities. However, RMS is an interesting starting point.

Table 2.1   Allowable CPU utilization based on number of tasks.

Number of Tasks	$n(2^{1/n} - 1)$
1	1.000
2	0.828
3	0.779
4	0.756
5	0.743
.	.
.	.
.	.
∞	0.693

2.18   Mutual Exclusion

The easiest way for tasks to communicate with each other is through shared data structures. This is especially easy when all tasks exist in a single address space and can reference global variables, pointers, buffers, linked lists, ring buffers, etc. Although sharing data simplifies the exchange of information, you must ensure that each task has exclusive access to the data to avoid contention and data corruption. The most common methods of obtaining exclusive access to shared resources are

- disabling interrupts,
- performing test-and-set operations,
- disabling scheduling, and
- using semaphores.

2.18.01   Disabling and Enabling Interrupts

The easiest and fastest way to gain exclusive access to a shared resource is by disabling and enabling interrupts, as shown in the pseudocode in Listing 2.3.

Listing 2.3   Disabling and enabling interrupts.

```
Disable interrupts;
Access the resource (read/write from/to variables);
Reenable interrupts;
```

μC/OS-II uses this technique (as do most, if not all, kernels) to access internal variables and data structures. In fact, μC/OS-II provides two macros that allow you to disable and then enable interrupts from your C code: OS_ENTER_CRITICAL() and OS_EXIT_CRITICAL(), respectively. You need to use these macros in tandem, as shown in Listing 2.4.

Listing 2.4   Using μC/OS-II macros to disable and enable interrupts.

```
void Function (void)
{
    OS_ENTER_CRITICAL();
    .
    .   /* You can access shared data in here */
    .
    OS_EXIT_CRITICAL();
}
```

You must be careful, however, not to disable interrupts for too long because this affects the response of your system to interrupts. This is known as *interrupt latency*. You should consider this method when you are changing or copying a few variables. Also, this is the only way that a task can share variables or data structures with an ISR. In all cases, you should keep interrupts disabled for as little time as possible.

If you use a kernel, you are basically allowed to disable interrupts for as much time as the kernel does without affecting interrupt latency. Obviously, you need to know how long the kernel will disable interrupts. Any good kernel vendor will provide you with this information. After all, if they sell a real-time kernel, time is important!

2.18.02   Test-And-Set

If you are not using a kernel, two functions could 'agree' that to access a resource, they must check a global variable and if the variable is 0, the function has access to the resource. To prevent the other function from accessing the resource, however, the first function that gets the resource simply sets the variable to 1. This is commonly called a *Test-And-Set* (or TAS) operation. Either the TAS operation must be performed indivisibly (by the processor) or you must disable interrupts when doing the TAS on the variable, as shown in Listing 2.5.

Listing 2.5   Using Test-And-Set to access a resource.

```
Disable interrupts;
if ('Access Variable' is 0) {
    Set variable to 1;
    Reenable interrupts;
    Access the resource;
    Disable interrupts;
    Set the 'Access Variable' back to 0;
    Reenable interrupts;
} else {
    Reenable interrupts;
    /* You don't have access to the resource, try back later; */
}
```

Some processors actually implement a TAS operation in hardware (e.g., the 68000 family of processors have the TAS instruction).

2.18.03   Disabling and Enabling the Scheduler

If your task is not sharing variables or data structures with an ISR, you can disable and enable scheduling, as shown in Listing 2.6 (using µC/OS-II as an example). In this case, two or more tasks can share data without the possibility of contention. You should note that while the scheduler is locked, interrupts are enabled, and if an interrupt occurs while in the critical section, the ISR is executed immediately. At the end of the ISR, the kernel always returns to the interrupted task, even if a higher priority task has been made ready to run by the ISR. The scheduler is invoked when OSSchedUnlock() is called to see if a higher priority task has been made ready to run by the task or an ISR. A context switch results if a higher priority task is ready to run. Although this method works well, you should avoid disabling the scheduler because it defeats the purpose of having a kernel in the first place. The next method should be chosen instead.

Listing 2.6   Accessing shared data by disabling and enabling scheduling.

```
void Function (void)
{
    OSSchedLock();
    .
    .      /* You can access shared data in here (interrupts are recognized) */
    .
    OSSchedUnlock();
}
```

2.18.04   Semaphores

The semaphore was invented by Edgser Dijkstra in the mid-1960s. It is a protocol mechanism offered by most multitasking kernels. Semaphores are used to

- control access to a shared resource (mutual exclusion),
- signal the occurrence of an event, and
- allow two tasks to synchronize their activities.

A semaphore is a key that your code acquires in order to continue execution. If the semaphore is already in use, the requesting task is suspended until the semaphore is released by its current owner. In other words, the requesting task says: "Give me the key. If someone else is using it, I am willing to wait for it!" There are two types of semaphores: *binary* semaphores and *counting* semaphores. As its name implies, a binary semaphore can only take two values: 0 or 1. A counting semaphore allows values between 0 and 255, 65535, or 4294967295, depending on whether the semaphore mechanism is implemented using 8, 16, or 32 bits, respectively. The actual size depends on the kernel used. Along with the semaphore's value, the kernel also needs to keep track of tasks waiting for the semaphore's availability.

Generally, only three operations can be performed on a semaphore: INITIALIZE (also called *CREATE*), WAIT (also called *PEND*), and SIGNAL (also called *POST*). The initial value of the semaphore must be provided when the semaphore is initialized. The waiting list of tasks is always initially empty.

A task desiring the semaphore will perform a WAIT operation. If the semaphore is available (the semaphore value is greater than 0), the semaphore value is decremented and the task continues execution. If the semaphore's value is 0, the task performing a WAIT on the semaphore is placed in a waiting list. Most kernels allow you to specify a timeout; if the semaphore is not available within a certain amount of time, the requesting task is made ready to run and an error code (indicating that a timeout has occurred) is returned to the caller.

A task releases a semaphore by performing a SIGNAL operation. If no task is waiting for the semaphore, the semaphore value is simply incremented. If any task is waiting for the semaphore, however, one of the tasks is made ready to run and the semaphore value is not incremented; the key is given to one of the tasks waiting for it. Depending on the kernel, the task that receives the semaphore is either

- the highest priority task waiting for the semaphore or
- the first task that requested the semaphore (First In First Out, or FIFO).

Some kernels have an option that allows you to choose either method when the semaphore is initialized. µC/OS-II only supports the first method. If the readied task has a higher priority than the current task (the task releasing the semaphore), a context switch occurs (with a preemptive kernel) and the higher priority task resumes execution; the current task is suspended until it again becomes the highest priority task ready to run.

Listing 2.7 shows how you can share data using a semaphore (in µC/OS-II). Any task needing access to the same shared data calls OSSemPend(), and when the task is done with the data, the task calls OSSemPost(). Both of these functions are described later. You should note that a semaphore is an object that needs to be initialized before it's used; for mutual exclusion, a semaphore is initialized to a value of 1. Using a semaphore to access shared data doesn't affect interrupt latency. If an ISR or the current task makes a higher priority task ready to run while accessing shared data, the higher priority task executes immediately.

Listing 2.7   Accessing shared data by obtaining a semaphore.

```
OS_EVENT *SharedDataSem;
void Function (void)
{
    INT8U err;
    OSSemPend(SharedDataSem, 0, &err);
    .
    .      /* You can access shared data in here (interrupts are recognized) */
    .
    OSSemPost(SharedDataSem);
}
```

Semaphores are especially useful when tasks share I/O devices. Imagine what would happen if two tasks were allowed to send characters to a printer at the same time. The printer would contain interleaved data from each task. For instance, the printout from Task 1 printing "I am Task 1!" and Task 2 printing "I am Task 2!" could result in:

I Ia amm T Tasask k1 !2!

In this case, use a semaphore and initialize it to 1 (i.e., a binary semaphore). The rule is simple: to access the printer each task first must obtain the resource's semaphore. Figure 2.10 shows tasks competing for a semaphore to gain exclusive access to the printer. Note that the semaphore is represented symbolically by a key, indicating that each task must obtain this key to use the printer.

Figure 2.10   Using a semaphore to get permission to access a printer.

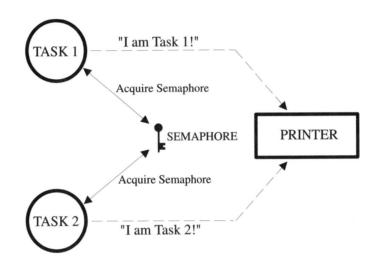

The above example implies that each task must know about the existence of the semaphore in order to access the resource. There are situations when it is better to encapsulate the semaphore. Each task would thus not know that it is actually acquiring a semaphore when accessing the resource. For example, an RS-232C port is used by multiple tasks to send commands and receive responses from a device connected at the other end (Figure 2.11).

The function CommSendCmd() is called with three arguments: the ASCII string containing the command, a pointer to the response string from the device, and finally, a timeout in case the device doesn't respond within a certain amount of time. The pseudocode for this function is shown in Listing 2.8.

Listing 2.8   Encapsulating a semaphore.

```
INT8U CommSendCmd(char *cmd, char *response, INT16U timeout)
{
    Acquire port's semaphore;
    Send command to device;
    Wait for response (with timeout);
    if (timed out) {
        Release semaphore;
        return (error code);
    } else {
```

Listing 2.8   Encapsulating a semaphore. (Continued)

```
        Release semaphore;
        return (no error);
    }
}
```

Each task that needs to send a command to the device has to call this function. The semaphore is assumed to be initialized to 1 (i.e., available) by the communication driver initialization routine. The first task that calls CommSendCmd() acquires the semaphore, proceeds to send the command, and waits for a response. If another task attempts to send a command while the port is busy, this second task is suspended until the semaphore is released. The second task appears simply to have made a call to a normal function that will not return until the function has performed its duty. When the semaphore is released by the first task, the second task acquires the semaphore and is allowed to use the RS-232C port.

Figure 2.11   Hiding a semaphore from tasks.

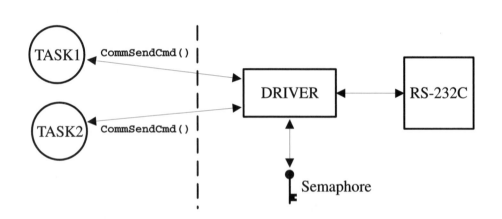

A counting semaphore is used when a resource can be used by more than one task at the same time. For example, a counting semaphore is used in the management of a buffer pool as shown in Figure 2.12. Assume that the buffer pool initially contains 10 buffers. A task would obtain a buffer from the buffer manager by calling BufReq(). When the buffer is no longer needed, the task would return the buffer to the buffer manager by calling BufRel(). The pseudocode for these functions is shown in Listing 2.9.

Listing 2.9   Buffer management using a semaphore.

```
BUF *BufReq(void)
{
    BUF *ptr;

    Acquire a semaphore;
    Disable interrupts;
    ptr          = BufFreeList;
    BufFreeList = ptr->BufNext;
    Enable interrupts;
    return (ptr);
}


void BufRel(BUF *ptr)
{
    Disable interrupts;
    ptr->BufNext = BufFreeList;
    BufFreeList  = ptr;
    Enable interrupts;
    Release semaphore;
}
```

Figure 2.12   Using a counting semaphore.

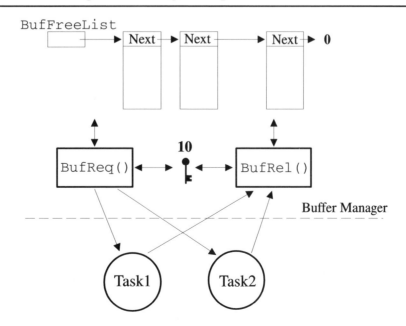

The buffer manager will satisfy the first 10 buffer requests because there are 10 keys. When all semaphores are used, a task requesting a buffer is suspended until a semaphore becomes available. Interrupts are disabled to gain exclusive access to the linked list (this operation is very quick). When a task is finished with the buffer it acquired, it calls BufRel() to return the buffer to the buffer manager; the buffer is inserted into the linked list before the semaphore is released. By encapsulating the interface to the buffer manager in BufReq() and BufRel(), the caller doesn't need to be concerned with the actual implementation details.

Semaphores are often overused. The use of a semaphore to access a simple shared variable is overkill in most situations. The overhead involved in acquiring and releasing the semaphore can consume valuable time. You can do the job just as efficiently by disabling and enabling interrupts (see section 2.18.01, Disabling and Enabling Interrupts). Suppose that two tasks are sharing a 32-bit integer variable. The first task increments the variable while the other task clears it. If you consider how long a processor takes to perform either operation, you will realize that you do not need a semaphore to gain exclusive access to the variable. Each task simply needs to disable interrupts before performing its operation on the variable and enable interrupts when the operation is complete. A semaphore should be used, however, if the variable is a floating-point variable and the microprocessor doesn't support floating point in hardware. In this case, the processing time involved in processing the floating-point variable could have affected interrupt latency if you had disabled interrupts.

2.19 Deadlock (or Deadly Embrace)

A deadlock, also called a *deadly embrace*, is a situation in which two tasks are each unknowingly waiting for resources held by the other. Assume task T1 has exclusive access to resource R1 and task T2 has exclusive access to resource R2. If T1 needs exclusive access to R2 and T2 needs exclusive access to R1, neither task can continue. They are deadlocked. The simplest way to avoid a deadlock is for tasks to

- acquire all resources before proceeding,
- acquire the resources in the same order, and
- release the resources in the reverse order.

Most kernels allow you to specify a timeout when acquiring a semaphore. This feature allows a deadlock to be broken. If the semaphore is not available within a certain amount of time, the task requesting the resource resumes execution. Some form of error code must be returned to the task to notify it that a timeout occurred. A return error code prevents the task from thinking it has obtained the resource. Deadlocks generally occur in large multitasking systems, not in embedded systems.

2.20 Synchronization

A task can be synchronized with an ISR (or another task when no data is being exchanged) by using a semaphore as shown in Figure 2.13. Note that, in this case, the semaphore is drawn as a flag to indicate that it is used to signal the occurrence of an event (rather than to ensure mutual exclusion, in which case it would be drawn as a key). When used as a synchronization mechanism, the semaphore is initialized to 0. Using a semaphore for this type of synchronization is called a *unilateral rendezvous*. A task initiates an I/O operation and waits for the semaphore. When the I/O operation is complete, an ISR (or another task) signals the semaphore and the task is resumed.

2

Figure 2.13 Synchronizing tasks and ISRs.

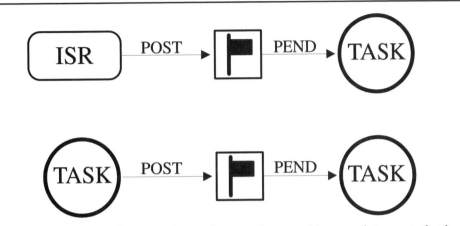

If the kernel supports counting semaphores, the semaphore would accumulate events that have not yet been processed. Note that more than one task can be waiting for an event to occur. In this case, the kernel could signal the occurrence of the event either to

- the highest priority task waiting for the event to occur or
- the first task waiting for the event.

Depending on the application, more than one ISR or task could signal the occurrence of the event.

Two tasks can synchronize their activities by using two semaphores, as shown in Figure 2.14. This is called a *bilateral rendezvous*. A bilateral rendezvous is similar to a unilateral rendezvous, except both tasks must synchronize with one another before proceeding.

For example, two tasks are executing as shown in Listing 2.10. When the first task reaches a certain point, it signals the second task [L2.10(1)] then waits for a return signal [L2.10(2)]. Similarly, when the second task reaches a certain point, it signals the first task [L2.10(3)] and waits for a return signal [L2.10(4)]. At this point, both tasks are synchronized with each other. A bilateral rendezvous cannot be performed between a task and an ISR because an ISR cannot wait on a semaphore.

Figure 2.14 Tasks synchronizing their activities.

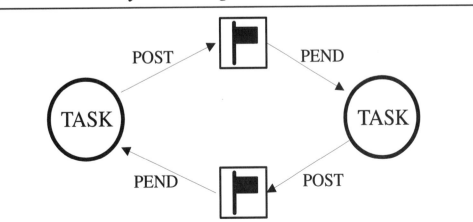

Listing 2.10   Bilateral rendezvous.

```
Task1()
{
    for (;;) {
        Perform operation;
        Signal task #2;                          (1)
        Wait for signal from task #2;            (2)
        Continue operation;
    }
}


Task2()
{
    for (;;) {
        Perform operation;
        Signal task #1;                          (3)
        Wait for signal from task #1;            (4)
        Continue operation;
    }
}
```

2.21 Event Flags

Event flags are used when a task needs to synchronize with the occurrence of multiple events. The task can be synchronized when any of the events have occurred. This is called disjunctive synchronization (logical OR). A task can also be synchronized when all events have occurred. This is called conjunctive synchronization (logical AND). Disjunctive and conjunctive synchronization are shown in Figure 2.15.

Common events can be used to signal multiple tasks, as shown in Figure 2.16. Events are typically grouped. Depending on the kernel, a group consists of 8, 16, or 32 events, each reprensnted by a bit. (mostly 32 bits, though). Tasks and ISRs can set or clear any event in a group. A task is resumed when all the events it requires are satisfied. The evaluation of which task will be resumed is performed when a new set of events occurs (i.e., during a SET operation).

Kernels supporting event flags offer services to SET event flags, CLEAR event flags, and WAIT for event flags (conjunctively or disjunctively). µC/OS-II does not currently support event flags.

Figure 2.15 Disjunctive and conjunctive synchronization.

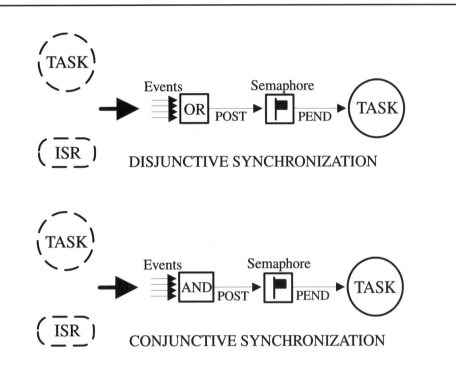

DISJUNCTIVE SYNCHRONIZATION

CONJUNCTIVE SYNCHRONIZATION

2.22 *Intertask Communication*

It is sometimes necessary for a task or an ISR to communicate information to another task. This information transfer is called *intertask communication*. Information may be communicated between tasks in two ways: through global data or by sending messages.

When using global variables, each task or ISR must ensure that it has exclusive access to the variables. If an ISR is involved, the only way to ensure exclusive access to the common variables is to disable interrupts. If two tasks are sharing data, each can gain exclusive access to the variables either by disabling and enabling interrupts or with the use of a semaphore (as we have seen). Note that a task can only communicate information to an ISR by using global variables. A task is not aware when a global variable is changed by an ISR, unless the ISR signals the task by using a semaphore or unless the task polls the contents of the variable periodically. To correct this situation, you should consider using either a *message mailbox* or a *message queue*.

Figure 2.16   Event flags.

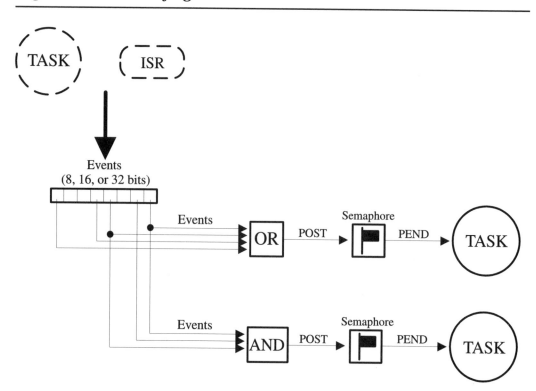

2.23 Message Mailboxes

Messages can be sent to a task through kernel services. A Message Mailbox, also called a message exchange, is typically a pointer-size variable. Through a service provided by the kernel, a task or an ISR can deposit a message (the pointer) into this mailbox. Similarly, one or more tasks can receive messages through a service provided by the kernel. Both the sending task and receiving task agree on what the pointer is actually pointing to.

A waiting list is associated with each mailbox in case more than one task wants to receive messages through the mailbox. A task desiring a message from an empty mailbox is suspended and placed on the waiting list until a message is received. Typically, the kernel allows the task waiting for a message to specify a timeout. If a message is not received before the timeout expires, the requesting task is made ready to run and an error code (indicating that a timeout has occurred) is returned to it. When a message is deposited into the mailbox, either the highest priority task waiting for the message is given the message (*priority based*) or the first task to request a message is given the message (*First-In-First-Out*, or FIFO). Figure 2.17 shows a task depositing a message into a mailbox. Note that the mailbox is represented by an I-beam and the timeout is represented by an hourglass. The number next to the hourglass represents the number of clock ticks (described later) the task will wait for a message to arrive.

Kernels typically provide the following mailbox services.

- Initialize the contents of a mailbox. The mailbox initially may or may not contain a message.

- Deposit a message into the mailbox (POST).

- Wait for a message to be deposited into the mailbox (PEND).

- Get a message from a mailbox if one is present, but do not suspend the caller if the mailbox is empty (ACCEPT). If the mailbox contains a message, the message is extracted from the mailbox. A return code is used to notify the caller about the outcome of the call.

Message mailboxes can also simulate binary semaphores. A message in the mailbox indicates that the resource is available, and an empty mailbox indicates that the resource is already in use by another task.

Figure 2.17   Message mailbox.

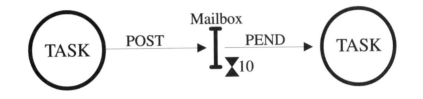

2.24 Message Queues

A message queue is used to send one or more messages to a task. A message queue is basically an array of mailboxes. Through a service provided by the kernel, a task or an ISR can deposit a message (the pointer) into a message queue. Similarly, one or more tasks can receive messages through a service provided by the kernel. Both the sending task and receiving task agree as to what the pointer is actually pointing to. Generally, the first message inserted in the queue will be the first message extracted from the queue (FIFO). In addition, to extract messages in a FIFO fashion, µC/OS-II allows a task to get messages Last-In-First-Out (LIFO).

As with the mailbox, a waiting list is associated with each message queue, in case more than one task is to receive messages through the queue. A task desiring a message from an empty queue is suspended and placed on the waiting list until a message is received. Typically, the kernel allows the task waiting for a message to specify a timeout. If a message is not received before the timeout expires, the requesting task is made ready to run and an error code (indicating a timeout has occurred) is returned to it. When a message is deposited into the queue, either the highest priority task or the first task to wait for the message is given the message. Figure 2.18 shows an ISR (Interrupt Service Routine) depositing a message into a queue. Note that the queue is represented graphically by a double I-beam. The "10" indicates the number of messages that can accumulate in the queue. A "0" next to the hourglass indicates that the task will wait forever for a message to arrive.

Kernels typically provide the message queue services listed below.

• Initialize the queue. The queue is always assumed to be empty after initialization.

• Deposit a message into the queue (POST).

• Wait for a message to be deposited into the queue (PEND).

• Get a message from a queue if one is present, but do not suspend the caller if the queue is empty (ACCEPT). If the queue contains a message, the message is extracted from the queue. A return code is used to notify the caller about the outcome of the call.

Figure 2.18   Message queue.

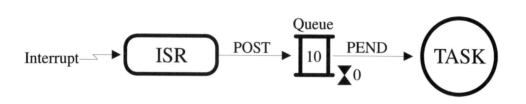

2.25   Interrupts

An interrupt is a hardware mechanism used to inform the CPU that an asynchronous event has occurred. When an interrupt is recognized, the CPU saves part (or all) of its context (i.e., registers) and jumps to a special subroutine called an *Interrupt Service Routine*, or ISR. The ISR processes the event, and upon completion of the ISR, the program returns to

• the background for a foreground/background system,

• the interrupted task for a non-preemptive kernel, or

• the highest priority task ready to run for a preemptive kernel.

Interrupts allow a microprocessor to process events when they occur. This prevents the microprocessor from continuously *polling* an event to see if it has occurred. Microprocessors allow interrupts to be ignored and recognized through the use of two special instructions: *disable interrupts* and *enable interrupts*, respectively. In a real-time environment, interrupts should be disabled as little as possible. Disabling interrupts affects interrupt latency (see section 2.26, Interrupt Latency) and may cause interrupts to be missed. Processors generally allow interrupts to be *nested*. This means that while servicing an interrupt, the processor will recognize and service other (more important) interrupts, as shown in Figure 2.19.

2.26   Interrupt Latency

Probably the most important specification of a real-time kernel is the amount of time interrupts are disabled. All real-time systems disable interrupts to manipulate critical sections of code and reenable interrupts when the critical section has executed. The longer interrupts are disabled, the higher the *interrupt latency*. Interrupt latency is given by Equation [2.2].

[2.2]   Maximum amount of time interrupts are disabled
               + Time to start executing the first instruction in the ISR

Figure 2.19   Interrupt nesting.

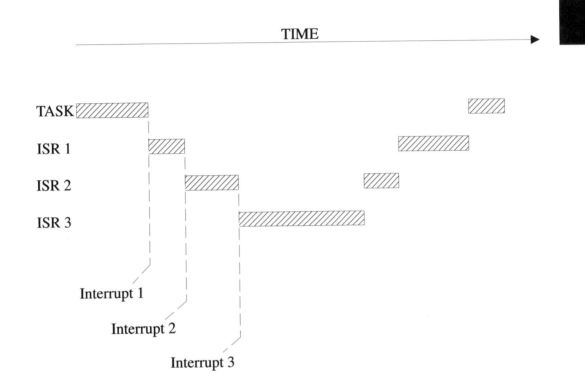

2.27 Interrupt Response

Interrupt response is defined as the time between the reception of the interrupt and the start of the user code that handles the interrupt. The interrupt response time accounts for all the overhead involved in handling an interrupt. Typically, the processor's context (CPU registers) is saved on the stack before the user code is executed.

For a foreground/background system, the user ISR code is executed immediately after saving the processor's context. The response time is given by Equation [2.3].

[2.3]   Interrupt latency + Time to save the CPU's context

For a non-preemptive kernel, the user ISR code is executed immediately after the processor's context is saved. The response time to an interrupt for a non-preemptive kernel is given by Equation [2.4].

[2.4]   Interrupt latency + Time to save the CPU's context

For a preemptive kernel, a special function provided by the kernel needs to be called. This function notifies the kernel that an ISR is in progress and allows the kernel to keep track of interrupt nesting. For μC/OS-II, this function is called OSIntEnter(). The response time to an interrupt for a preemptive kernel is given by Equation [2.5].

[2.5]   Interrupt latency
                     + Time to save the CPU's context
                     + Execution time of the kernel ISR entry function

A system's worst case interrupt response time is its only response. Your system may respond to interrupts in 50μs 99 percent of the time, but if it responds to interrupts in 250μs the other 1 percent, you must assume a 250μs interrupt response time.

2.28   Interrupt Recovery

Interrupt recovery is defined as the time required for the processor to return to the interrupted code. Interrupt recovery in a foreground/background system simply involves restoring the processor's context and returning to the interrupted task. Interrupt recovery is given by Equation [2.6].

[2.6]   Time to restore the CPU's context
                     + Time to execute the return from interrupt instruction

As with a foreground/background system, interrupt recovery with a non-preemptive kernel (Equation [2.7]) simply involves restoring the processor's context and returning to the interrupted task.

[2.7]   Time to restore the CPU's context
                     + Time to execute the return from interrupt instruction

For a preemptive kernel, interrupt recovery is more complex. Typically, a function provided by the kernel is called at the end of the ISR. For μC/OS-II, this function is called OSIntExit() and allows the kernel to determine if all interrupts have nested. If they have nested (i.e., a return from interrupt would return to task-level code), the kernel determines if a higher priority task has been made ready to run as a result of the ISR. If a higher priority task is ready to run as a result of the ISR, this task is resumed. Note that, in this case, the interrupted task will resume only when it again becomes the highest priority task ready to run. For a preemptive kernel, interrupt recovery is given by Equation [2.8].

[2.8]   Time to determine if a higher priority task is ready
                     + Time to restore the CPU's context of the highest priority task
                     + Time to execute the return from interrupt instruction

2.29   Interrupt Latency, Response, and Recovery

Figures 2.20 through 2.22 show the interrupt latency, response, and recovery for a foreground/background system, a non-preemptive kernel, and a preemptive kernel, respectively.

You should note that for a preemptive kernel, the exit function either decides to return to the interrupted task [F2.22(A)] or to a higher priority task that the ISR has made ready to run [F2.22(B)]. In the later case, the execution time is slightly longer because the kernel has to perform a context switch. I made the difference in execution time somewhat to scale assuming μC/OS-II on an Intel 80186 processor (see Table 9.3, Execution times of μC/OS-II services on 33MHz 80186). This allows you to see the cost (in execution time) of switching context.

2.30 ISR Processing Time

Although ISRs should be as short as possible, there are no absolute limits on the amount of time for an ISR. One cannot say that an ISR must always be less than 100μs, 500μs, or 1ms. If the ISR code is the most important code that needs to run at any given time, it could be as long as it needs to be. In most cases, however, the ISR should recognize the interrupt, obtain data or a status from the interrupting device, and signal a task to perform the actual processing. You should also consider whether the over-head involved in signaling a task is more than the processing of the interrupt. Signaling a task from an ISR (i.e., through a semaphore, a mailbox, or a queue) requires some processing time. If processing your interrupt requires less than the time required to signal a task, you should consider processing the interrupt in the ISR itself and possibly enabling interrupts to allow higher priority interrupts to be recognized and serviced.

Figure 2.20   Interrupt latency, response, and recovery (foreground/background).

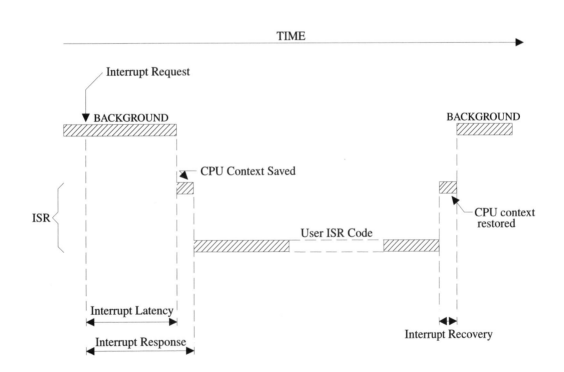

2.31 Nonmaskable Interrupts (NMIs)

Sometimes, an interrupt must be serviced as quickly as possible and cannot afford to have the latency imposed by a kernel. In these situations, you may be able to use the *Nonmaskable Interrupt* (NMI) pro-vided on most microprocessors. Because the NMI cannot be disabled, interrupt latency, response, and recovery are minimal. The NMI is generally reserved for drastic measures such as saving important information during a power down. If, however, your application doesn't have this requirement, you could use the NMI to service your most time-critical ISR. The following equations show how to determine the interrupt latency [2.9], response [2.10], and recovery [2.11], respectively, of an NMI.

[2.9]   Time to execute longest instruction + Time to start executing the NMI ISR

[2.10]   Interrupt latency + Time to save the CPU's context

[2.11]   Time to restore the CPU's context
               + Time to execute the return from interrupt instruction

I have used the NMI in an application to respond to an interrupt that could occur every 150µs. The processing time of the ISR took from 80 to 125µs, and the kernel I used disabled interrupts for about 45µs. As you can see, if I had used maskable interrupts, the ISR could have been late by 20µs.

When you are servicing an NMI, you cannot use kernel services to signal a task because NMIs cannot be disabled to access critical sections of code. However, you can still pass parameters to and from the NMI. Parameters passed must be global variables and the size of these variables must be read or written indivisibly; that is, not as separate byte read or write instructions.

Figure 2.21   Interrupt latency, response, and recovery (non-preemptive kernel).

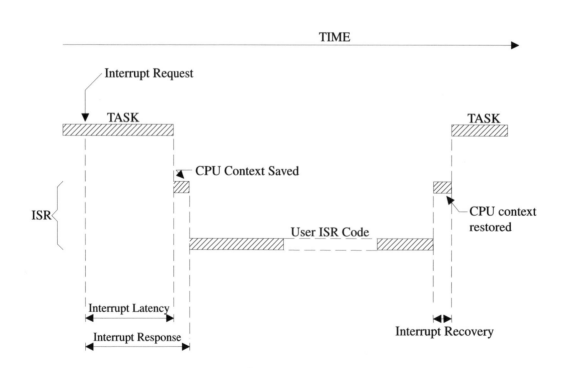

Figure 2.22   *Interrupt latency, response, and recovery (preemptive kernel).*

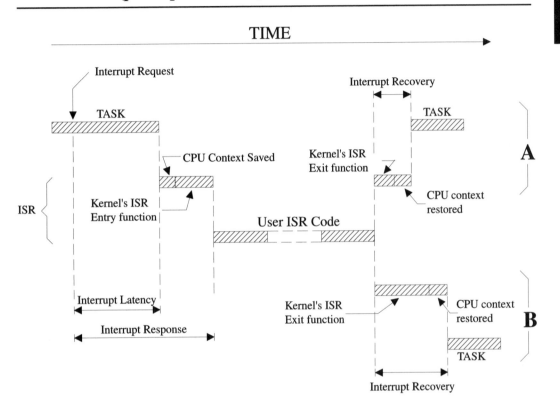

NMIs can be disabled by adding external circuitry, as shown in Figure 2.23. Assuming that both the interrupt and the NMI are positive-going signals, a simple AND gate is inserted between the interrupt source and the processor's NMI input. Interrupts are disabled by writing a 0 to an output port. You wouldn't want to disable interrupts to use kernel services, but you could use this feature to pass parameters (i.e., larger variables) to and from the ISR and a task.

Figure 2.23   *Disabling nonmaskable interrupts.*

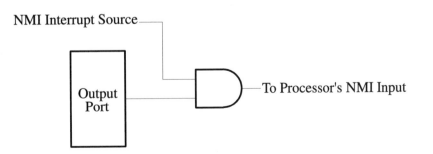

Now, suppose that the NMI service routine needs to signal a task every 40 times it executes. If the NMI occurs every 150μs, a signal would be required every 6ms (40 x 150μs). From a NMI ISR, you cannot use the kernel to signal the task, but you could use the scheme shown in Figure 2.24. In this case, the NMI service routine would generate a hardware interrupt through an output port (i.e., bring an output high). Since the NMI service routine typically has the highest priority and interrupt nesting is typically not allowed while servicing the NMI ISR, the interrupt would not be recognized until the end of the NMI service routine. At the completion of the NMI service routine, the processor would be interrupted to service this hardware interrupt. This ISR would clear the interrupt source (i.e., bring the port output low) and post to a semaphore that would wake up the task. As long as the task services the semaphore well within 6ms, your deadline would be met.

Figure 2.24   Signaling a task from a nonmaskable interrupt.

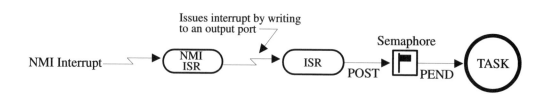

2.32 Clock Tick

A *clock tick* is a special interrupt that occurs periodically. This interrupt can be viewed as the system's heartbeat. The time between interrupts is application specific and is generally between 10 and 200ms. The clock tick interrupt allows a kernel to delay tasks for an integral number of clock ticks and to provide timeouts when tasks are waiting for events to occur. The faster the tick rate, the higher the overhead imposed on the system.

All kernels allow tasks to be delayed for a certain number of clock ticks. The resolution of delayed tasks is one clock tick; however, this does not mean that its accuracy is one clock tick.

Figures 2.25 through 2.27 are timing diagrams showing a task delaying itself for one clock tick. The shaded areas indicate the execution time for each operation being performed. Note that the time for each operation varies to reflect typical processing, which would include loops and conditional statements (i.e., if/else, switch, and ?:). The processing time of the Tick ISR has been exaggerated to show that it too is subject to varying execution times.

Case 1 (Figure 2.25) shows a situation where higher priority tasks and ISRs execute prior to the task, which needs to delay for one tick. As you can see, the task attempts to delay for 20ms but because of its priority, actually executes at varying intervals. This causes the execution of the task to *jitter*.

Figure 2.25   Delaying a task for one tick (Case 1).

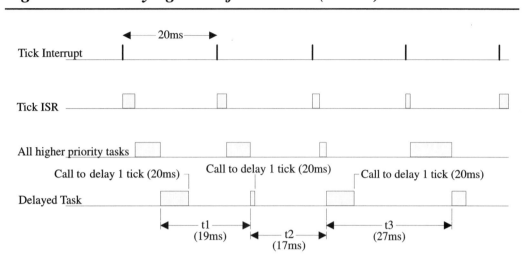

Case 2 (Figure 2.26) shows a situation where the execution times of all higher priority tasks and ISRs are slightly less than one tick. If the task delays itself just before a clock tick, the task will execute again almost immediately! Because of this, if you need to delay a task at least one clock tick, you must specify one extra tick. In other words, if you need to delay a task for at least five ticks, you must specify six ticks!

Figure 2.26   Delaying a task for one tick (Case 2).

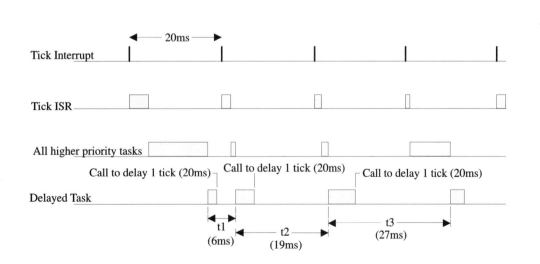

Case 3 (Figure 2.27) shows a situation in which the execution times of all higher priority tasks and ISRs extend beyond one clock tick. In this case, the task that tries to delay for one tick actually executes two ticks later and misses its deadline. This might be acceptable in some applications, but in most cases it isn't.

These situations exist with all real-time kernels. They are related to CPU processing load and possibly incorrect system design. Here are some possible solutions to these problems:

- Increase the clock rate of your microprocessor.
- Increase the time between tick interrupts.
- Rearrange task priorities.
- Avoid using floating-point math (if you must, use single precision).
- Get a compiler that performs better code optimization.
- Write time-critical code in assembly language.
- If possible, upgrade to a faster microprocessor in the same family; that is, 8086 to 80186, 68000 to 68020, etc.

Regardless of what you do, jitter will always occur.

Figure 2.27   Delaying a task for one tick (Case 3).

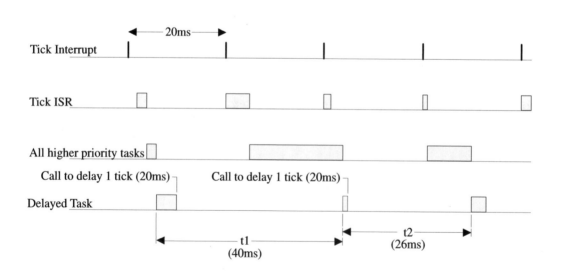

2.33 Memory Requirements

If you are designing a foreground/background system, the amount of memory required depends solely on your application code. With a multitasking kernel, things are quite different. To begin with, a kernel requires extra code space (ROM). The size of the kernel depends on many factors. Depending on the features provided by the kernel, you can expect anywhere from 1 to 100Kb. A minimal kernel for an 8-bit CPU that provides only scheduling, context switching, semaphore management, delays, and time-outs should require about 1 to 3Kb of code space. The total code space is given by Equation [2.12].

[2.12]   Application code size + Kernel code size

Because each task runs independently of the others, it must be provided with its own stack area (RAM). As a designer, you must determine the stack requirement of each task as closely as possible (this is sometimes a difficult undertaking). The stack size must not only account for the task require- ments (local variables, function calls, etc.), it must also account for maximum interrupt nesting (saved registers, local storage in ISRs, etc.). Depending on the target processor and the kernel used, a separate stack can be used to handle all interrupt-level code. This is a desirable feature because the stack require- ment for each task can be substantially reduced. Another desirable feature is the ability to specify the stack size of each task on an individual basis (µC/OS-II permits this). Conversely, some kernels require that all task stacks be the same size. All kernels require extra RAM to maintain internal variables, data structures, queues, etc. The total RAM required if the kernel does not support a separate interrupt stack is given by Equation [2.13].

[2.13]   Application code requirements
           + Data space (i.e., RAM) needed by the kernel
           + SUM(task stacks + MAX(ISR nesting))

If the kernel supports a separate stack for interrupts, the total RAM required is given by Equation [2.14].

[2.14]   Application code requirements
           + Data space (i.e., RAM) needed by the kernel
           + SUM(task stacks)
           + MAX(ISR nesting)

Unless you have large amounts of RAM to work with, you need to be careful how you use the stack space. To reduce the amount of RAM needed in an application, you must be careful how you use each task's stack for

- large arrays and structures declared locally to functions and ISRs,
- function (i.e., subroutine) nesting,
- interrupt nesting,
- library functions stack usage, and
- function calls with many arguments.

To summarize, a multitasking system requires more code space (ROM) and data space (RAM) than a foreground/background system. The amount of extra ROM depends only on the size of the kernel, and the amount of RAM depends on the number of tasks in your system.

2.34 Advantages and Disadvantages of Real-Time Kernels

A real-time kernel, also called a *Real-Time Operating System*, or *RTOS*, allows real-time applications to be designed and expanded easily; functions can be added without requiring major changes to the soft- ware. The use of an RTOS simplifies the design process by splitting the application code into separate tasks. With a preemptive RTOS, all time-critical events are handled as quickly and as efficiently as pos- sible. An RTOS allows you to make better use of your resources by providing you with valuable ser- vices, such as semaphores, mailboxes, queues, time delays, timeouts, etc.

You should consider using a real-time kernel if your application can afford the extra requirements: extra cost of the kernel, more ROM/RAM, and 2 to 4 percent additional CPU overhead.

The one factor I haven't mentioned so far is the cost associated with the use of a real-time kernel. In some applications, cost is everything and would preclude you from even considering an RTOS.

There are currently about 80+ RTOS vendors. Products are available for 8-, 16-, 32-, and even 64-bit microprocessors. Some of these packages are complete operating systems and include not only the real-time kernel but also an input/output manager, windowing systems (display), a file system, networking, language interface libraries, debuggers, and cross-platform compilers. The cost of an RTOS varies from $70 to well over $30,000. The RTOS vendor may also require royalties on a per-target-system basis. This is like buying a chip from the RTOS vendor that you include with each unit sold. The RTOS vendors call this *silicon software*. The royalty fee varies between $5 to about $250 per unit. Like any other software package these days, you also need to consider the maintenance cost, which can set you back another $100 to $5,000 per year!

2.35 Real-Time Systems Summary

Table 2.2 summarizes the three types of real-time systems: foreground/background, non-preemptive kernel, and preemptive kernel.

Table 2.2   Real-time systems summary.

	Foreground/ Background	Non-Preemptive Kernel	Preemptive Kernel
Interrupt latency (Time)	MAX(Longest instruction, User int. disable) + Vector to ISR	MAX(Longest instruction, User int. disable, Kernel int. disable) + Vector to ISR	MAX(Longest instruction, User int. disable, Kernel int. disable) + Vector to ISR
Interrupt response (Time)	Int. latency + Save CPU's context	Int. latency + Save CPU's context	Interrupt latency + Save CPU's context + Kernel ISR entry function
Interrupt recovery (Time)	Restore background's context + Return from int.	Restore task's context + Return from int.	Find highest priority task + Restore highest priority task's context + Return from interrupt
Task response (Time)	Background	Longest task + Find highest priority task + Context switch	Find highest priority task + Context switch
ROM size	Application code	Application code + Kernel code	Application code + Kernel code
RAM size	Application code	Application code + Kernel RAM + SUM(Task stacks + MAX(ISR stack))	Application code + Kernel RAM + SUM(Task stacks + MAX(ISR stack))
Services available?	Application code must provide	Yes	Yes

2.36 Bibliography

Allworth, Steve T. 1981. *Introduction To Real-Time Software Design.* New York: Springer-Verlag. ISBN 0-387-91175-8.

Bal Sathe, Dhananjay. 1988. Fast Algorithm Determines Priority. *EDN* (India), September, p. 237.

Comer, Douglas. 1984.*Operating System Design, The XINU Approach.* Englewood Cliffs, New Jersey: Prentice-Hall. ISBN 0-13-637539-1.

Deitel, Harvey M. and Michael S. Kogan. 1992. *The Design Of OS/2.* Reading, Massachusetts: Addison-Wesley. ISBN 0-201-54889-5.

Ganssle, Jack G. 1992. *The Art of Programming Embedded Systems.* San Diego: Academic Press. ISBN 0-122-748808.

Gareau, Jean L. 1998. Embedded x86 Programming: Protected Mode. *Embedded Systems Programming*, April, p. 80–93.

Halang, Wolfgang A. and Alexander D. Stoyenko. 1991. *Constructing Predictable Real Time Systems.* Norwell, Massachusetts: Kluwer Academic Publishers Group. ISBN 0-7923-9202-7.

Hunter & Ready. 1986. *VRTX Technical Tips.* Palo Alto, California: Hunter & Ready.

Hunter & Ready. 1983. *Dijkstra Semaphores, Application Note.* Palo Alto, California: Hunter & Ready.

Hunter & Ready. 1986. *VRTX and Event Flags.* Palo Alto, California: Hunter & Ready.

Intel Corporation. 1986. *iAPX 86/88, 186/188 User's Manual: Programmer's Reference.* Santa Clara, California: Intel Corporation.

Kernighan, Brian W. and Dennis M. Ritchie. 1988. *The C Programming Language,* 2nd edition. Englewood Cliffs, New Jersey: Prentice Hall. ISBN 0-13-110362-8.

Klein, Mark H., Thomas Ralya, Bill Pollak, Ray Harbour Obenza, and Michael Gonzlez. 1993. *A Practioner's Handbook for Real-Time Analysis: Guide to Rate Monotonic Analysis for Real-Time Systems.* Norwell, Massachusetts: Kluwer Academic Publishers Group. ISBN 0-7923-9361-9.

Labrosse, Jean J. 1992. µC/OS, The Real-Time Kernel. Lawrence, Kansas: R&D Publications. ISBN 0-87930-444-8.

Laplante, Phillip A. 1992. *Real-Time Systems Design and Analysis, An Engineer's Handbook.* Piscataway, New Jersey: IEEE Computer Society Press. ISBN 0-780-334000.

Lehoczky, John, Lui Sha, and Ye Ding. 1989. The Rate Monotonic Scheduling Algorithm: Exact Characterization and Average Case Behavior. In: *Proceedings of the IEEE Real-Time Systems Symposium.,* Los Alamitos, California. Piscataway, New Jersey: IEEE Computer Society, p. 166–171.

Madnick, E. Stuart and John J. Donovan. 1974. *Operating Systems.* New York: McGraw-Hill. ISBN 0-07-039455-5.

Ripps, David L. 1989. *An Implementation Guide To Real-Time Programming.* Englewood Cliffs, New Jersey: Yourdon Press. ISBN 0-13-451873-X.

Savitzky, Stephen R. 1985. *Real-Time Microprocessor Systems.* New York: Van Nostrand Reinhold. ISBN 0-442-28048-3.

Wood, Mike and Tom Barrett . 1990. A Real-Time Primer. *Embedded Systems Programming*, February, p. 20–28.

Chapter 3

Keyboards

A large number of embedded products, such as microwave ovens, FAX machines, copiers, laser print-ers, Point Of Sale (POS) terminals, Programmable Logic Controls (PLCs), and so on, rely on a key-board or keypad interface for user input. The keyboard might be used to input numerical data as well as to select the operating mode of the controlling device. As an embedded system designer, you are always concerned with the cost of your products. Chips are currently available to perform keyboard scanning, but a software approach to keyboard scanning has the benefit of reducing the recurring cost of a system and requires very little CPU overhead.

In this chapter, I will describe how a microprocessor can scan a keyboard, and I will also provide you with a complete, portable *m x n* matrix keyboard scanning module. The module can scan any key-board matrix arrangement up to an 8x8 matrix, but can easily be modified to handle a larger number of keys. The matrix keyboard module code is an important building block for embedded systems. The key-board module presented in this chapter has the following features:

- Scans any keyboard arrangement from a 3x3 to an 8x8 key matrix.
- Provides buffering (user configurable buffer size).
- Supports auto-repeat.
- Keeps track of how long a key has been pressed.
- Allows up to three Shift keys.

All you need to do to use this module is to write three simple hardware interface functions and set the value of 17 #define constants. The keyboard module assumes the presence of a real-time kernel but can easily be modified to work in a foreground/background environment.

3.00 *Keyboard Basics*

A momentary contact switch is typically used in a keyboard, and a closure can easily be detected by a microprocessor using the simple circuit shown in Figure 3.1. The pull-up resistor provides a logic 1 when the switch is opened and a logic 0 when the switch is closed. Unfortunately, switches are not per-fect in that they do not generate a crisp 1 or 0 when they are pressed or released. Although a contact may appear to close firmly and quickly, at the fast running speed of a microprocessor, the action is comparatively slow. As the contact closes, the contact bounces like a ball. This bouncing effect produces multiple pulses as shown in Figure 3.1. The duration of the bounce typically will last between 5 and 30 mS. If multiple keys are needed, each switch can be connected to its own input port on the microprocessor. As the number of switches increases, however, this method quickly begins to use up all the input ports.

Figure 3.1   Keyboard switch.

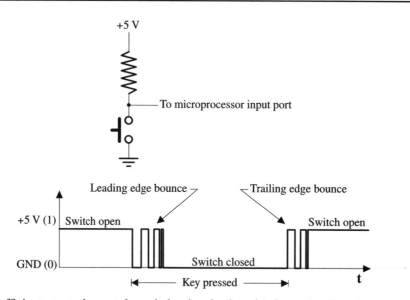

The most efficient way to lay out the switches in a keyboard (when more than *five* keys are needed) is to form a two-dimensional matrix as shown in Figure 3.2. The most optimum arrangement (where I/O lines are concerned) occurs when there are as many rows as columns, that is, a square matrix. A momentary contact switch (push button) is placed at the intersection of each row and column. The number of keys needed in the matrix is obviously application dependent. Each row is driven by a bit of an output port, while each column is pulled up by a resistor and fed to a bit on an input port.

Figure 3.2   Keyboard matrix.

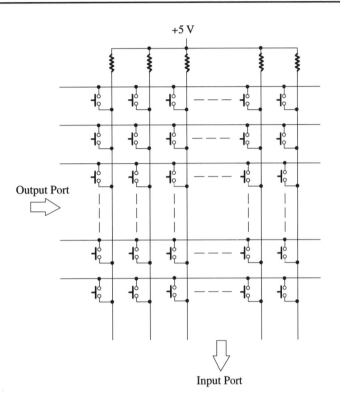

Keyboard scanning is the process of having the microprocessor look at the keyboard matrix at a regular interval to see if a key has been pressed. Once the processor determines that a key has been pressed, the keyboard scanning software filters out the bounce and determines which of the keys was pressed. Each key is assigned a unique identifier called a *scan code*. The scan code is used by your application to determine what action is to be taken based on the key pressed. In other words, the scan code tells your application which key was pressed.

Pressing (accidentally or deliberately) more than one key at a time is called *rollover*. Any algorithm that can correctly recognize that a new key has been pressed — even though *n-1* keys are already pressed — is said to have *n-key rollover* capability. The matrix keyboard module presented in this chapter does not implement an *n-key rollover* algorithm because of the extra code required. The code presented here is intended for small embedded systems where user input would occur one keystroke after the other. Such systems typically do not require full-featured keyboards like the ones found on terminals or computer systems.

3.01 Matrix Keyboard Scanning Algorithm

During initialization, all rows (output port) are forced low (see Figure 3.2). When no key is pressed, all columns (input port) read high. Any key closure will cause one of the columns to go low. To see if a key has been pressed, the microprocessor only needs to see if any of the input lines are low. Once the microprocessor has detected that a key has been pressed, it needs to find out which key it was. This process is quite simple. The microprocessor outputs a low on only one of the rows. If it finds a 0 on the input port, the microprocessor knows that the key closure occurred on the selected row. Conversely, if the input port had all highs, the key pressed was not on that row and the microprocessor selects the next row, repeating the process until it finds the row. Once the row has been identified, the specific column of the pressed key can be established by locating the position of the single low bit on the input port. The time required for the microprocessor to perform these steps is very short compared to the minimum switch closure time and it is thus assumed that the key will remain pressed during that interval.

To filter through the bouncing problem, the microprocessor samples the keyboard at regular intervals, typically between 20 mS and 100 mS (called the *debounce period*) depending on the bounce characteristics of the switches being used.

The scan code of the key pressed is typically placed in a buffer until the application is ready to process a keystroke. Buffering is a handy feature because it prevents losing keystrokes when the application cannot process them as they occur. The size of the buffer depends on your application requirements. A buffer size of 10 keystrokes is a good starting point. The buffer is generally implemented as a circular queue. When a key is pressed, the scan code is placed at the next empty location in the queue. When your application obtains a scan code from the keyboard module, the scan code is extracted from the oldest location in the queue. If the queue is full, any further keystrokes are lost.

Another nice feature is what is called *auto-repeat* or *typematic*. Auto-repeat allows the scan code of a key pressed to be repeatedly inserted into the buffer for as long as you press the key or until the buffer fills up. Auto-repeat capability is nice to have if you plan on incrementing or decrementing the value of a parameter (i.e., a variable) without having to continuously press and release the key. The timing diagram of Figure 3.3 shows how auto-repeat works. The scan code of the key pressed is inserted in the buffer as soon as the closure is detected. If the key is held down longer than the *auto-repeat start delay*, the scan code is again inserted in the buffer. From then on, if the key remains pressed, the scan code will be inserted in the buffer every *auto-repeat delay*.

Figure 3.3   *Auto-repeat.*

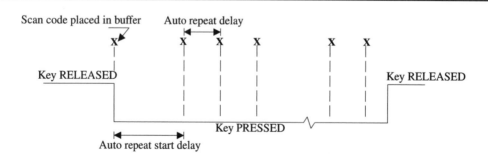

By also telling you how long the key has been pressed, your application can speed up the process of incrementing or decrementing the value of a parameter based on how long the key has been pressed.

To reduce the recurring cost of your system, you can assign multiple functions to each key. To access the alternate function of each key, you can either assign a *prefix key* (like calculators) or provide one or more Shift keys. With a prefix key, you access the alternate function by pressing the prefix key followed by the desired key. To execute another alternate function, you generally have to press the prefix key again. With a Shift key, you access the alternate function by first pressing and holding down the Shift key and then pressing the desired key. In both cases, the keyboard scanning code can keep track of the operation and provide your application with a unique scan code for each type of key pressed. The matrix keyboard module supports the second method and allows you to have up to three Shift keys. Note that you can still use the prefix keys with the keyboard module except that your user interface software will have to keep track of them.

3.02 Matrix Keyboard Module

The source code for the matrix keyboard module is found in the \SOFTWARE\BLOCKS\KEY_MN\SOURCE directory. The source code is found in the files KEY.C and KEY.H. The source code is shown in Listing 3.1 (KEY.C) and Listing 3.2 (KEY.H). As a convention, all functions and variables related to the keyboard module start with Key while all #define constants start with KEY_.

The code allows you to scan a keyboard having any number of rows and columns up to an 8x8 matrix. Rows are driven by an output port (up to 8 bits). The module assumes that rows are populated starting with bit 0 on the output port. Columns are fed to an input port (up to 8 bits). As with the rows, columns must be populated starting with bit 0. You must sacrifice column inputs if your application requires Shift keys. The module can accommodate up to three Shift keys. Shift keys must be populated starting with bit 7 of the input port. In other words, your first Shift key should be placed on bit 7 of the input port, the next one, on bit 6, and the third on bit 5.

The module in Listing 3.1 and 3.2 has been configured and tested assuming the keyboard layout shown in Figure 3.4: a 4-row by 6-column keyboard matrix with two Shift keys. Each key in the matrix has a scan code associated with it (see Figure 3.4). When no Shift key is pressed, the scan code for a key is between 0 and 23 (incl.). When the SHIFT1 key is pressed, the scan code for each is the number shown in Figure 3.4 plus 24. Similarly, if the SHIFT2 key is pressed, 48 is added to the scan codes in Figure 3.4. (See Table 3.1).

3

Figure 3.4   Keyboard matrix.

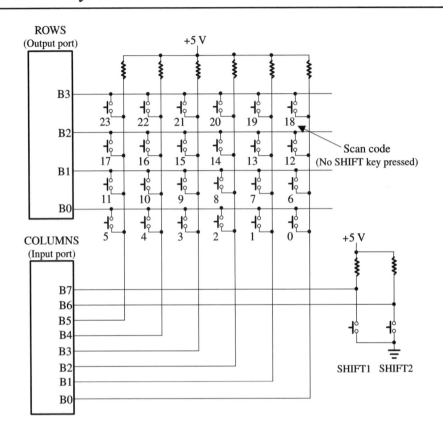

Table 3.1   Scan codes for keyboard shown in Figure 3.4.

Scan code	Shift key(s) pressed
0..23	None
24..47	Shift1
48..71	Shift2
72..95	Shift1 and Shift2

3.03 Internals

Figure 3.5 shows a flow diagram of the matrix keyboard module. To use this module, all you need to do is to adapt three hardware interface functions to your environment and change the value of 17 #define constants. As shown in Figure 3.5, the code assumes the presence of a real-time kernel. The keyboard scanning module only makes use of two kernel services: semaphores and time delays. You should refer to Listing 3.1 and 3.2 for the following description. A single task, KeyScanTask(), is responsible for scanning the keyboard. `KeyScanTask()` is created when your application calls `KeyInit()`. Once created, `KeyScanTask()` executes every `KEY_SCAN_TASK_DLY` milliseconds. `KEY_SCAN_TASK_DLY` should be set to produce a scan rate between 10 and 30 Hz (rate in Hertz is `1000 / KEY_SCAN_TASK_DLY`).

Figure 3.5 Matrix keyboard driver flow diagram.

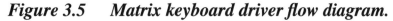

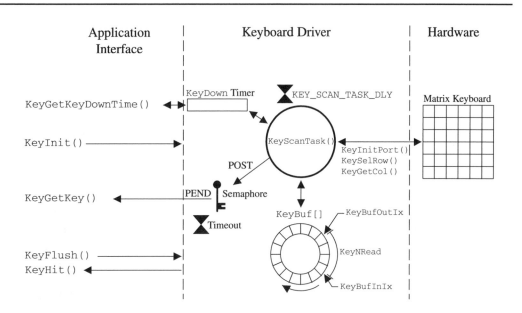

The simplest method I have found to scan a keyboard and implement all the features described previously is to build a simple state machine as shown in Figure 3.6. The state machine is executed every debounce period. Only one of the four states is executed every `KEY_SCAN_TASK_DLY` milliseconds.

Figure 3.6   Matrix keyboard driver state machine.

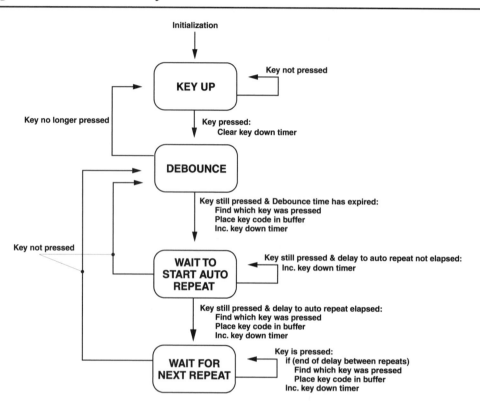

Initially, the state machine is in the KEY_STATE_UP state. When a key is pressed, the state of the state machine changes to KEY_STATE_DEBOUNCE, which will execute KEY_SCAN_TASK_DLY milliseconds later. Notice that the operating system's (i.e., µC/OS-II) function OSTimeDlyHMSM() provides a convenient way to debounce and scan the keyboard at a regular interval.

After the delay, KeyScanTask() executes the code in the KEY_STATE_DEBOUNCE state, which again checks to see if the key is pressed. The state machine returns to the KEY_STATE_UP state if the key is released. If the key is still pressed, however, the scan code is found by calling KeyDecode() and inserted in the circular buffer through KeyBufIn(). KeyBufIn() discards the scan code if the buffer is already full. KeyBufIn() also signals the keyboard semaphore, allowing your application to obtain the scan code of the key through KeyGetKey(). The state machine is then changed to the KEY_STATE_RPT_START_DLY state.

The auto-repeat function will engage if the key is pressed for more than KEY_RPT_START_DLY scan times. In this case, the scan code is inserted in the buffer and the state is changed to the KEY_STATE_RPT_DLY state. If the key is no longer pressed, the state of the state machine is changed to the KEY_STATE_DEBOUNCE state to debounce the released key.

After a scan period, KeyScanTask() executes the code in the KEY_STATE_RPT_DLY state, where the scan code for a pressed key will be inserted into the buffer every KEY_RPT_DLY scan times. As with the other states, debouncing will be required if the key is released.

3.04 Interface Functions

Figure 3.7 shows a block diagram of the matrix keyboard module. Your application knows about the keyboard module only through five functions: KeyFlush(), KeyGetKey(), KeyGetKeyDownTime(), KeyHit(), and KeyInit().

Figure 3.7   Matrix keyboard driver block diagram.

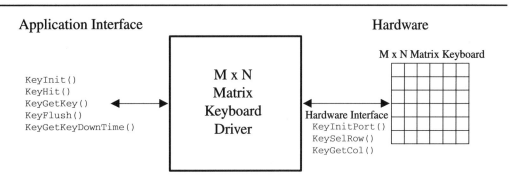

3

KeyFlush()

`void KeyFlush(void);`

The matrix keyboard module buffers user keystrokes until they are consumed by your application. In some instances, it may be useful to flush the buffer and start with fresh user input. In other words, you may want to throw away previously accumulated keystrokes and start with an empty keyboard buffer. You can accomplish this by calling KeyFlush().

Arguments

None

Return Value

None

Notes/Warnings

None

Example

```
void Task (void *pdata)
{
    .
    .
    .
    for (;;) {
        .
        KeyFlush();                    /* Clear the keyboard buffer    */
        .
        .
    }
}
```

KeyGetKey()

INT8U KeyGetKey(INT16U to);

KeyGetKey() is called by your application to obtain a scan code from the keyboard module. If a key has not been pressed, the calling task will be suspended until the user presses a key or until a user-specified timeout expires; the timeout is passed as an argument to KeyGetKey(). If a timeout occurs, KeyGetKey() returns 0xFF.

3

Arguments

to is a user specified time out specified in 'clock ticks'. To wait for ever for a key press, specify a timeout of 0.

Return Value

The scan code corresponding to the key pressed or 0xFF if the specified timeout period expires. The scan code returned by KeyGetKey() depends on whether or not any of the Shift keys are pressed, as shown in .

Notes/Warnings

This function will suspend the calling task until a key is pressed.

Example

```
void Task (void *pdata)
{
    INT8U scancode;



    .
    .

    for (;;) {
        scancode = KeyGetKey(10);       /* Wait for key to be pressed */
                                        /* ... up to 10 ticks         */

        .
        .

    }
}
```

KeyGetKeyDownTime()

INT16U KeyGetKeyDownTime(void);

KeyGetKeyDownTime() returns the amount of time (in milliseconds) that a key has been pressed. This function is useful to speed up the process of incrementing or decrementing the value of a parameter based on how long a key has been pressed.

The key down time is not cleared when the pressed key is released. Instead, the key down time is reset only when the next key is pressed. In other words, you can always obtain the amount of time that the last key was pressed.

Arguments

None

Return Value

The amount of time that the current key is being pressed.

Notes/Warnings

The first edition of this book returned the time the key was pressed in number of clock ticks instead of milliseconds. You will thus have to change your code if you used the previous version of this function.

Example

```c
void Task (void *pdata)
{
    INT16U time;
    .
    .
    for (;;) {
        .
        time = KeyGetKeyDownTime();      /* See how long last key was pressed    */
        .
        .
    }
}
```

KeyHit()

BOOLEAN KeyHit(void);

KeyHit() allows your application to determine if a key has been pressed. Unlike KeyGetKey(), KeyHit() does not suspend the caller. KeyHit() immediately returns TRUE if a key was pressed and FALSE otherwise.

3

Arguments

None

Return Value

TRUE is a key is available from the keyboard buffer.
FALSE if no key has been pressed.

Notes/Warnings

None

Example

```
void Task (void *pdata)
{
    INT8U scancode;




    .

    .

    for (;;) {
        if (KeyHit()) {             /* See if a key has been pressed */
            scancode = KeyGetKey(0); /* Yes, get scan code            */
        }
        .

        .

    }
}
```

KeyInit()

void KeyInit(void);

KeyInit() is the initialization code for the module and it must be called before you invoke any of the other functions. KeyInit() is responsible for initializing internal variables used by the module, initializing the hardware ports, and creating a task that will be responsible for scanning the keyboard.

Arguments

None

Return Value

None

Notes/Warnings

None

Example

```
void main (void)
{
    .
    .
    KeyInit();        /* Initialize the keyboard handler */
    .
    .
}
```

3.05 Configuration

Configuration of the matrix keyboard module code involves changing the value of 17 #defines and adapting three hardware-interface functions to your environment. The #defines are found in KEY.H (section: User Defined Constants) and are also found in CFG.H. The #defines are fully described in KEY.H. You should typically assign a low task priority to keyboard scanning.

WARNING:

In the previous edition of this book, you needed to specify KEY_SCAN_TASK_DLY in number of ticks between execution of KeyScanTask(). Because µC/OS-II provides a more convenient function (i.e., OSTimeDlyHMSM()) to specify the task execution period in hours, minutes, seconds, and milliseconds — KEY_SCAN_TASK_DLY now specifies the scan period in milliseconds instead of ticks.

WARNING

In the previous edition of this book, KEY_SCAN_TASK_STK_SIZE specified the size of the stack for KeyScanTask() in number of bytes. µC/OS-II assumes the stack is specified in stack width elements.

To make this module as portable as possible, access to hardware ports has been isolated into three functions: KeyInitPort(), KeySelRow(), and KeyGetCol(). The matrix keyboard module can be adapted to just about any environment as long as you write these functions as described.

KeyInitPort() is responsible for initializing the I/O ports used for the rows and columns. I tested the code using an Intel 82C55A PPI (Programmable Peripheral Interface). KeyInitPort() is called by KeyInit().

KeySelRow() is used to select rows. KeySelRow() expects a single argument that can either be KEY_ALL_ROWS (to force all rows low) or a number between 0 and 7 (to force a specific row low).

KeyGetCol() reads and returns the complement of the columns input port (a 1 indicates a key pressed).

3.06 How to Use the Matrix Keyboard Module

Let's suppose that your application needs a keyboard, as shown in Figure 3.8. This keyboard should look somewhat familiar except for the four function keys: F1 to F4.

Before you can use any of the keyboard module's services, you must call KeyInit():

```
void main(void)
{
    .
    OSInit();                 /* Initialize the O.S. (mC/OS-II)    */
    .
    .
    KeyInit();                /* Initialize the keyboard module    */
    .
    .
    OSStart();                /* Start multitasking (mC/OS-II)     */
}
```

Figure 3.8   Using the keyboard module.

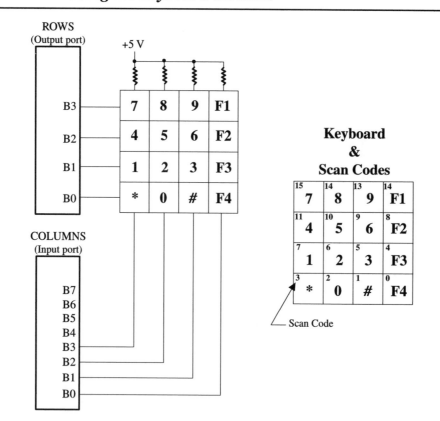

Once multitasking has started, the keyboard will be scanned at the rate defined by KEY_SCAN_TASK_DLY. At this point, your application task (typically some type of user interface) will call one of the four keyboard module services: KeyGetKey(), KeyHit(), KeyFlush(), or KeyGetKeyDownTime().

In the following code, the user interface task calls KeyGetKey() by specifying a timeout of 0. In this case, the user interface will be suspended until a key is pressed. When a key is pressed, KeyGetKey() returns the scan code of the key pressed. For example, if you pressed the 8 key, the scan code returned by KeyGetKey() would be 14 (see Figure 3.8).

```
void UserIFTask (void *data)
{
    INT8U key;


    data = data;
    for (;;) {
        key = KeyGetKey(0);              /* Wait for user input (no timeout) */
        switch (key) {

            .

            .

            .

        }
    }
}
```

You can map scan codes to anything you want by defining a lookup table:

```
char UserKeyMapTbl[] = {
    'A',                    /* F4 key           */
    '#',                    /* #  key           */
    '0',                    /* 0  key           */
    '*',                    /* *  key           */
    'B',                    /* F3 key           */
    '3',                    /* 3  key           */
    '2',                    /* 2  key           */
    '1',                    /* 1  key           */
    'C',                    /* F2 key           */
    '6',                    /* 6  key           */
    '5',                    /* 5  key           */
    '4',                    /* 4  key           */
    'D',                    /* F1 key           */
    '9',                    /* 9  key           */
    '8',                    /* 8  key           */
    '7'                     /* 7  key           */
};
```

The user interface code would now look as shown following this paragraph. With UserKeyMapTbl[], the 8 key would now be returned to your application as ASCII 8 or, '8', the # would be returned as ASCII '#', etc.

```
void UserIFTask (void *data)
{
    INT8U code;
    char  key;

    data = data;
    for (;;) {
        code = KeyGetKey(0);            /* Wait for user input    */
        key  = UserKeyMapTbl[code];     /* Get ASCII value of key */
        switch (key) {
            .
            .
            .
        }
    }
}
```

One of the disadvantages of the user interface code shown previously is that the user interface code is suspended until a key is pressed. If your user interface also needs to display run-time information, you can run the user interface code at a regular rate and poll the keyboard module:

```
void UserIFTask (void *data)
{
    INT8U code;
    char  key;

    data = data;
    for (;;) {
        OSTimeDlyHMSM(???);                  /* Delay user I/F        */
        if (KeyHit()) {                      /* See if key was pressed */
            code = KeyGetKey(0);             /* Get user input        */
            key  = UserKeyMapTbl[code];      /* Convert to ASCII key  */
            switch (key) {
                .
                .
                .
            }
        }
        /* User interface display functions */
    }
}
```

3.07 Bibliography

Dybowski, John
"Negotiating a Keyboard Interface"
The Computer Applications Journal, October/November 1992, p.88–93

Lipovski, G. J.
Single- and Multiple-Chip Microcomputer Interfacing
Englewood Cliffs, NJ
Prentice Hall

Texas Instruments
TMS7000 Keyboard Interface (SPNA003)
Houston, TX
Texas Instruments, 1985

Zaks, Rodnay
Microprocessors, from Chips to Systems
Berkeley, CA
Sybex

3

Listing 3.1   KEY.C

```
/*
********************************************************************************************
*                           Embedded Systems Building Blocks
*                           Complete and Ready-to-Use Modules in C
*
*                           Matrix Keyboard Driver
*
*                     (c) Copyright 1999, Jean J. Labrosse, Weston, FL
*                                    All Rights Reserved
*
* Filename   : KEY.C
* Programmer : Jean J. Labrosse
********************************************************************************************
*                                      DESCRIPTION
*
*     The keyboard is assumed to be a matrix having 4 rows by 6 columns.  However, this code works for any
* matrix arrangements up to an 8 x 8 matrix.  By using from one to three of the column inputs, the driver
* can support "SHIFT" keys.  These keys are: SHIFT1, SHIFT2 and SHIFT3.
*
*     Your application software must declare (see KEY.H):
*
*     KEY_BUF_SIZE              Size of the KEYBOARD buffer
*
*     KEY_MAX_ROWS             The maximum number of rows    on the keyboard
*     KEY_MAX_COLS             The maximum number of columns on the keyboard
*
*     KEY_RPT_DLY              Number of scan times before auto repeat executes the function again
*     KEY_RPT_START_DLY        Number of scan times before auto repeat function engages
*
*     KEY_SCAN_TASK_DLY        The number of milliseconds between keyboard scans
*     KEY_SCAN_TASK_PRIO       Sets the priority of the keyboard scanning task
*     KEY_SCAN_TASK_STK_SIZE   The size of the keyboard scanning task stack
*
*     KEY_SHIFT1_MSK           The mask which determines which column input handles the SHIFT1 key
*                                   (A 0x00 indicates that a SHIFT1 key is not present)
*     KEY_SHIFT1_OFFSET        The scan code offset to add when the SHIFT1 key is pressed
*
*     KEY_SHIFT2_MSK           The mask which determines which column input handles the SHIFT2 key
*                                   (A 0x00 indicates that an SHIFT2 key is not present)
*     KEY_SHIFT2_OFFSET        The scan code offset to add when the SHIFT2 key is pressed
*
*     KEY_SHIFT3_MSK           The mask which determines which column input handles the SHIFT3 key
*                                   (A 0x00 indicates that a SHIFT3 key is not present)
*     KEY_SHIFT3_OFFSET        The scan code offset to add when the SHIFT3 key is pressed
*
*
*     KEY_PORT_ROW             The port address of the keyboard matrix ROWs
*     KEY_PORT_COL             The port address of the keyboard matrix COLUMNs
*     KEY_PORT_CW              The port address of the keyboard I/O ports control word
*
*     KeyInitPort, KeySelRow() and KeyGetCol() are the only three hardware specific functions.  This has
*     been done to localize the interface to the hardware in only these two functions and thus make is
*     easier to adapt to your application.
********************************************************************************************
*/

/*$PAGE*/
```

Listing 3.1 (continued)   KEY.C

```
/*
********************************************************************************************
*                                   INCLUDE FILES
********************************************************************************************
*/

#include "includes.h"

/*
********************************************************************************************
*                                  LOCAL CONSTANTS
********************************************************************************************
*/

#define KEY_STATE_UP                 1      /* Key scanning states used in KeyScan()              */
#define KEY_STATE_DEBOUNCE           2
#define KEY_STATE_RPT_START_DLY      3
#define KEY_STATE_RPT_DLY            4

/*
********************************************************************************************
*                                  GLOBAL VARIABLES
********************************************************************************************
*/

static  INT8U     KeyBuf[KEY_BUF_SIZE];      /* Keyboard buffer                                    */
static  INT8U     KeyBufInIx;                /* Index into key buf where next scan code will be inserted*/
static  INT8U     KeyBufOutIx;               /* Index into key buf where next scan code will be removed */
static  INT16U    KeyDownTmr;                /* Counts how long key has been pressed               */
static  INT8U     KeyNRead;                  /* Number of keys read from the keyboard              */

static  INT8U     KeyRptStartDlyCtr;         /* Number of scan times before auto repeat is started */
static  INT8U     KeyRptDlyCtr;              /* Number of scan times before auto repeat executes again */

static  INT8U     KeyScanState;              /* Current state of key scanning function             */

static  OS_STK    KeyScanTaskStk[KEY_SCAN_TASK_STK_SIZE];  /* Keyboard scanning task stack         */

static  OS_EVENT *KeySemPtr;                                /* Pointer to keyboard semaphore        */

/*
********************************************************************************************
*                              LOCAL FUNCTION PROTOTYPES
********************************************************************************************
*/

static  void      KeyBufIn(INT8U code);      /* Insert scan code into keyboard buffer              */
static  INT8U     KeyDecode(void);           /* Get scan code from current key pressed             */
static  BOOLEAN   KeyIsKeyDown(void);        /* See if key has been pressed                        */
static  void      KeyScanTask(void *data);   /* Keyboard scanning task                             */

/*$PAGE*/
```

Listing 3.1 (continued)   KEY.C

```
/*
*********************************************************************************************
*                            INSERT KEY CHARACTER INTO KEYBOARD BUFFER
*
* Description : This function inserts a key character into the keyboard buffer
* Arguments   : code    is the keyboard scan code to insert into the buffer
* Returns     : none
*********************************************************************************************
*/

static  void  KeyBufIn (INT8U code)
{
    OS_ENTER_CRITICAL();                     /* Start of critical section of code, disable ints   */
    if (KeyNRead < KEY_BUF_SIZE) {           /* Make sure that we don't overflow the buffer       */
        KeyNRead++;                          /* Increment the number of keys read                 */
        KeyBuf[KeyBufInIx++] = code;         /* Store the scan code into the buffer               */
        if (KeyBufInIx >= KEY_BUF_SIZE) {    /* Adjust index to the next scan code to put in buffer*/
            KeyBufInIx = 0;
        }
        OS_EXIT_CRITICAL();                  /* End of critical section of code                   */
        OSSemPost(KeySemPtr);                /* Signal sem if scan code inserted in the buffer    */
    } else {                                 /* Buffer is full, key scan code is lost             */
        OS_EXIT_CRITICAL();                  /* End of critical section of code                   */
    }
}

/*$PAGE*/
```

Listing 3.1 (continued) **KEY.C**

```
/*
*********************************************************************************************
*                                    DECODE KEYBOARD
*
* Description : This function is called to determine the key scan code of the key pressed.
* Arguments   : none
* Returns     : the key scan code
*********************************************************************************************
*/

static  INT8U  KeyDecode (void)
{
    INT8U   col;
    INT8U   row;
    INT8U   offset;
    BOOLEAN done;
    INT8U   col_id;
    INT8U   msk;


    done = FALSE;
    row  = 0;
    while (row < KEY_MAX_ROWS && !done) {       /* Find out in which row key was pressed    */
        KeySelRow(row);                          /* Select a row                             */
        if (KeyIsKeyDown()) {                    /* See if key is pressed in this row        */
            done = TRUE;                         /* We are done finding the row              */
        } else {
            row++;                               /* Select next row                          */
        }
    }
    col    = KeyGetCol();                         /* Read columns                             */
    offset = 0;                                   /* No SHIFT1, SHIFT2 or SHIFT3 key pressed  */
    if (col & KEY_SHIFT1_MSK) {                   /* See if SHIFT1 key was also pressed       */
        offset += KEY_SHIFT1_OFFSET;
    }
    if (col & KEY_SHIFT2_MSK) {                   /* See if SHIFT2 key was also pressed       */
        offset += KEY_SHIFT2_OFFSET;
    }
    if (col & KEY_SHIFT3_MSK) {                   /* See if SHIFT3 key was also pressed       */
        offset += KEY_SHIFT3_OFFSET;
    }
    msk    = 0x01;                                /* Set bit mask to scan for the column      */
    col_id =    0;                               /* Set column value (0..7)                  */
    done   = FALSE;
    while (col_id < KEY_MAX_COLS && !done) {      /* Go through all columns                    */
        if (col & msk) {                          /* See if key was pressed in this columns   */
            done  = TRUE;                         /* Done, i has column value of the key (0..7) */
        } else {
            col_id++;
            msk <<= 1;
        }
    }
    return (row * KEY_MAX_COLS + offset + col_id);  /* Return scan code                       */
}

/*$PAGE*/
```

Listing 3.1 (continued)   KEY.C

```
/*
*********************************************************************************************
*                                    FLUSH KEYBOARD BUFFER
*
* Description : This function clears the keyboard buffer
* Arguments   : none
* Returns     : none
*********************************************************************************************
*/

void  KeyFlush (void)
{
    while (KeyHit()) {                       /* While there are keys in the buffer...      */
        KeyGetKey(0);                        /* ... extract the next key from the buffer   */
    }
}



/*
*********************************************************************************************
*                                         GET KEY
*
* Description : Get a keyboard scan code from the keyboard driver.
* Arguments   : 'to'    is the amount of time KeyGetKey() will wait (in number of ticks) for a key to be
*                       pressed.  A timeout of '0' means that the caller is willing to wait forever for
*                       a key to be pressed.
* Returns     : != 0xFF  is the key scan code of the key pressed
*               == 0xFF  indicates that there is no key in the buffer within the specified timeout
*********************************************************************************************
*/

INT8U  KeyGetKey (INT16U to)
{
    INT8U code;
    INT8U err;


    OSSemPend(KeySemPtr, to, &err);          /* Wait for a key to be pressed               */
    OS_ENTER_CRITICAL();                     /* Start of critical section of code, disable ints  */
    if (KeyNRead > 0) {                      /* See if we have keys in the buffer          */
        KeyNRead--;                          /* Decrement the number of keys read          */
        code = KeyBuf[KeyBufOutIx];          /* Get scan code from the buffer              */
        KeyBufOutIx++;
        if (KeyBufOutIx >= KEY_BUF_SIZE) {   /* Adjust index into the keyboard buffer      */
            KeyBufOutIx = 0;
        }
        OS_EXIT_CRITICAL();                  /* End of critical section of code            */
        return (code);                       /* Return the scan code of the key pressed    */
    } else {
        OS_EXIT_CRITICAL();                  /* End of critical section of code            */
        return (0xFF);                       /* No scan codes in the buffer, return -1     */
    }
}

/*$PAGE*/
```

Listing 3.1 (continued) **KEY.C**

```
/*
***************************************************************************************************
*                                GET HOW LONG KEY HAS BEEN PRESSED
*
* Description : This function returns the amount of time the key has been pressed.
* Arguments   : none
* Returns     : key down time in 'milliseconds'
***************************************************************************************************
*/

INT32U  KeyGetKeyDownTime (void)
{
    INT16U tmr;


    OS_ENTER_CRITICAL();
    tmr = KeyDownTmr;
    OS_EXIT_CRITICAL();
    return (tmr * KEY_SCAN_TASK_DLY);
}

/*$PAGE*/
/*
***************************************************************************************************
*                                SEE IF ANY KEY IN BUFFER
*
* Description : This function checks to see if a key was pressed
* Arguments   : none
* Returns     : TRUE    if a key has been pressed
*               FALSE   if no key pressed
***************************************************************************************************
*/

BOOLEAN  KeyHit (void)
{
    BOOLEAN hit;


    OS_ENTER_CRITICAL();
    hit = (BOOLEAN)(KeyNRead > 0) ? TRUE : FALSE;
    OS_EXIT_CRITICAL();
    return (hit);
}
```

3

Listing 3.1 (continued)   KEY.C

```
/*
*********************************************************************************************
*                                    KEYBOARD INITIALIZATION
*
* Description: Keyboard initialization function.  KeyInit() must be called before calling any other of
*             the user accessible functions.
* Arguments  : none
* Returns    : none
*********************************************************************************************
*/

void  KeyInit (void)
{
    KeySelRow(KEY_ALL_ROWS);                     /* Select all row                                  */
    KeyScanState = KEY_STATE_UP;                 /* Keyboard should not have a key pressed          */
    KeyNRead     = 0;                            /* Clear the number of keys read                   */
    KeyDownTmr   = 0;
    KeyBufInIx   = 0;                            /* Key codes inserted at  the beginning of the buffer */
    KeyBufOutIx  = 0;                            /* Key codes removed from the beginning of the buffer */
    KeySemPtr    = OSSemCreate(0);               /* Initialize the keyboard semaphore               */
    KeyInitPort();                               /* Initialize I/O ports used in keyboard driver    */
    OSTaskCreate(KeyScanTask, (void *)0, &KeyScanTaskStk[KEY_SCAN_TASK_STK_SIZE], KEY_SCAN_TASK_PRIO);
}

/*$PAGE*/
/*
*********************************************************************************************
*                                    SEE IF KEY PRESSED
*
* Description : This function checks to see if a key is pressed
* Arguments   : none
* Returns     : TRUE   if a key is     pressed
*               FALSE  if a key is not pressed
* Note        : (1 << KEY_MAX_COLS) - 1   is used as a mask to isolate the column inputs (i.e. mask off
*                                         the SHIFT keys).
*********************************************************************************************
*/

static  BOOLEAN  KeyIsKeyDown (void)
{
    if (KeyGetCol() & ((1 << KEY_MAX_COLS) - 1)) {    /* Key not pressed if 0                       */
        OS_ENTER_CRITICAL();
        KeyDownTmr++;                                 /* Update key down counter                    */
        OS_EXIT_CRITICAL();
        return (TRUE);
    } else {
        return (FALSE);
    }
}

/*$PAGE*/
```

Listing 3.1 (continued) KEY.C

```
/*
*********************************************************************************************
*                                    KEYBOARD SCANNING TASK
*
* Description : This function contains the body of the keyboard scanning task.  The task should be
*               assigned a low priority.  The scanning period is determined by KEY_SCAN_TASK_DLY.
* Arguments   : 'data'   is a pointer to data passed to task when task is created (NOT USED).
* Returns     : KeyScanTask() never returns.
* Notes       : - An auto repeat of the key pressed will be executed after the key has been pressed for
*                 more than KEY_RPT_START_DLY scan times.  Once the auto repeat has started, the key will
*                 be repeated every KEY_RPT_DLY scan times as long as the key is pressed.  For example,
*                 if the scanning of the keyboard occurs every 50 mS and KEY_RPT_START_DLY is set to 40
*                 and KEY_RPT_DLY is set to 2, then the auto repeat function will engage after 2 seconds
*                 and will repeat every 100 mS (10 times per second).
*********************************************************************************************
*/

/*$PAGE*/

static  void  KeyScanTask (void *data)
{
    INT8U code;
```

Listing 3.1 *(continued)*   `KEY.C`

```c
    data = data;                                        /* Avoid compiler warning (uC/OS-II req.)   */
    for (;;) {
        OSTimeDlyHMSM(0, 0, 0, KEY_SCAN_TASK_DLY);      /* Delay between keyboard scans              */
        switch (KeyScanState) {
            case KEY_STATE_UP:                          /* See if need to look for a key pressed     */
                if (KeyIsKeyDown()) {                   /* See if key is pressed                      */
                    KeyScanState = KEY_STATE_DEBOUNCE;  /* Next call we will have debounced the key   */
                    KeyDownTmr   = 0;                    /* Reset key down timer                      */
                }
                break;

            case KEY_STATE_DEBOUNCE:                    /* Key pressed, get scan code and buffer     */
                if (KeyIsKeyDown()) {                   /* See if key is pressed                      */
                    code            = KeyDecode();      /* Determine the key scan code                */
                    KeyBufIn(code);                     /* Input scan code in buffer                  */
                    KeyRptStartDlyCtr = KEY_RPT_START_DLY;/* Start delay to auto-repeat function      */
                    KeyScanState      = KEY_STATE_RPT_START_DLY;
                } else {
                    KeySelRow(KEY_ALL_ROWS);            /* Select all row                            */
                    KeyScanState      = KEY_STATE_UP;   /* Key was not pressed after all!             */
                }
                break;

            case KEY_STATE_RPT_START_DLY:
                if (KeyIsKeyDown()) {                   /* See if key is still pressed                */
                    if (KeyRptStartDlyCtr > 0) {        /* See if we need to delay before auto rpt    */
                        KeyRptStartDlyCtr--;            /* Yes, decrement counter to start of rpt     */
                        if (KeyRptStartDlyCtr == 0) {   /* If delay to auto repeat is completed ...   */
                            code            = KeyDecode();/* Determine the key scan code              */
                            KeyBufIn(code);             /* Input scan code in buffer                  */
                            KeyRptDlyCtr = KEY_RPT_DLY; /* Load delay before next repeat              */
                            KeyScanState = KEY_STATE_RPT_DLY;
                        }
                    }
                } else {
                    KeyScanState = KEY_STATE_DEBOUNCE;  /* Key was not pressed after all              */
                }
                break;

            case KEY_STATE_RPT_DLY:
                if (KeyIsKeyDown()) {                   /* See if key is still pressed                */
                    if (KeyRptDlyCtr > 0) {             /* See if we need to wait before repeat key   */
                        KeyRptDlyCtr--;                 /* Yes, dec. wait time to next key repeat     */
                        if (KeyRptDlyCtr == 0) {        /* See if it's time to repeat key             */
                            code            = KeyDecode();/* Determine the key scan code              */
                            KeyBufIn(code);             /* Input scan code in buffer                  */
                            KeyRptDlyCtr = KEY_RPT_DLY; /* Reload delay counter before auto repeat    */
                        }
                    }
                } else {
                    KeyScanState = KEY_STATE_DEBOUNCE;  /* Key was not pressed after all              */
                }
                break;
        }
    }
}

/*$PAGE*/
```

Listing 3.1 (continued)   KEY.C

```
/*
**********************************************************************************************
*                                    READ COLUMNS
*
* Description : This function is called to read the column port.
* Arguments   : none
* Returns     : the complement of the column port thus, ones are keys pressed
**********************************************************************************************
*/

#ifndef CFG_C
INT8U  KeyGetCol (void)
{
    return (~inp(KEY_PORT_COL));                   /* Complement columns (ones indicate key is pressed)  */
}
#endif




/*
**********************************************************************************************
*                                  INITIALIZE I/O PORTS
**********************************************************************************************
*/

#ifndef CFG_C
void  KeyInitPort (void)
{
    outp(KEY_PORT_CW, 0x82);                       /* Initialize 82C55: A=OUT, B=IN (COLS), C=OUT (ROWS) */
}
#endif




/*
**********************************************************************************************
*                                    SELECT A ROW
*
* Description : This function is called to select a row on the keyboard.
* Arguments   : 'row'  is the row number (0..7) or KEY_ALL_ROWS
* Returns     : none
* Note        : The row is selected by writing a LOW.
**********************************************************************************************
*/

#ifndef CFG_C
void  KeySelRow (INT8U row)
{
    if (row == KEY_ALL_ROWS) {
        outp(KEY_PORT_ROW, 0x00);                  /* Force all rows LOW                                 */
    } else {
        outp(KEY_PORT_ROW, ~(1 << row));           /* Force desired row LOW                              */
    }
}
#endif
```

Listing 3.2   *KEY.H*

```
/*
*********************************************************************************************
*                             Embedded Systems Building Blocks
*                             Complete and Ready-to-Use Modules in C
*
*                             Matrix Keyboard Driver
*
*                       (c) Copyright 1999, Jean J. Labrosse, Weston, FL
*                                    All Rights Reserved
*
* Filename   : KEY.H
* Programmer : Jean J. Labrosse
*********************************************************************************************
*                             USER DEFINED CONSTANTS
*
* Note: These #defines would normally reside in your application specific code.
*********************************************************************************************
*/

#ifndef  CFG_H
#define  KEY_BUF_SIZE            10        /* Size of the KEYBOARD buffer                        */

#define  KEY_PORT_ROW            0x0312    /* The port address of the keyboard matrix ROWs       */
#define  KEY_PORT_COL            0x0311    /* The port address of the keyboard matrix COLUMNs    */
#define  KEY_PORT_CW             0x0313    /* The port address of the I/O ports control word     */

#define  KEY_MAX_ROWS            4         /* The maximum number of rows   on the keyboard       */
#define  KEY_MAX_COLS            6         /* The maximum number of columns on the keyboard      */

#define  KEY_RPT_DLY             2         /* Number of scan times before auto repeat executes again */
#define  KEY_RPT_START_DLY       10        /* Number of scan times before auto repeat function engages*/

#define  KEY_SCAN_TASK_DLY       50        /* Number of milliseconds between keyboard scans      */
#define  KEY_SCAN_TASK_PRIO      50        /* Set priority of keyboard scan task                 */
#define  KEY_SCAN_TASK_STK_SIZE  1024      /* Size of keyboard scan task stack                   */

#define  KEY_SHIFT1_MSK          0x80      /* The SHIFT1 key is on bit B7 of the column input port  */
                                           /*     (A 0x00 indicates that a SHIFT1 key is not present) */
#define  KEY_SHIFT1_OFFSET       24        /* The scan code offset to add when SHIFT1 is pressed  */

#define  KEY_SHIFT2_MSK          0x40      /* The SHIFT2 key is on bit B6 of the column input port  */
                                           /*     (A 0x00 indicates that an SHIFT2 key is not present)*/
#define  KEY_SHIFT2_OFFSET       48        /* The scan code offset to add when SHIFT2 is pressed  */

#define  KEY_SHIFT3_MSK          0x00      /* The SHIFT3 key is on bit B5 of the column input port  */
                                           /*     (A 0x00 indicates that a SHIFT3 key is not present) */
#define  KEY_SHIFT3_OFFSET       0         /* The scan code offset to add when SHIFT3 is pressed  */
#endif


#define  KEY_ALL_ROWS            0xFF      /* Select all rows (i.e. all rows LOW)                 */
```

Listing 3.2 (continued)   KEY.H

```
/*
********************************************************************************************************
*                                   FUNCTION PROTOTYPES
********************************************************************************************************
*/

void    KeyFlush(void);              /* Flush the keyboard buffer                                    */
INT8U   KeyGetKey(INT16U to);        /* Get a key scan code from driver if one is present, -1 else   */
INT32U  KeyGetKeyDownTime(void);     /* Get how long key has been pressed (in milliseconds)          */
BOOLEAN KeyHit(void);                /* See if a key has been pressed (TRUE if so, FALSE if not)     */
void    KeyInit(void);               /* Initialize the keyboard handler                              */


void    KeyInitPort(void);           /* Initialize I/O ports                                         */
INT8U   KeyGetCol(void);             /* Read COLUMNs                                                 */
void    KeySelRow(INT8U row);        /* Select a ROW                                                 */
```

3

## Chapter 4

Multiplexed LED Displays

A large number of embedded systems offer some form of display device to convey information to the user. The display can consist of anything from a light indicating that power is on, to a complex graphical display showing a representation of the process. Simple control systems can be equipped with complex displays while more complex systems can offer limited information to its user; there are no set rules as to how much information has to be displayed or how it has to be presented. The world of information display is becoming extremely complex, especially when you consider new technologies such as virtual reality.

In this chapter, I will take a very modest position and describe how to interface to LED (Light Emitting Diode) displays. Specifically, I provide you with a module that allows you to control up to 64 multiplexed LEDs. The LEDs can either be seven-segment digits or discrete devices. The module presented allows you to:

- Display limited ASCII characters using seven-segment digits.
- Display numbers.
- Turn ON or OFF individual (discrete) LEDs.

4.00   LED Displays

The *Light Emitting Diode*, or *LED*, is a semiconductor device that produces visible light when a current flows through it as shown in the schematic of Figure 4.1. The intensity of the LED is proportional to the current flowing through the LED. LEDs that produce either RED, YELLOW, GREEN, or BLUE light are now commonly available. The most common color for LEDs is RED, while BLUE LEDs have just been available in the past few years.

Figure 4.1   *Turning ON an LED.*

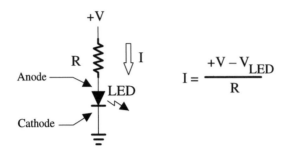

As shown in Figure 4.2, a microprocessor can easily control one or more LEDs by using an output port. LEDs are turned on by writing a 0 to the appropriate bit position of the port. Here, I assume that the port can sink the current required for each LED.

Figure 4.2   *Controlling LEDs with a microprocessor.*

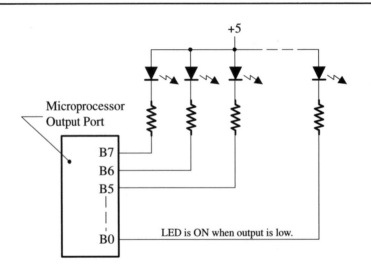

Numbers can be displayed by using what are called *seven-segment LED displays* as shown in Figure 4.3. Two types of seven-segment LED displays are available: common anode and common cathode. Figure 4.2 shows a common anode arrangement, while Figure 4.3 shows a common cathode arrangement.

Figure 4.3   Common cathode seven-segment LED display.

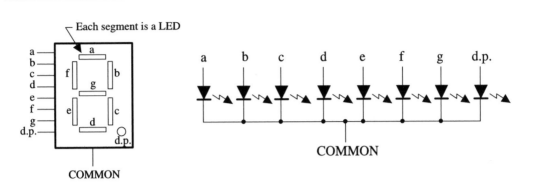

Controlling LEDs using output ports becomes expensive when the number of digits in a display increases. Fortunately, LEDs can be *multiplexed*. Multiplexing simply consists of connecting the LEDs in a matrix as shown in Figure 4.4 and sending the information for each digit in succession. Each digit must be updated very quickly to give the impression that all digits are turned on at the same time. Flickering will occur if the digit update rate is too low. Updating all digits at a rate of about 60 to 100 times per second will produce good results. Multiplexing is not restricted to seven-segment LED displays. The matrix shown in Figure 4.4 also includes discrete LEDs which can be used to display status information. For example, if the display is used in an automobile, the status LEDs can indicate whether the number being displayed represents engine RPM, vehicle speed, odometer reading, trip odometer, etc. Because of the high refresh rate needed to avoid flickering, multiplexing consumes a fair amount of CPU time.

Figure 4.4   Multiplexing LEDs.

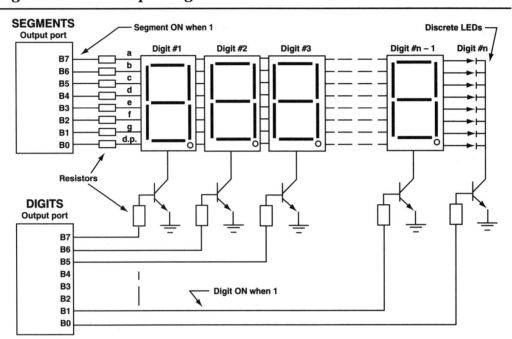

If you need additional seven-segment digits or discrete LEDs, you can add one or more 8-bit ports. The additional port(s) can be used to control more DIGITS or SEGMENTS. Adding DIGIT ports will increase the CPU overhead but will not increase the current consumption of your system. Similarly, you can add SEGMENT ports if you prefer to reduce the overhead on the CPU. In this case, however, you will be increasing the current consumption. The software presented in this chapter can be easily adapted for either situation.

If the LED display matrix needs to be located some appreciable distance from the microprocessor, you might consider using a hardware approach. In this case, a hardware solution might be less expensive, especially if you consider the cost of the connectors and cables needed to bring the control signals to the display. The Maxim 7219 should be considered in this case. The Maxim 7219 is outlined by Jeff Bachiochi in the article, "Seven-Segment LEDs Live ON" (see "Bibliography" on page 148). Using of the Maxim 7219 would eliminate the need for a multiplexing ISR (thus reducing the CPU overhead) but the segment manipulation functions would still be applicable.

4.01 Multiplexed LED Display Module

The source code for the multiplexed LED display module code is found in the \SOFT-WARE\BLOCKS\LED\SOURCE directory. The source code is found in three files: LED.C (Listing 4.1), LED.H (Listing 4.2), and LED_IA.ASM (Listing 4.3). As a convention, all functions and variables related to the display module start with Disp while all #defines constants start with DISP_.

The code allows you to multiplex up to 64 LEDs (using two 8-bit output ports). The LEDs can be either be seven-segment displays, discrete LEDs, or any combination of both. The module can easily be changed if you need to add more seven-segment digits or discrete LEDs.

4.02 Internals

The software provided does not require the presence of a real-time kernel. LED_IA.ASM, however, increments the global variable OSIntNesting and calls OSIntExit(). OSIntNesting is used to notify µC/OS-II that an ISR has started and OSIntExit() is used to noitfy µC/OS-II that the ISR has completed. If you are not planning on using µC/OS-II in your application, you may delete these two lines.

Implementing multiplexing in software is fairly straightforward, as shown in Figure 4.5. Here, I assume you have less than eight digits (including status indicators). You will need a hardware timer that will generate interrupts at a rate of about:

DISP_N_DIG x 60 (Hz)

4

Figure 4.5   LED multiplexing (block diagram).

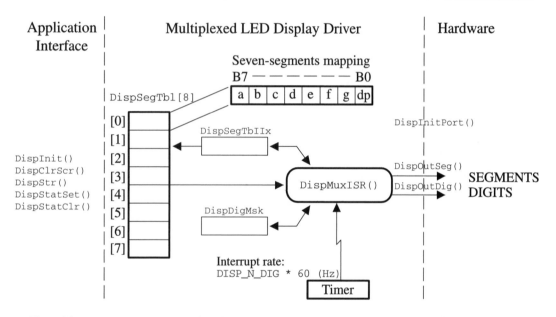

The table DispSegTbl[] contains the segment pattern for the corresponding digit (a one indicates that the segment will be turned on). The first entry in DispSegTbl[] contains the segment patterns for the leftmost digit. DispSegTblIx is an index into the segment table that will point to the next digit to be displayed. DispDigMsk is a mask used to select the next digit to be displayed. Note that only one of the digits can be selected at any given time. The pseudocode for the ISR is:

```
void DispMuxISR (void)
{
    Save CPU registers;
    Clear timer interrupt source;
    Turn OFF the segments of the current digit;
    Select the next digit to display;
    Output the segments pattern for the digit to display;
    Restore CPU registers;
    Return from interrupt;
}
```

You should implement `DispMuxISR()` in assembly language to reduce CPU utilization. I tested a C version of `DispMuxISR()` on an Intel 80386 running at 16 MHz. `DispMuxISR()` was using up 7 percent of the processor's time. Imagine how much time the C version of `DispMuxISR()` would use on an 8-bit CPU!

`DispMuxISR()` turns OFF the segments of the current digit before selecting the next digit. This very important step is taken to prevent what is called *ghosting*. If the segments were not turned OFF before the next digit is selected, the segments of the previous digit would appear briefly on the newly selected digit. `DispMuxISR()` is only concerned with updating the display at the desired refresh rate. How the segment patterns got into `DispSegTbl[]` is the responsibility of task-level code, specifically, the application interface functions.

Conversion of decimal or hexadecimal numbers to seven-segment patterns is very straightforward when using a lookup table, as shown in Figure 4.6. The number to convert is used as an index into `DispHexToSegTbl[]`. Note that a limited number of alphabetical characters can also be displayed using seven-segments. `DispASCIItoSegTbl[]`, shown in Listing 4.1, provides an ASCII to seven-segment conversion table. Note that the table starts with ASCII ' ' (i.e., 0x20) and ends with ASCII 'z' (0x7A). To obtain the seven-segment pattern of an ASCII character, you must index the table after subtracting 0x20 from the desired ASCII character as follows:

```
seg = DispASCIItoSegTbl[c -  0x20];
```

Figure 4.6   *Hexadecimal to seven-segments lookup table.*

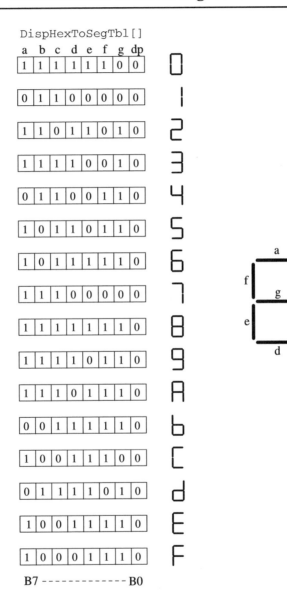

The ASCII to seven-segments table is very useful when you combine it with standard library functions such as `itoa()`, `ltoa()`, `sprintf()`, etc. For example, you can easily display numbers (converted to ASCII with `itoa()`) using the function `DispStr()` as:

```
void DispStr (char *s, INT8U dig)
{
    INT8U stat;


    while (*s && dig < DISP_N_SS) {
        Disable Interrupts;
        stat               = DispSegTbl[dig] & 0x01;
        DispSegTbl[dig++]  = DispASCIItoSegTbl[*s++ -  0x20] | stat;
        Enable Interrupts;
    }
}
```

DispStr() needs to set the seven-segment pattern without changing the state of the status bit (i.e., bit 0) because a DispSegTbl[] entry contains both the pattern for a seven-segment digit and a status bit. This is why I mask off the upper seven bits in order to isolate the state of the status. The bit pattern for the ASCII character is then merged with the status information (ORed). Interrupts are disabled when a DispSegTbl[] entry is changed because DispSegTbl[] is a critical section. DISP_N_SS defines the number of seven-segment digits in the display. Seven-segment display patterns are also assumed to be in DispSegTbl[0] through DispSegTbl[DISP_N_SS - 1].

4.03 Interface Functions

Figure 4.7 shows a block diagram of the multiplexed LED display module. Your application interfaces to the module through five functions: DispInit(), DispClrScr(), DispStr(), DispStatSet(), and DispStatClr().

Figure 4.7   LED multiplexing driver block diagram.

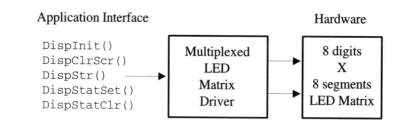

DispClrScr()

void DispClrScr(void);

DispClrScr() is called by your application to clear (i.e., turn off) the display. In other words, DispClrScr() blanks the display.

Arguments

None

Return Value

None

Notes/Warnings

None

Example

```
void Task (void *pdata)
{
    .
    .
    .
    for (;;) {
        .
        DispClrScr();        /* Clear everything on the display */
        .
        .
    }
}
```

DispInit()

void DispInit(void);

DispInit() is the initialization code for the module and must be invoked before any of the other functions. DispInit() is responsible for initializing internal variables used by the module and initializing the hardware ports.

Arguments

None

Return Value

None

Notes/Warnings

None

Example

```
void main (void)
{
    .
    .
    OSInit();
    .
    DispInit();
    .
    .
    OSStart();
}
```

DispStatClr()

void DispStatClr(INT8U dig, INT8U seg);

DispStatClr() is used to turn off a single LED. This function is the complement to DispStatSet(). This function is useful when some of the LEDs are used as status indicators or decimal points for numerical data.

Arguments

dig specifies the digit that will get its segment cleared.

seg specifies the specific segment to set. seg corresponds to the bit position in the digit as follows:

- 0 sets segment dp (bit 0)
- 1 sets segment g (bit 1)
- 2 sets segment f (bit 2)
- 3 sets segment e (bit 3)
- 4 sets segment d (bit 4)
- 5 sets segment c (bit 5)
- 6 sets segment b (bit 6)
- 7 sets segment a (bit 7)

Return Value

None

Notes/Warnings

You can #define status indicators and icons to make your code clearer.

Example

```
void Task (void *pdata)
{
    .
    .
    for (;;) {
        .
        DispStatClr(5, USER_TRIP_ODOMETER_ICON);
        .
        .
    }
}
```

4

DispStatSet()

void DispStatSet(INT8U dig, INT8U seg);

DispStatSet() is used to turn on a single LED. This function is useful when some of the LEDs are used as status indicators or decimal points for numerical data.

Arguments

dig specifies the digit that will get its segment set.

seg specifies the specific segment to set. seg corresponds to the bit position in the digit as follows:

- 0 sets segment dp (bit 0)
- 1 sets segment g (bit 1)
- 2 sets segment f (bit 2)
- 3 sets segment e (bit 3)
- 4 sets segment d (bit 4)
- 5 sets segment c (bit 5)
- 6 sets segment b (bit 6)
- 7 sets segment a (bit 7)

Return Value

None

Notes/Warnings

You can #define status indicators and icons to make your code clearer.

Example

```
void Task (void *pdata)
{
    .
    .
    .
    for (;;) {
        .
        DispStatSet(5, USER_TRIP_ODOMETER_ICON);
        .
        .
    }
}
```

DispStr()

void DispStr(INT8U dig, char *s);

DispStr() is called to display an ASCII string. Not all ASCII characters can be displayed using a seven-segment display. Because of this, you must be careful in the selection of messages to display.

Arguments

dig is the starting position where the ASCII string will be displayed (0 is the first 7-segment digit, 1 is the second digit, etc.).

s is a pointer to the ASCII string. The length of the ASCII string must not exceed the number of seven-segment digits. For example, DispStr(2, "Hello") will display the string as HELLo starting at the third seven-segments digit. Because of the limitation of seven-segments, only the last character would appear in lower case (you should display "HELLO" instead).

Return Value

None

Notes/Warnings

None

Example

```
void Task (void *pdata)
{
    .
    .
    for (;;) {
        .
        DispStr(2, "HELLO");
        .
        .
    }
}
```

4

4.04 Configuration

Configuring the multiplexed LED display module is fairly straightforward.

1. You need to change the value of four #defines. The #defines are found and described in LED.H and are also found in CFG.H.

2. You need to adapt three hardware interface functions to your environment. To make this module as portable as possible, access to hardware ports has been encapsulated into three functions: DispInitPort(), DispOutSeg(), and DispOutDig() (described in the following paragraphs).

3. You will need a hardware timer that will interrupt the CPU at the desired multiplexing rate. The interrupt should vector to DispMuxISR() which is defined in LED_IA.ASM.

DispInitPort() is responsible for initializing the output ports used for the segment and the digit outputs. The code assumed two 8-bit latches such as the 74HC573. Initialization thus consists only of turning off all the segment and digit outputs. I assumed 74HC573s over an 82C55 because of the higher current drive capability of the 74HC573. DispInitPort() is called by DispInit().

DispOutSeg() is used to output the segments while DispOutDig() is used to select the current digit to display. Both functions are called by the multiplexing ISR handler, DispMuxHandler().

To reduce the ISR processing time, the multiplexing ISR code should be written entirely in assembly language and DispOutSeg() and DispOutDig() should be integrated in the ISR. The C code is very inefficient and would not be used in an actual implementation, however, the C code is portable.

4.05 How to Use the Multiplexed LED Display Module

Let's suppose you have a four-digit LED display and four annunciator lights as shown in Figure 4.8.

Figure 4.8   Multiplexing LEDs.

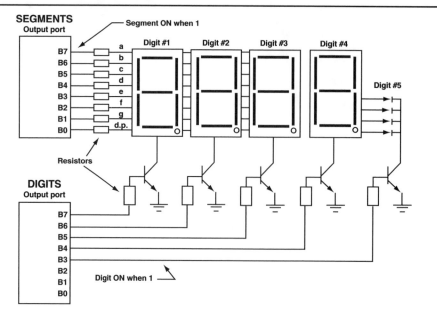

As shown, you must call `DispInit()` before you can use any of the multiplexed LED module's services:

```
void main (void)
{

    .
    .
    DispInit();
    .
    .

}
```

Your application can use the services provided by the multiplexed LED module immediately after `DispInit()`. Display multiplexing will start as soon as you enable interrupts. Your display should be blank because `DispInit()` clears the display buffer `DispSegTbl[]`. You can display the speed as follows:

```
void UserDispSpeed (void)
{
    char s[5];


    DispClrScr();               /* Erase what was being displayed   */
    sprintf(s, "%4d", Speed);   /* Format the speed into ASCII ...   */
    DispStr(0, s);              /* ... and display                   */
    DispStatSet(4, 1);          /* Turn ON Speed indicator           */
}
```

Similarly, you can display the current value of the trip odometer, as shown following this paragraph. Note that the trip odometer is displayed as ###.# and thus, we also need to turn ON the decimal point:

```
void UserDispTripOdometer (void)
{
    char s[5];

                                    /* Note: Display as ###.#          */
    DispClrScr();                   /* Erase what was being displayed */
    sprintf(s, "%4d", TripOdometer); /* Format trip odo. to ASCII ...   */
    DispStr(0, s);                  /* ... and display                 */
    DispStatSet(4, 2);              /* Turn ON trip odo. indicator     */
    DispStatSet(2, 0);              /* Turn ON decimal point           */
}
```

4.06 Bibliography

Artusi, Daniel
"LED display drivers interface to uCs on just three I/O lines"
EDN, November 14, 1985, p259–265

Bachiochi, Jeff
"Seven-Segment LEDs Live On"
The Computer Applications Journal, March 1993, p60–66

Cantrell, Tom
"Smart LEDs: The Hard Way, the Soft Way, and the Right Way"
The Computer Applications Journal, February 1993, p62–67

The Hewlett-Packard Applications Engineering Staff
Optoelectronics Applications Manual
McGraw-Hill Book Company, 1977, ISBN 0-07-028605-1

Listing 4.1   LED.C

```
/*
*********************************************************************************************************
*                                    Embedded Systems Building Blocks
*                                   Complete and Ready-to-Use Modules in C
*
*                                      Multiplexed LED Display Driver
*
*                              (c) Copyright 1999, Jean J. Labrosse, Weston, FL
*                                            All Rights Reserved
*
* Filename   : LED.C
* Programmer : Jean J. Labrosse
*********************************************************************************************************
*                                              DESCRIPTION
*
* This module provides an interface to a multiplexed "8 segments x N digits" LED matrix.
*
* To use this driver:
*
*     1) You must define (LED.H):
*
*        DISP_N_DIG          The total number of digits to display (up to 8)
*        DISP_N_SS           The total number of seven-segment digits in the display (up to 8)
*        DISP_PORT_DIG       The address of the DIGITS   output port
*        DISP_PORT_SEG       The address of the SEGMENTS output port
*
*     2) You must allocate a hardware timer which will interrupt the CPU at a rate of at least:
*
*        DISP_N_DIG * 60   (Hz)
*
*        The timer interrupt must vector to DispMuxISR (defined in LED_IA.ASM).  You MUST write the
*        code to clear the interrupt source.  The interrupt source must be cleared either in DispMuxISR
*        or in DispMuxHandler().
*
*     3) Adapt DispInitPort(), DispOutSeg() and DispOutDig() for your environment.
*********************************************************************************************************
*/

/*$PAGE*/
```

Listing 4.1 *(continued)*   LED.C

```
/*
*********************************************************************************
*                                  INCLUDE FILES
*********************************************************************************
*/

#include "includes.h"

/*
*********************************************************************************
*                                 LOCAL VARIABLES
*********************************************************************************
*/

static   INT8U  DispDigMsk;                /* Bit mask used to point to next digit to display      */
static   INT8U  DispSegTbl[DISP_N_DIG];    /* Segment pattern table for each digit to display      */
static   INT8U  DispSegTblIx;              /* Index into DispSegTbl[] for next digit to display     */

/*$PAGE*/
```

Listing 4.1 (continued)   LED.C

```
/*
*********************************************************************************************
*                          ASCII to SEVEN-SEGMENT conversion table
*                                                                    a
*                                                                  ------
*                                                              f |      | b
*                                                                |  g   |
* Note: The segments are mapped as follows:                      ------
*                                                              e |      | c
*       a    b    c    d    e    f    g                          |  d   |
*      --   --   --   --   --   --   --   --                     ------
*      B7   B6   B5   B4   B3   B2   B1   B0
*********************************************************************************************
*/

const INT8U DispASCIItoSegTbl[] = {     /* ASCII to SEVEN-SEGMENT conversion table              */
    0x00,                               /* ' '                                                  */
    0x00,                               /* '!', No seven-segment conversion for exclamation point */
    0x44,                               /* '"', Double quote                                    */
    0x00,                               /* '#', Pound sign                                      */
    0x00,                               /* '$', No seven-segment conversion for dollar sign     */
    0x00,                               /* '%', No seven-segment conversion for percent sign    */
    0x00,                               /* '&', No seven-segment conversion for ampersand       */
    0x40,                               /* ''', Single quote                                    */
    0x9C,                               /* '(', Same as '['                                     */
    0xF0,                               /* ')', Same as ']'                                     */
    0x00,                               /* '*', No seven-segment conversion for asterix         */
    0x00,                               /* '+', No seven-segment conversion for plus sign       */
    0x00,                               /* ',', No seven-segment conversion for comma           */
    0x02,                               /* '-', Minus sign                                      */
    0x00,                               /* '.', No seven-segment conversion for period          */
    0x00,                               /* '/', No seven-segment conversion for slash           */
    0xFC,                               /* '0'                                                  */
    0x60,                               /* '1'                                                  */
    0xDA,                               /* '2'                                                  */
    0xF2,                               /* '3'                                                  */
    0x66,                               /* '4'                                                  */
    0xB6,                               /* '5'                                                  */
    0xBE,                               /* '6'                                                  */
    0xE0,                               /* '7'                                                  */
    0xFE,                               /* '8'                                                  */
    0xF6,                               /* '9'                                                  */
    0x00,                               /* ':', No seven-segment conversion for colon           */
    0x00,                               /* ';', No seven-segment conversion for semi-colon      */
    0x00,                               /* '<', No seven-segment conversion for less-than sign   */
    0x12,                               /* '=', Equal sign                                      */
    0x00,                               /* '>', No seven-segment conversion for greater-than sign */
    0xCA,                               /* '?', Question mark                                   */
    0x00,                               /* '@', No seven-segment conversion for commercial at-sign */
    0xEE,                               /* 'A'                                                  */
    0x3E,                               /* 'B', Actually displayed as 'b'                       */
    0x9C,                               /* 'C'                                                  */
    0x7A,                               /* 'D', Actually displayed as 'd'                       */
    0x9E,                               /* 'E'                                                  */
    0x8E,                               /* 'F'                                                  */
/*$PAGE*/
```

Listing 4.1 (continued)   LED.C

```
    0xBC,                    /* 'G', Actually displayed as 'g'                           */
    0x6E,                    /* 'H'                                                      */
    0x60,                    /* 'I', Same as '1'                                         */
    0x78,                    /* 'J'                                                      */
    0x00,                    /* 'K', No seven-segment conversion                         */
    0x1C,                    /* 'L'                                                      */
    0x00,                    /* 'M', No seven-segment conversion                         */
    0x2A,                    /* 'N', Actually displayed as 'n'                           */
    0xFC,                    /* 'O', Same as '0'                                         */
    0xCE,                    /* 'P'                                                      */
    0x00,                    /* 'Q', No seven-segment conversion                         */
    0x0A,                    /* 'R', Actually displayed as 'r'                           */
    0xB6,                    /* 'S', Same as '5'                                         */
    0x1E,                    /* 'T', Actually displayed as 't'                           */
    0x7C,                    /* 'U'                                                      */
    0x00,                    /* 'V', No seven-segment conversion                         */
    0x00,                    /* 'W', No seven-segment conversion                         */
    0x00,                    /* 'X', No seven-segment conversion                         */
    0x76,                    /* 'Y'                                                      */
    0x00,                    /* 'Z', No seven-segment conversion                         */
    0x00,                    /* '[',                                                     */
    0x00,                    /* '\', No seven-segment conversion                         */
    0x00,                    /* ']',                                                     */
    0x00,                    /* '^', No seven-segment conversion                         */
    0x00,                    /* '_', Underscore                                          */
    0x00,                    /* '`', No seven-segment conversion for reverse quote       */
    0xFA,                    /* 'a'                                                      */
    0x3E,                    /* 'b'                                                      */
    0x1A,                    /* 'c'                                                      */
    0x7A,                    /* 'd'                                                      */
    0xDE,                    /* 'e'                                                      */
    0x8E,                    /* 'f', Actually displayed as 'F'                           */
    0xBC,                    /* 'g'                                                      */
    0x2E,                    /* 'h'                                                      */
    0x20,                    /* 'i'                                                      */
    0x78,                    /* 'j', Actually displayed as 'J'                           */
    0x00,                    /* 'k', No seven-segment conversion                         */
    0x1C,                    /* 'l', Actually displayed as 'L'                           */
    0x00,                    /* 'm', No seven-segment conversion                         */
    0x2A,                    /* 'n'                                                      */
    0x3A,                    /* 'o'                                                      */
    0xCE,                    /* 'p', Actually displayed as 'P'                           */
    0x00,                    /* 'q', No seven-segment conversion                         */
    0x0A,                    /* 'r'                                                      */
    0xB6,                    /* 's', Actually displayed as 'S'                           */
    0x1E,                    /* 't'                                                      */
    0x38,                    /* 'u'                                                      */
    0x00,                    /* 'v', No seven-segment conversion                         */
    0x00,                    /* 'w', No seven-segment conversion                         */
    0x00,                    /* 'x', No seven-segment conversion                         */
    0x76,                    /* 'y', Actually displayed as 'Y'                           */
    0x00                     /* 'z', No seven-segment conversion                         */
};

/*$PAGE*/
```

Listing 4.1 (continued)   LED.C

```
/*
***********************************************************************************************
*                           HEXADECIMAL to SEVEN-SEGMENT conversion table
*                                                                      a
*                                                                    ------
*                                                                  f |    | b
*                                                                    | g  |
* Note: The segments are mapped as follows:                         ------
*                                                                  e |    | c
*       a    b    c    d    e    f    g                              | d  |
*       --   --   --   --   --   --   --   --                       ------
*       B7   B6   B5   B4   B3   B2   B1   B0
***********************************************************************************************
*/

const INT8U DispHexToSegTbl[] = {      /* HEXADECIMAL to SEVEN-SEGMENT conversion table              */
    0xFC,                              /* '0'                                                         */
    0x60,                              /* '1'                                                         */
    0xDA,                              /* '2'                                                         */
    0xF2,                              /* '3'                                                         */
    0x66,                              /* '4'                                                         */
    0xB6,                              /* '5'                                                         */
    0xBE,                              /* '6'                                                         */
    0xE0,                              /* '7'                                                         */
    0xFE,                              /* '8'                                                         */
    0xF6,                              /* '9'                                                         */
    0xEE,                              /* 'A'                                                         */
    0x3E,                              /* 'B', Actually displayed as 'b'                              */
    0x9C,                              /* 'C'                                                         */
    0x7A,                              /* 'D', Actually displayed as 'd'                              */
    0x9E,                              /* 'E'                                                         */
    0x8E                               /* 'F'                                                         */
};

/*$PAGE*/
```

Listing 4.1 (continued)   LED.C

```
/*
*********************************************************************************************************
*                                        CLEAR THE DISPLAY
*
* Description: This function is called to clear the display.
* Arguments  : none
* Returns    : none
*********************************************************************************************************
*/

void  DispClrScr (void)
{
    INT8U i;


    for (i = 0; i < DISP_N_DIG; i++) {            /* Clear the screen by turning OFF all segments       */
        OS_ENTER_CRITICAL();
        DispSegTbl[i] = 0x00;
        OS_EXIT_CRITICAL();
    }
}

/*$PAGE*/

/*
*********************************************************************************************************
*                                    DISPLAY DRIVER INITIALIZATION
*
* Description : This function initializes the display driver.
* Arguments   : None.
* Returns     : None.
*********************************************************************************************************
*/

void  DispInit (void)
{
    DispInitPort();                     /* Initialize I/O ports used in display driver                 */
    DispDigMsk   = 0x80;
    DispSegTblIx = 0;
    DispClrScr();                       /* Clear the Display                                           */
}

/*$PAGE*/
```

Listing 4.1 (continued)   LED.C

```
/*
*********************************************************************************************
*                              DISPLAY NEXT SEVEN-SEGMENT DIGIT
*
* Description: This function is called by DispMuxISR() to output the segments and select the next digit
*              to be multiplexed.  DispMuxHandler() is called by DispMuxISR() defined in LED_IA.ASM
* Arguments  : none
* Returns    : none
* Notes      : - You MUST supply the code to clear the interrupt source.  Note that with some
*                microprocessors (i.e. Motorola's MC68HC11), you must clear the interrupt source before
*                enabling interrupts.
*********************************************************************************************
*/

void  DispMuxHandler (void)
{
                                              /* Insert code to CLEAR INTERRUPT SOURCE here        */

    DispOutSeg(0x00);                         /* Turn OFF segments while changing digits           */
    DispOutDig(DispDigMsk);                   /* Select next digit to display                      */
    DispOutSeg(DispSegTbl[DispSegTblIx]);     /* Output digit's seven-segment pattern              */
    if (DispSegTblIx == (DISP_N_DIG - 1)) {   /* Adjust index to next seven-segment pattern        */
        DispSegTblIx =    0;                   /* Index into first segments pattern                 */
        DispDigMsk   = 0x80;                   /* 0x80 will select the first seven-segment digit    */
    } else {
        DispSegTblIx++;
        DispDigMsk >>= 1;                     /* Select next digit                                 */
    }
}

/*$PAGE*/
```

Listing 4.1 (continued)   LED.C

```
/*
*********************************************************************************************
*                                    CLEAR STATUS SEGMENT
*
* Description: This function is called to turn OFF a single segment on the display.
* Arguments  : dig   is the position of the digit where the segment appears (0..DISP_N_DIG-1)
*              bit   is the segment bit to turn OFF (0..7)
* Returns    : none
*********************************************************************************************
*/

void  DispStatClr (INT8U dig, INT8U bit)
{
    OS_ENTER_CRITICAL();
    DispSegTbl[dig] &= ~(1 << bit);
    OS_EXIT_CRITICAL();
}


/*
*********************************************************************************************
*                                     SET STATUS SEGMENT
*
* Description: This function is called to turn ON a single segment on the display.
* Arguments  : dig   is the position of the digit where the segment appears (0..DISP_N_DIG-1)
*              bit   is the segment bit to turn ON (0..7)
* Returns    : none
*********************************************************************************************
*/

void  DispStatSet (INT8U dig, INT8U bit)
{
    OS_ENTER_CRITICAL();
    DispSegTbl[dig] |= 1 << bit;
    OS_EXIT_CRITICAL();
}

/*$PAGE*/
```

Listing 4.1 (continued)   LED.C

```
/*
*********************************************************************************************************
*                           DISPLAY ASCII STRING ON SEVEN-SEGMENT DISPLAY
*
* Description: This function is called to display an ASCII string on the seven-segment display.
* Arguments  : dig   is the position of the first digit where the string will appear:
*                         0 for the first  seven-segment digit.
*                         1 for the second seven-segment digit.
*                         .   .    .     .     .      .      .
*                         .   .    .     .     .      .      .
*                         DISP_N_SS - 1 is the last seven-segment digit.
*              s      is the ASCII string to display
* Returns    : none
* Notes      : - Not all ASCII characters can be displayed on a seven-segment display.  Consult the
*                ASCII to seven-segment conversion table DispASCIItoSegTbl[].
*********************************************************************************************************
*/

void  DispStr (INT8U dig, char *s)
{
    INT8U stat;


    while (*s && dig < DISP_N_SS) {
        OS_ENTER_CRITICAL();
        stat            = DispSegTbl[dig] & 0x01;                   /* Save state of B0 (i.e. status) */
        DispSegTbl[dig++] = DispASCIItoSegTbl[*s++ - 0x20] | stat;
        OS_EXIT_CRITICAL();
    }
}

/*$PAGE*/
```

4

Listing 4.1 *(continued)*   LED.C

```c
#ifndef CFG_C
/*
*********************************************************************************************
*                                  I/O PORTS INITIALIZATION
*
* Description: This is called by DispInit() to initialize the output ports used in the LED multiplexing.
* Arguments  : none
* Returns    : none
* Notes      : 74HC573  8 bit latches are used for both the segments and digits outputs.
*********************************************************************************************
*/

void  DispInitPort (void)
{
    outp(DISP_PORT_SEG, 0x00);             /* Turn OFF segments                            */
    outp(DISP_PORT_DIG, 0x00);             /* Turn OFF digits                              */
}


/*
*********************************************************************************************
*                                     DIGIT output
*
* Description: This function outputs the digit selector.
* Arguments  : msk    is the mask used to select the current digit.
* Returns    : none
*********************************************************************************************
*/

void  DispOutDig (INT8U msk)
{
    outp(DISP_PORT_DIG, msk);
}


/*
*********************************************************************************************
*                                    SEGMENTS output
*
* Description: This function outputs seven-segment patterns.
* Arguments  : seg     is the seven-segment pattern to output
* Returns    : none
*********************************************************************************************
*/

void  DispOutSeg (INT8U seg)
{
    outp(DISP_PORT_SEG, seg);
}
#endif
```

Listing 4.2   LED.H

```
/*
********************************************************************************************************
*                                    Embedded Systems Building Blocks
*                                    Complete and Ready-to-Use Modules in C
*
*                                    Multiplexed LED Display Driver
*
*                              (c) Copyright 1999, Jean J. Labrosse, Weston, FL
*                                         All Rights Reserved
*
* Filename   : LED.H
* Programmer : Jean J. Labrosse
********************************************************************************************************
*/

/*
********************************************************************************************************
*                                             CONSTANTS
********************************************************************************************************
*/

#ifndef  CFG_H
#define  DISP_PORT_DIG    0x0301            /* Port address of DIGITS   output                      */
#define  DISP_PORT_SEG    0x0300            /* Port address of SEGMENTS output                      */

#define  DISP_N_DIG          8             /* Total number of digits (including status indicators) */
#define  DISP_N_SS           7             /* Total number of seven-segment digits                 */
#endif

/*
********************************************************************************************************
*                                        FUNCTION PROTOTYPES
********************************************************************************************************
*/

void  DispClrScr(void);
void  DispInit(void);
void  DispMuxHandler(void);
void  DispMuxISR(void);
void  DispStr(INT8U dig, char *s);
void  DispStatClr(INT8U dig, INT8U bit);
void  DispStatSet(INT8U dig, INT8U bit);

/*
********************************************************************************************************
*                                        FUNCTION PROTOTYPES
*                                          HARDWARE SPECIFIC
********************************************************************************************************
*/

void  DispInitPort(void);
void  DispOutDig(INT8U msk);
void  DispOutSeg(INT8U seg);
```

Listing 4.3   LED_IA.ASM

```
;********************************************************************************
;                          Embedded Systems Building Blocks
;                          Complete and Ready-to-Use Modules in C
;
;                              Multiplexed LED Display Driver
;                                   LED Multiplex ISR
;                               Intel 80x86 (LARGE MODEL)
;
;                          (c) Copyright 1999, Jean J. Labrosse, Weston, FL
;                                   All Rights Reserved
;
; File : LED_IA.ASM
; By   : Jean J. Labrosse
;********************************************************************************

          PUBLIC  _DispMuxISR

          EXTRN   _DispMuxHandler:FAR
          EXTRN   _OSIntExit:FAR
          EXTRN   _OSIntNesting:BYTE

.MODEL    LARGE
.CODE
.186


;********************************************************************************
;                     OUTPUT NEXT SEGMENTS PATTERN TO LED DISPLAY MATRIX
;                                   void DispMuxISR(void)
;********************************************************************************

_DispMuxISR PROC FAR
;
          PUSHA                             ; Save processor's context
          PUSH    ES
          PUSH    DS
;
          INC     BYTE PTR _OSIntNesting    ; Notify uC/OS-II of ISR
          CALL    FAR  PTR _DispMuxHandler  ; Call C routine to handle multiplexing
          CALL    FAR  PTR _OSIntExit       ; Exit through uC/OS-II scheduler
;
          POP     DS                        ; Restore processor's context
          POP     ES
          POPA
;
          IRET                              ; Return to interrupted code
;
_DispMuxISR ENDP

          END
```

## Chapter 5

Character LCD Modules

In this chapter, I provide you with a software module that will allow you to interface with character LCD (Liquid Crystal Display) modules. This software package works with just about any character module based on the Hitachi HD44780 Dot Matrix LCD Controller & Driver. The module allows you to:

- Control LCD modules containing up to 80 characters.
- Display ASCII characters.
- Display ASCII strings.
- Define up to eight symbols based on a 5x7 dot matrix.
- Display bargraphs.

5.00 Liquid Crystal Displays

Liquid Crystal Displays (LCDs) are a passive display technology. This means that LCDs do not emit light but instead manipulate ambient light. By manipulating this light, LCDs can display images using very little power. This characteristic has made LCDs the preferred technology whenever low power consumption is critical. An LCD is basically a reflective part. It needs ambient light to reflect back to a user's eyes. In applications where ambient light is low or nonexistent, a light source can be placed behind the LCD. This is known as *backlighting*.

Backlighting can be accomplished by either using *electroluminescent* (EL) or LED light sources. EL backlights are very thin and lightweight and produce a very even light source. EL backlights for LCDs are available in a variety of colors with white being the most popular. EL backlights consume very little power but require high voltages (80 to 100 Vac). EL backlights also have a limited life of about 2,000 to 3,000 hours. LEDs are used for backlighting and are primarily used for character modules. LEDs offer a much longer life (at least 50,000 hours) and are brighter than ELs. Unfortunately, LEDs consume more power than ELs. LEDs are typically mounted in an array directly behind the display. LEDs come in a variety of colors but yellow-green LEDs are the most common.

Controlling LCDs is a little bit trickier than controlling LEDs. LCDs are almost always controlled with dedicated hardware. Figure 5.1 shows the three types of LCDs currently available:

1. *Custom* displays with individual segment controls (similar to LED displays). LCDs lend themselves very well to custom displays, as shown in Figure 5.1. You can design a display with just about any type of annunciation. Where software is concerned, these types of displays are similar to LED displays because each segment is controlled individually.

2. *Alphanumeric* or *character* displays. These types of displays are currently available in *modules*. A module contains the LCD and the drive electronics. Character displays are composed of one to four lines of 16 to 40 character blocks. Each character block consists of a 5x8 dot matrix that is used to display any ASCII character and a limited number of symbols.

3. *Full graphics* displays. As with character displays, full graphics displays are available in modules. Graphic modules offer the greatest flexibility in formatting data on the display. They allow for text, graphics, pictures, or any combinations of these. Because character size is defined by software, graphic modules allow any language or character font. Limitations are driven by the resolution. Graphic modules are organized in rows (horizontal) and columns (vertical) of pixels. Each pixel is addressed individually, which allows any pixel to be ON or OFF. Graphics displays are available in a wide variety of configurations from 64x32 to 640x480 pixels (columns x rows). From a software point of view, interfacing with graphics displays is at least an order of magnitude more complex than interfacing with the other two types of displays. I will not be covering graphics displays in this book.

Figure 5.1   Types of LCDs.

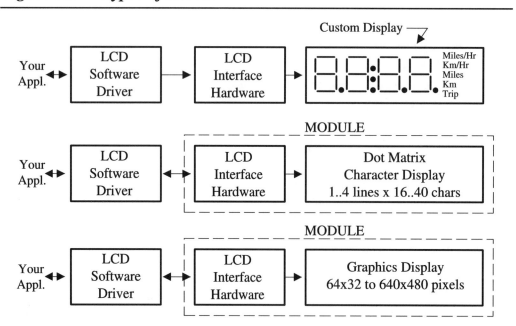

5.01 Character LCD Modules

A character module contains the LCD and the drive electronics. Character displays are composed of one to four lines each having between 16 and 40 character blocks. Each character block consists of a 5x8 dot matrix which is used to display any ASCII character and a limited number of symbols. In this chapter, I will be providing a software interface module for character display modules. Character modules are finding their way into a large number of embedded systems such as:

- air conditioners
- audio amplifiers
- FAX machines
- laser printers
- medical equipment
- security systems
- telephones

Because of their popularity, character modules are available from an increasing number of manufacturers, including:

- Densitron Corporation
- Optrex Corp.
- Seiko Instruments
- Stanley Electric

Character modules generally have at least one thing in common: they pretty much all use the Hitachi HD44780 LCD module controller. A subset of the Hitachi HD44780 data sheet can be found on the CD-ROM, 44780.pdf. The HD44780 can interface directly with any 4- or 8-bit data bus, draws very little current (less than 1 mA), is fully ASCII-compatible, can display up to 80 characters, and contains eight user-programmable 5x8 symbols. The good news is that, where software is concerned, once a display module is written, it can be used with just about any module based on the HD44780.

The hardware interface of an LCD module is quite straightforward. LCD modules can generally interface directly with most microprocessor buses either as an I/O device or a memory mapped I/O. The HD44780 has a 500 nS (nano-second) access time. Connecting the LCD module on the microprocessor bus is economical but becomes problematic if the display is located some distance from the microprocessor bus. In this case, parallel I/O ports can be used to interface with the LCD module, as shown in Figure 5.2. Here, I used an Intel 82C55 Programmable Peripheral Interface (PPI) controller. As shown in Figure 5.2, only 11 parallel output lines are required to interface to the LCD module. Eight of the lines are used for data transfer while the other three are used as control lines for the LCD module.

Figure 5.2   Interfacing to an LCD module.

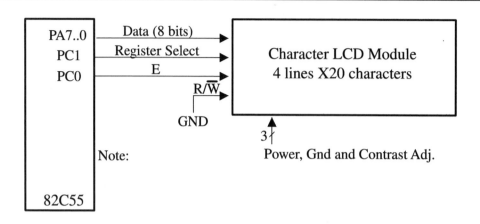

The HD44780 takes a certain amount of time to process commands or data sent to it. The Hitachi data sheet provides you with the maximum amount of time required for each type of data transfer. Because of this, the software can simply send a command or data and wait at least the amount of time specified before sending the next command or data. Note that the HD44780 itself allows the microprocessor to read a BUSY status. The BUSY status can be read by the microprocessor to determine if the HD44780 is ready to accept another command or more data. If you can, you should make use of the BUSY capability of the HD44780 because this provides you with a true indication that the HD44780 is ready to accept another command or more data. As a precaution, however, you should still provide a timeout loop to prevent hanging up the microprocessor in case of a malfunction with the interface electronics. Unless the LCD module is directly connected to the microprocessor bus, implementing read capability with parallel I/O ports is more costly. Note that the 82C55 does have a bidirectional mode but is more complex to use. This is why the circuit shown is implemented with output ports only instead of a bidirectional data port and three control lines (i.e., RS, E, and R/W).

The interface circuit is simplified by choosing to have the CPU wait between commands and data. It turns out that this scheme also makes the software easier to write. Waiting is done using a software loop. You might be thinking that software loops should be avoided because they are not accurate. Well, in this case, accuracy is not required. All you need to do is wait at least the amount of time specified by Hitachi before sending the next command or data. A software loop also doesn't affect responsiveness to asynchronous events since interrupts are enabled while in the loop. (Besides, how else would you wait just 40 μS with a low end processor?)

With the hardware interface shown, the LCD module appears as two write-only registers (note that the R/W line is always low). The first write register is called the *data register* (when RS is high) while the other write register is called the *instruction register* (when RS is low). The software presented in this chapter calls the instruction register the control register. Characters to display are written to the data register. The *control register* allows the software to control the operating mode of the module: clear the display, set the position of the cursor, turn the display ON or OFF, etc.

5.02   Character LCD Module, Internals

The source code for the LCD module is found in the \SOFTWARE\BLOCKS\LCD\SOURCE directory. The source code is found in files LCD.C (Listing 5.1) and LCD.H (Listing 5.2). As a convention, all functions and variables related to the display module start with Disp while all #defines constants start with DISP_.

The code allows you to interface to just about any LCD module based on the Hitachi HD44780 LCD module controller. At first view, you might think that writing a software module for an LCD module is a trivial task. This is not quite the case because the HD44780 has its quirks. The HD44780 was originally designed for a 40 characters by 2 lines display (40x2) and thus has internal memory to hold 80 characters. The first 40 characters are stored at memory locations[1] 0x80 through 0xA7 (128 to 167) while the next 40 characters are stored at memory locations 0xC0 through 0xE7 (192 to 231)! Tables 5.1 through 5.4 show the memory mapping for different LCD module configurations. The addresses are shown in decimal and are actually based at 0x80. That is, address 00 actually corresponds to 0x80, address 64 is actually 0xC0 (i.e., 0x80 + 64), etc.

Table 5.1 shows the memory organization for 16-character displays. Notice how the 16 characters by 1 line module appears as a two-line display. This is done by the LCD module manufacturers to reduce the cost of their product by fully using the drive capability of the HD44780.

Table 5.1   16-character LCD modules.

16 Characters x 1 lines															
00	01	02	03	04	05	06	07	64	65	66	67	68	69	70	71
16 Characters x 2 lines															
00	01	02	03	04	05	06	07	08	09	10	11	12	13	14	15
64	65	66	67	68	69	70	71	72	73	74	75	76	77	78	79
16 Characters x 4 lines															
00	01	02	03	04	05	06	07	08	09	10	11	12	13	14	15
64	65	66	67	68	69	70	71	72	73	74	75	76	77	78	79
16	17	18	19	20	21	22	23	24	25	26	27	28	29	30	31
80	81	82	83	84	85	86	87	88	89	90	91	92	93	94	95

Table 5.2 shows the memory organization for 20-character displays. Again, the single-line display appears as a two-line module.

Table 5.2   20-character LCD modules.

20 Characters x 1 lines																			
00	01	02	03	04	05	06	07	08	09	64	65	66	67	68	69	70	71	72	73
20 Characters x 2 lines																			
00	01	02	03	04	05	06	07	08	09	10	11	12	13	14	15	16	17	18	19
64	65	66	67	68	69	70	71	72	73	74	75	76	77	78	79	80	81	82	83

1. Memory locations inside the HD44780 chip.

Table 5.2   20-character LCD modules.

20 Characters x 4 lines																			
00	01	02	03	04	05	06	07	08	09	10	11	12	13	14	15	16	17	18	19
64	65	66	67	68	69	70	71	72	73	74	75	76	77	78	79	80	81	82	83
20	21	22	23	24	25	26	27	28	29	30	31	32	33	34	35	36	37	38	39
84	85	86	87	88	89	90	91	92	93	94	95	96	97	98	99	100	101	102	103

Table 5.3 shows the memory organization for 24-character displays. As with the 16- and 20-character displays, the single-line display appears as a two-line module.

Table 5.3   24-character LCD modules.

24-Characters x 1 line																							
00	01	02	03	04	05	06	07	08	09	10	11	64	65	66	67	68	69	70	71	72	73	74	75
24-Characters x 2 lines																							
00	01	02	03	04	05	06	07	08	09	10	11	12	13	14	15	16	17	18	19	20	21	22	23
64	65	66	67	68	69	70	71	72	73	74	75	76	77	78	79	80	81	82	83	84	85	86	87

Table 5.4 shows the memory organization for 40-character displays. As with the other module configurations, the single-line display appears as a two-line module. Note that each line of a 40-character display is shown broken down into two separate lines; the second line is offset from the first. This has been done to avoid reducing the character font in order to fit within the width of the page.

Table 5.4   40-character LCD modules.

40 Characters x 1 line																							
00	01	02	03	04	05	06	07	08	09	10	11	12	13	14	15	16	17	18	19				
				64	65	66	67	68	69	70	71	72	73	74	75	76	77	78	79	80	81	82	83
40 Characters x 2 lines																							
00	01	02	03	04	05	06	07	08	09	10	11	12	13	14	15	16	17	18	19				
				20	21	22	23	24	25	26	27	28	29	30	31	32	33	34	35	36	37	38	39
64	65	66	67	68	69	70	71	72	73	74	75	76	77	78	79	80	81	82	83				
				84	85	86	87	88	89	90	91	92	93	94	95	96	97	98	99	100	101	102	103

The software module presented in this book will support any LCD module that is organized as shown in Tables 5.1 through 5.4. The software was actually tested with an Optrex DMC20434. Table 5.5 shows a list of available LCD module configurations and their manufacturer's part numbers.

Table 5.5   LCD module configurations available.

#Lines	#Characters	Densitron P/N	Optrex P/N	Seiko P/N	Stanley P/N	FEMA P/N
1	16	LM4020	DMC16117A	M1641	GMD1610	MDL1611
2	16	LM4222	DMC16207	M1632	GMD1620	MDL1621
4	16	LM4443	DMC16433	M1614	GMD1640	-
1	20	LM432	-	-	-	-
2	20	LM4261	DMC20215	L2012	GMD2020	MDL2021
4	20	LM4821	DMC20434	L2014	GMD2040	MDL2041
1	24	LM413	DMC24138	-	-	MDL2411
2	24	LM4227	DMC24227	L2432	GMD2420	MDL2421
1	40	LM414	-	L4041	-	MDL4011
2	40	LM4218	DMC40218	L4042	GMD4020	MDL4021

5

5.03   Interface Functions

Figure 5.3 shows a block diagram of the LCD module. Your application knows about the display only through the interface functions provided.

Figure 5.3   LCD module driver block diagram.

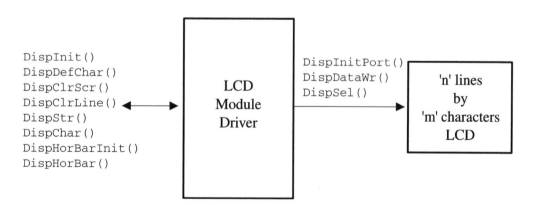

The module assumes the presence of a real-time kernel because it requires a semaphore and time delay services. The display module makes use of a binary semaphore to prevent multiple tasks from accessing the display at the same time. Use of the semaphore is encapsulated in the code, and thus, your application doesn't have to worry about it.

DispChar()

void DispChar(INT8U row, INT8U col, char c);

DispChar() allows you to display a single character anywhere on the display.

Arguments

row and **col** will specify the coordinates (row, col) where the character will appear. rows (i.e., lines) are numbered from 0 to DispMaxRows − 1, and columns are numbered from 0 to DispMaxCols − 1.

c is the character to display. The Hitachi HD44780 allows you to specify up to eight characters or symbols numbered from 0 to 7 (i.e., its identification). You display a user-defined character or symbol by calling DispChar(), the row/column position, and the character or symbol's identification number.

Return Value

None

Notes/Warnings

None

Example

```
void Task (void *pdata)
{
    .
    .
    .
    for (;;) {
        .
        DispChar(1, 3, '$');/* Display '$' on second row, 4th character */
        .
        .
    }
}
```

DispClrLine()

void DispClrLine(INT8U line);

DispClrLine() allows your application to clear one of the LCD module's lines. The line is basically filled with the ASCII character ' ' (i.e., 0x20).

Arguments

line is the line (i.e., row) to clear. Note that lines are numbered from 0 to DispMaxRows − 1.

Return Value

None

Notes/Warnings

None

Example

```
void Task (void *pdata)
{
   .
   .
   for (;;) {
      .
      DispClrLine(0);        /* Clear the first line of the display */
      .
      .
   }
}
```

5

DispClrScr()

void DispClrScr(void);

DispClrScr() allows you to clear the screen. The cursor is positioned on the top leftmost character. The screen is basically filled with the ASCII character ' ' (i.e., 0x20).

Arguments

None

Return Value

None

Notes/Warnings

None

Example

```
void Task (void *pdata)
{
    .
    .
    for (;;) {
        .
        DispClrScr();        /* Clear everything on the display */
        .
        .
    }
}
```

DispDefChar()

void DispDefChar(INT8U id, INT8U *pat);

DispDefChar() allows you to define up to eight custom 5x8 pixel characters or symbols. This is one of the most powerful features of the LCD modules because it allows you to create graphics such as icons, bargraphs, arrows, etc.

Figure 5.4 shows how to define a character or a symbol. The 5x8 pixel matrix is organized as a bit-map table. The first entry of the table corresponds to pixels for the first row, the second entry, the pixels for the second row, etc. A pixel is turned ON when its corresponding bit is set (i.e., 1).

Figure 5.4   Defining characters, or symbols.

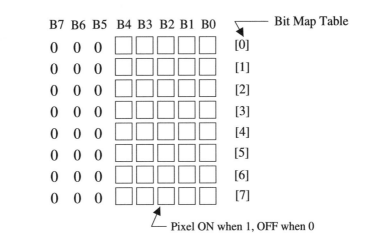

All you need to do to define a new character or symbol is to declare an initialized array of INT8Us containing eight entries and call DispDefChar().

Arguments

id specifies an identification number for the new character or symbol (a number between 0 and 7). The identification number will be used to actually display the new character or symbol.

pat is a pointer to the bitmap table which defines what the character or symbol will look like.

Return Value

None

Notes/Warnings

None

Example

```
const INT8U DispRightArrowChar[] = {
    0x08, 0x0C, 0x0E, 0x1F, 0x1F, 0x0E, 0x0C, 0x08
}


void Task (void *pdata)
{
    .
    .
    for (;;) {
        .
        DispDefChar(0, &DispRightArrowChar[0]); /* Define arrow char. */
        .
        .
    }
}
```

Figure 5.5 shows examples of bitmaps to create arrows and other symbols. Once symbols are created, you can display them by calling DispChar().

Figure 5.5   Symbol examples.

DispHorBar()

void DispHorBar(INT8U row, INT8U col, INT8U val);

You can use the LCD module to create remarkably high quality bargraphs. The linear bargraph is an excellent trend indicator and can greatly enhance operator feedback. Depending on the size of the module, many bargraphs can be simultaneously displayed. The LCD module software allows you to display bargraphs of any size anywhere on the screen.

DispHorBar() is used to display horizontal bars anywhere on the screen.

Figure 5.6 also shows that a 16xN-character display can produce bargraphs with up to 80 bars (16 x 5 bars per character block). In Figure 5.6, I started the bargraph on the first column on a 16xN-character display. Once scaled, bargraphs can represent just about anything. For example, the 38 bars shown in Figure 5.6 can represent 47.5 percent (38 bars = 47.5/100 \ 80), 100.7 degrees if the bargraph is used to represent temperatures from 0 to 212 degrees, etc.

Figure 5.6   Bargraphs with 16-character displays.

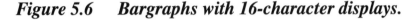

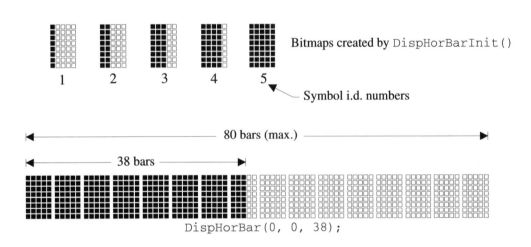

DispHorBar(0, 0, 38);

Arguments

row and **col** will specify the coordinates (row, col) where the first character in the bargraph will appear. rows (i.e., lines) are numbered from 0 to DispMaxRows − 1, and columns are numbered from 0 to DispMaxCols − 1.

val is the number of bars you want to have turned on (a number between 0 to 80 in this example).

Return Value

None

Notes/Warnings

Before you can use `DispHorBar()`, you *must* call `DispHorBarInit()` which defines 5 characters used for bargraphs.

Example

You could actually use fewer bars and display the actual value next to the bargraph, as shown in Figure 5.7. In this example, I am displaying 100.7 degrees (28 bars) on a scale of 0 to 212 degrees (60 bars).

Figure 5.7   Bargraph with value.

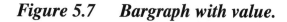

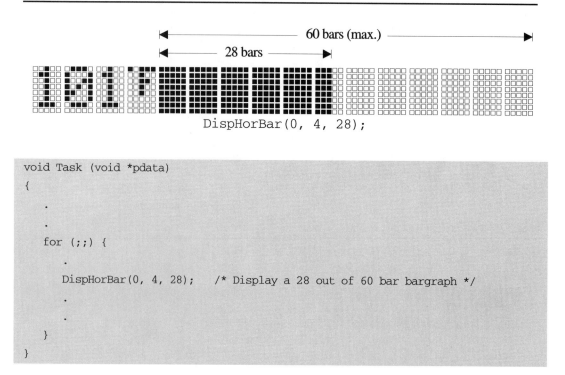

```
DispHorBar(0, 4, 28);
```

```c
void Task (void *pdata)
{
    .
    .
    for (;;) {
        .
        DispHorBar(0, 4, 28);    /* Display a 28 out of 60 bar bargraph */
        .
        .
    }
}
```

DispHorBarInit()

void DispHorBarInit(void);

DispHorBarInit() defines five special symbols with identification numbers 1 through 5 as shown in Figure 5.6. You must be call before you use DispHorBar(). You only need to call DispHorBarInit() once unless you intend to redefine the symbol identifiers dynamically for other purposes.

Arguments

None

Return Value

None

Notes/Warnings

Because DispHorBarInit() defines the five symbols shown in Figure 5.6, you must use other identification numbers (i.e., 0, 6, and 7) for your own symbols.

Example

```
void Task (void *pdata)
{
   .
   .
   .
   DispHorBarInit();           /* Initialize the bargraph capability  */
   for (;;) {
      .
      DispHorBar(0, 4, 28);    /* Display a 28 out of 60 bar bargraph */
      .
      .
      .
   }
}
```

DispInit()

void DispInit(INT8U maxrows, INT8U maxcols);

DispInit() is the initialization code for the module and *must* be invoked before any of the other functions. DispInit() assumes that multitasking has started because it uses services provided by the real-time kernel.

DispInit() initializes the hardware, creates the semaphore, and sets the operating mode of the LCD module.

Arguments

maxrows is the LCD module's maximum number of rows (lines), and **maxcols** is the maximum number of columns (characters per line).

Return Value

None

Notes/Warnings

None

Example

You should call DispInit() from your user interface task as follows:

```
void UserIFTask (void *data)
{
   DispInit(4, 20);          /* Initialize the 4x20 LCD display */
   for (;;) {
       .
       .
       User interface code;
       .
       .
   }
}
```

DispStr()

void DispStr(INT8U row, INT8U col, char *s);

DispStr() allows you to display ASCII strings anywhere on the display. You can easily display either integer or floating-point numbers using the standard library functions itoa(), ltoa(), sprintf(), etc. Of course, you should ensure that these functions are reentrant if you are using them in a multitasking environment.

Arguments

row and **col** will specify the coordinates (row, col) where the first character of the ASCII string will appear. Note that rows (i.e., lines) are numbered from 0 to DispMaxRows − 1. Similarly, columns are numbered from 0 to DispMaxCols − 1. The upper-left corner is coordinate 0, 0.

s is a pointer to the ASCII string. The displayed string will be truncated if the string is longer than the available space on the specified line.

Return Value

None

Notes/Warnings

None

Example

```
void UserIFTask(void *data)
{
    DispInit(4, 20);            /* Initialize the 4x20 LCD display */
    for (;;) {

        .

        .

        DispStr(0, 0, "Hello World");

        .

        .

    }
}
```

5

5.04 LCD Module Display, Configuration

Configuring the LCD display module is quite straightforward.

1. You need to change the value of three #defines. The #defines are found and described in LCD.H and also in CFG.H. DISP_DLY_CNTS is used to adjust delays between sending commands or data to the HD44780. You will need to change this constant so that a delay of at least 40 µS occurs between writes to the HD44780.

2. You need to adapt three hardware interface functions to your environment. To make this module as portable as possible, access to hardware ports has been encapsulated into the following functions: DispInitPort(), DispDataWr(), and DispSel() (described as follows).

    DispInitPort() is responsible for initializing the output ports used to interface with the LCD module. I used an Intel 82C55 PPI to verify the code. DispInitPort() is called by DispInit().

    DispDataWr() is used to write a single byte to the LCD module. Depending on the state of the RS line (see Figure 5.2), the byte will be either sent to the data (RS is 1) or control register (RS is 0).

    Changing the state of the RS line is the responsibility of the function DispSel(). DispSel() is called by the LCD display module with one argument that can either be set to DISP_SEL_CMD_REG or DISP_SEL_DATA_REG.

5.05 LCD Module Manufacturers

Densitron Corporation
2039 HW 11
Camden, SC 29020
(803) 432-5008

Hitachi America, Ltd.
Electron Tube Division
3850 Holcomd Bridge Rd.
Norcross, GA 30092
(404) 409-3000

Optrex Corp.
23399-T Commerce Drive
Suite B-8
Farmington Hills, MI 48335
(313) 471-6220

Seiko Instruments USA, Inc.
Electronic Components Division
2990 West Lomita Blvd.
Torrance, CA 90505
(310) 517-7829

Stanley Electric
2660 Barranca Parkway
Irvine, CA 92714
(714) 222-0777

Listing 5.1   LCD.C

```
/*
*********************************************************************************************
*                          Embedded Systems Building Blocks
*                          Complete and Ready-to-Use Modules in C
*
*                          LCD Display Module Driver
*
*                 (c) Copyright 1999, Jean J. Labrosse, Weston, FL
*                               All Rights Reserved
*
* Filename   : LCD.C
* Programmer : Jean J. Labrosse
*********************************************************************************************
*                                   DESCRIPTION
*
*
* This module provides an interface to an alphanumeric display module.
*
* The current version of this driver supports any  alphanumeric LCD module based on the:
*     Hitachi HD44780 DOT MATRIX LCD controller.
*
* This driver supports LCD displays having the following configuration:
*
*     1 line  x 16 characters     2 lines x 16 characters     4 lines x 16 characters
*     1 line  x 20 characters     2 lines x 20 characters     4 lines x 20 characters
*     1 line  x 24 characters     2 lines x 24 characters
*     1 line  x 40 characters     2 lines x 40 characters
*********************************************************************************************
*/


/*$PAGE*/
```

Listing 5.1 (continued)   LCD.C

```
/*
*********************************************************************************************
*                                        INCLUDE FILES
*********************************************************************************************
*/

#include "includes.h"

/*
*********************************************************************************************
*                                       LOCAL CONSTANTS
*********************************************************************************************
*/

                                    /* --------------------- HD44780 COMMANDS -------------------- */
#define  DISP_CMD_CLS       0x01  /* Clr display : clears display and returns cursor home          */
#define  DISP_CMD_FNCT      0x3B  /* Function Set: Set 8 bit data length, 1/16 duty, 5x8 dots      */
#define  DISP_CMD_MODE      0x06  /* Entry mode  : Inc. display data address when writing          */
#define  DISP_CMD_ON_OFF    0x0C  /* Disp ON/OFF : Display ON, cursor OFF and no BLINK character   */

/*
*********************************************************************************************
*                                       LOCAL VARIABLES
*********************************************************************************************
*/

static   INT8U      DispMaxCols;     /* Maximum number of columns (i.e. characters per line)      */
static   INT8U      DispMaxRows;     /* Maximum number of rows for the display                    */
static   OS_EVENT   *DispSem;        /* Semaphore used to access display functions                */


static   INT8U      DispBar1[] = {0x10, 0x10, 0x10, 0x10, 0x10, 0x10, 0x10, 0x10};
static   INT8U      DispBar2[] = {0x18, 0x18, 0x18, 0x18, 0x18, 0x18, 0x18, 0x18};
static   INT8U      DispBar3[] = {0x1C, 0x1C, 0x1C, 0x1C, 0x1C, 0x1C, 0x1C, 0x1C};
static   INT8U      DispBar4[] = {0x1E, 0x1E, 0x1E, 0x1E, 0x1E, 0x1E, 0x1E, 0x1E};
static   INT8U      DispBar5[] = {0x1F, 0x1F, 0x1F, 0x1F, 0x1F, 0x1F, 0x1F, 0x1F};

/*
*********************************************************************************************
*                                   LOCAL FUNCTION PROTOTYPES
*********************************************************************************************
*/

static   void       DispCursorSet(INT8U row, INT8U col);

/*$PAGE*/
```

Listing 5.1 (continued)   LCD.C

```
/*
*********************************************************************************************
*                                  DISPLAY A CHARACTER
*
* Description : This function is used to display a single character on the display device
* Arguments   : 'row'   is the row    position of the cursor in the LCD Display
*                       'row' can be a value from 0 to 'DispMaxRows - 1'
*               'col'   is the column position of the cursor in the LCD Display
*                       'col' can be a value from 0 to 'DispMaxCols - 1'
*               'c'     is the character to write to the display at the current ROW/COLUMN position.
* Returns     : none
*********************************************************************************************
*/

void  DispChar (INT8U row, INT8U col, char c)
{
    INT8U err;


    if (row < DispMaxRows && col < DispMaxCols) {
        OSSemPend(DispSem, 0, &err);           /* Obtain exclusive access to the display       */
        DispCursorSet(row, col);               /* Position cursor at ROW/COL                   */
        DispSel(DISP_SEL_DATA_REG);
        DispDataWr(c);                         /* Send character to display                    */
        OSSemPost(DispSem);                    /* Release access to display                    */
    }
}


/*
*********************************************************************************************
*                                     CLEAR LINE
*
* Description : This function clears one line on the LCD display and positions the cursor at the
*               beginning of the line.
* Arguments   : 'line'  is the line number to clear and can take the value
*                       0 to 'DispMaxRows - 1'
* Returns     : none
*********************************************************************************************
*/

void  DispClrLine (INT8U line)
{
    INT8U i;
    INT8U err;


    if (line < DispMaxRows) {
        OSSemPend(DispSem, 0, &err);           /* Obtain exclusive access to the display       */
        DispCursorSet(line, 0);                /* Position cursor at begin of the line to clear*/
        DispSel(DISP_SEL_DATA_REG);            /* Select the LCD Display DATA register         */
        for (i = 0; i < DispMaxCols; i++) {    /* Write ' ' into all column positions of that line */
            DispDataWr(' ');                   /* Write an ASCII space at current cursor position   */
        }
        DispCursorSet(line, 0);                /* Position cursor at begin of the line to clear */
        OSSemPost(DispSem);                    /* Release access to display                    */
    }
}

/*$PAGE*/
```

Listing 5.1 (continued)   LCD.C

```
/*
*********************************************************************************************
*                                    CLEAR THE SCREEN
*
* Description : This function clears the display
* Arguments   : none
* Returns     : none
*********************************************************************************************
*/

void  DispClrScr (void)
{
    INT8U err;


    OSSemPend(DispSem, 0, &err);        /* Obtain exclusive access to the display               */
    DispSel(DISP_SEL_CMD_REG);          /* Select the LCD display command register              */
    DispDataWr(DISP_CMD_CLS);           /* Send command to LCD display to clear the display     */
    OSTimeDly(2);                       /* Delay at least  2 mS (2 ticks ensures at least this much)  */
    OSSemPost(DispSem);                 /* Release access to display                            */
}

/*$PAGE*/
```

Listing 5.1 *(continued)*   LCD.C

```
/*
*********************************************************************************************************
*                                          POSITION THE CURSOR (Internal)
*
* Description : This function positions the cursor into the LCD buffer
* Arguments   : 'row'   is the row    position of the cursor in the LCD Display
*                       'row' can be a value from 0 to 'DispMaxRows - 1'
*               'col'   is the column position of the cursor in the LCD Display
*                       'col' can be a value from 0 to 'DispMaxCols - 1'
* Returns     : none
*********************************************************************************************************
*/

static  void  DispCursorSet (INT8U row, INT8U col)
{
    DispSel(DISP_SEL_CMD_REG);                              /* Select LCD display command register     */
    switch (row) {
        case 0:
             if (DispMaxRows == 1) {                        /* Handle special case when only one line  */
                 if (col < (DispMaxCols >> 1)) {
                     DispDataWr(0x80 + col);                /* First  half of the line starts at 0x80  */
                 } else {                                   /* Second half of the line starts at 0xC0  */
                     DispDataWr(0xC0 + col - (DispMaxCols >> 1));
                 }
             } else {
                 DispDataWr(0x80 + col);                    /* Select LCD's display line 1             */
             }
             break;

        case 1:
             DispDataWr(0xC0 + col);                        /* Select LCD's display line 2             */
             break;

        case 2:
             DispDataWr(0x80 + DispMaxCols + col);          /* Select LCD's display line 3             */
             break;

        case 3:
             DispDataWr(0xC0 + DispMaxCols + col);          /* Select LCD's display line 4             */
             break;
    }
}


/*$PAGE*/
```

5

Listing 5.1 *(continued)*   LCD.C

```
/*
*********************************************************************************************
*                                  DEFINE CHARACTER
*
* Description : This function defines the dot pattern for a character.
* Arguments   : 'id'    is the identifier for the desired dot pattern.
*               'pat'   is a pointer to an 8 BYTE array containing the dot pattern.
* Returns     : None.
*********************************************************************************************
*/

void  DispDefChar (INT8U id, INT8U *pat)
{
    INT8U err;
    INT8U i;


    OSSemPend(DispSem, 0, &err);            /* Obtain exclusive access to the display      */
    DispSel(DISP_SEL_CMD_REG);              /* Select command register                     */
    DispDataWr(0x40 + (id << 3));           /* Set address of CG RAM                       */
    DispSel(DISP_SEL_DATA_REG);             /* Select the data register                    */
    for (i = 0; i < 8; i++) {
        DispDataWr(*pat++);                 /* Write pattern into CG RAM                   */
    }
    OSSemPost(DispSem);                     /* Release access to display                   */
}

/*$PAGE*/
```

Listing 5.1 *(continued)* LCD.C

```
/*
*********************************************************************************************
*                                    DUMMY FUNCTION
*
* Description : This function doesn't do anything.  It is used to act like a NOP (i.e. No Operation) to
*               waste a few CPU cycles and thus, act as a short delay.
* Arguments   : none
* Returns     : none
*********************************************************************************************
*/

void  DispDummy (void)
{
}


/*
*********************************************************************************************
*                                 DISPLAY A HORIZONTAL BAR
*
* Description : This function allows you to display horizontal bars (bar graphs) on the LCD module.
* Arguments   : 'row'    is the row    position of the cursor in the LCD Display
*                        'row' can be a value from 0 to 'DispMaxRows - 1'
*                'val'    is the value of the horizontal bar.  This value cannot exceed:
*                             DispMaxCols * 5
* Returns     : none
* Notes       : To use this function, you must first call DispHorBarInit()
*********************************************************************************************
*/

void  DispHorBar (INT8U row, INT8U col, INT8U val)
{
    INT8U i;
    INT8U full;
    INT8U fract;
    INT8U err;


    full  = val / 5;                     /* Find out how many 'full' blocks to turn ON                    */
    fract = val % 5;                     /* Compute portion of block                                      */
    if (row < DispMaxRows && (col + full - 1) < DispMaxCols) {
        OSSemPend(DispSem, 0, &err);     /* Obtain exclusive access to the display                        */
        i = 0;                           /* Set counter to limit column to maximum allowable column */
        DispCursorSet(row, col);         /* Position cursor at beginning of the bar graph                 */
        DispSel(DISP_SEL_DATA_REG);
        while (full > 0) {               /* Write all 'full' blocks                                       */
            DispDataWr(5);               /* Send custom character #5 which is full block                  */
            i++;                         /* Increment limit counter                                       */
            full--;
        }
        if (fract > 0) {
            DispDataWr(fract);           /* Send custom character # 'fract' (i.e. portion of block) */
        }
        OSSemPost(DispSem);              /* Release access to display                                     */
    }
}

/*$PAGE*/
```

Listing 5.1 (continued)   LCD.C

```
/*
*********************************************************************************************
*                                  INITIALIZE HORIZONTAL BAR
*
* Description : This function is used to initialize the bar graph capability of this module.  You must
*              call this function prior to calling DispHorBar().
* Arguments   : none
* Returns     : none
*********************************************************************************************
*/

void  DispHorBarInit (void)
{
    DispDefChar(1, &DispBar1[0]);
    DispDefChar(2, &DispBar2[0]);
    DispDefChar(3, &DispBar3[0]);
    DispDefChar(4, &DispBar4[0]);
    DispDefChar(5, &DispBar5[0]);
}


/*
*********************************************************************************************
*                                  DISPLAY DRIVER INITIALIZATION
*
* Description : This function initializes the display driver.
* Arguments   : maxrows      specifies the number of lines on the display (1 to 4)
*               maxcols      specified the number of characters per line
* Returns     : None.
* Notes       : - DispInit() MUST be called only when multitasking has started.  This is because
*                 DispInit() requires time delay services from the operating system.
*               - DispInit() MUST only be called once during initialization.
*********************************************************************************************
*/

void  DispInit (INT8U maxrows, INT8U maxcols)
{
    DispInitPort();                         /* Initialize I/O ports used in display driver             */
    DispMaxRows = maxrows;
    DispMaxCols = maxcols;
    DispSem     = OSSemCreate(1);           /* Create display access semaphore                         */

                                            /* INITIALIZE THE DISPLAY MODULE                           */
    DispSel(DISP_SEL_CMD_REG);              /* Select command register.                                */
    OSTimeDlyHMSM(0, 0, 0, 50);             /* Delay more than 15 mS after power up (50 mS should be enough)*/
    DispDataWr(DISP_CMD_FNCT);              /* Function Set: Set 8 bit data length, 1/16 duty, 5x8 dots */
    OSTimeDly(2);                           /* Busy flag cannot be checked yet!                        */
    DispDataWr(DISP_CMD_FNCT);              /* The above command is sent four times!                   */
    OSTimeDly(2);                           /*    This is recommended by Hitachi in the HD44780 data sheet */
    DispDataWr(DISP_CMD_FNCT);
    OSTimeDly(2);
    DispDataWr(DISP_CMD_FNCT);
    OSTimeDly(2);

    DispDataWr(DISP_CMD_ON_OFF);            /* Disp ON/OFF: Display ON, cursor OFF and no BLINK character */
    DispDataWr(DISP_CMD_MODE);              /* Entry mode: Inc. display data address when writing      */
    DispDataWr(DISP_CMD_CLS);               /* Send command to LCD display to clear the display        */
    OSTimeDly(2);                           /* Delay at least  2 mS (2 ticks ensures at least this much) */
}

/*$PAGE*/
```

Listing 5.1 (continued)   LCD.C

```
/*
*********************************************************************************************************
*                                         DISPLAY AN ASCII STRING
*
* Description : This function is used to display an ASCII string on a line of the LCD display
* Arguments   : 'row'   is the row    position of the cursor in the LCD Display
*                       'row' can be a value from 0 to 'DispMaxRows - 1'
*               'col'   is the column position of the cursor in the LCD Display
*                       'col' can be a value from 0 to 'DispMaxCols - 1'
*               's'     is a pointer to the string to write to the display at
*                       the desired row/col.
* Returns     : none
*********************************************************************************************************
*/

void  DispStr (INT8U row, INT8U col, char *s)
{
    INT8U i;
    INT8U err;


    if (row < DispMaxRows && col < DispMaxCols) {
        OSSemPend(DispSem, 0, &err);         /* Obtain exclusive access to the display              */
        DispCursorSet(row, col);             /* Position cursor at ROW/COL                          */
        DispSel(DISP_SEL_DATA_REG);
        i = col;                             /* Set counter to limit column to maximum allowable column */
        while (i < DispMaxCols && *s) {      /* Write all chars within str + limit to DispMaxCols   */
            DispDataWr(*s++);                /* Send character to LCD display                       */
            i++;                             /* Increment limit counter                            */
        }
        OSSemPost(DispSem);                  /* Release access to display                          */
    }
}

/*$PAGE*/
```

5

Listing 5.1 (continued)   LCD.C

```
/*
********************************************************************************************
*                               WRITE DATA TO DISPLAY DEVICE
*
* Description : This function sends a single BYTE to the display device.
* Arguments   : 'data'  is the BYTE to send to the display device
* Returns     : none
* Notes       : You will need to adjust the value of DISP_DLY_CNTS (LCD.H) to produce a delay between
*               writes of at least 40 uS.  The display I used for the test actually required a delay of
*               80 uS!  If characters seem to appear randomly on the screen, you might want to increase
*               the value of DISP_DLY_CNTS.
********************************************************************************************
*/

#ifndef CFG_C
void  DispDataWr (INT8U data)
{
    INT8U  dly;


    outp(DISP_PORT_DATA, data);            /* Write data to display module                 */
    outp(DISP_PORT_CMD,  0x01);            /* Set E   line HIGH                            */
    DispDummy();                           /* Delay about 1 uS                             */
    outp(DISP_PORT_CMD,  0x00);            /* Set E   line LOW                             */
    for (dly = DISP_DLY_CNTS; dly > 0; dly--) {  /* Delay for at least 40 uS               */
        DispDummy();
    }
}
#endif


/*
********************************************************************************************
*                             INITIALIZE DISPLAY DRIVER I/O PORTS
*
* Description : This initializes the I/O ports used by the display driver.
* Arguments   : none
* Returns     : none
********************************************************************************************
*/

#ifndef CFG_C
void  DispInitPort (void)
{
    outp(DISP_PORT_CMD, 0x82);            /* Set to Mode 0: A are output, B are inputs, C are outputs     */
}
#endif
```

Listing 5.1 (continued)   LCD.C

```
/*
*********************************************************************************************
*                             SELECT COMMAND OR DATA REGISTER
*
* Description : This function read a BYTE from the display device.
* Arguments   : none
*********************************************************************************************
*/

#ifndef CFG_C
void  DispSel (INT8U sel)
{
    if (sel == DISP_SEL_CMD_REG) {
        outp(DISP_PORT_CMD, 0x02);      /* Select the command register (RS low)          */
    } else {
        outp(DISP_PORT_CMD, 0x03);      /* Select the data    register (RS high)         */
    }
}
#endif
```

5

Listing 5.2   LCD.H

```
/*
*********************************************************************************************************
*                                     Embedded Systems Building Blocks
*                                    Complete and Ready-to-Use Modules in C
*
*                                          LCD Display Module Driver
*
*                               (c) Copyright 1999, Jean J. Labrosse, Weston, FL
*                                             All Rights Reserved
*
* Filename   : LCD.H
* Programmer : Jean J. Labrosse
*********************************************************************************************************
*/

/*
*********************************************************************************************************
*                                               CONSTANTS
*********************************************************************************************************
*/

#ifndef  CFG_H
#define  DISP_DLY_CNTS           8          /* Number of iterations to delay for 40 uS (software loop) */

#define  DISP_PORT_CMD        0x0303        /* Address of the Control Word (82C55) to control RS & E    */
#define  DISP_PORT_DATA       0x0300        /* Port address of the DATA port of the LCD module          */
#endif

#define  DISP_SEL_CMD_REG        0
#define  DISP_SEL_DATA_REG       1

/*
*********************************************************************************************************
*                                         FUNCTION PROTOTYPES
*********************************************************************************************************
*/

void  DispChar(INT8U row, INT8U col, char c);
void  DispClrLine(INT8U line);
void  DispClrScr(void);
void  DispDefChar(INT8U id, INT8U *pat);
void  DispDummy(void);
void  DispHorBar(INT8U row, INT8U col, INT8U val);
void  DispHorBarInit(void);
void  DispInit(INT8U maxrows, INT8U maxcols);
void  DispStr(INT8U row, INT8U col, char *s);

/*
*********************************************************************************************************
*                                         FUNCTION PROTOTYPES
*                                           HARDWARE SPECIFIC
*********************************************************************************************************
*/

void  DispDataWr(INT8U data);
void  DispInitPort(void);
void  DispSel(INT8U sel);
```

Chapter 6

Time-of-Day Clock

The management of time is important in many microprocessor-based embedded systems. For instance, what would VCRs (Video Cassette Recorders) be without clock/calendars to schedule the recording of television programs?

In this chapter, I will describe how I implemented a Y2K-compliant clock/calendar module. The clock/calendar module offers the following features:

- Maintains hours, minutes, and seconds.

- Contains a calendar which keeps track of: month, day, year (including leap-years), and day-of-week.

- Allows your application to obtain a *timestamp* to mark the occurrence of events. A timestamp is the current date and time packed into a 32-bit integer.

6.00 Clocks/Calendars

A clock/calendar is a useful module for an embedded system. If you need a clock/calendar, you have to decide whether to implement it in hardware or software.

Clock/calendar chips are readily available and most can directly interface with microprocessors. These chips accurately maintain the time-of-day, and some chips even provide a built-in calendar. Some chips include a battery and can continue to keep track of date and time even when power is removed from the unit. Clock/calendar chips generally require a crystal, which further increases the recurring cost of your system. Clock/calendar chips are manufactured by a large number of semiconductor companies such as Motorola, National Semiconductor, Maxim, Dallas Semiconductor, etc. Just because you have a clock/calendar chip doesn't mean you don't need to write any software. Your application software will still need to:

- program the clock/calendar chip with the correct date and time,

- program any alarm clock functions, and

- read the current date and time.

A software-maintained clock/calendar is the best solution when your application cannot afford the extra cost associated with a clock/calendar chip, a battery, and an extra crystal. A software-implemented clock/calendar module can offer most of the benefits of a hardware approach (except that it can't maintain date and time when power is removed). A software approach requires very little ROM, RAM, and CPU time and does not add recurring cost to your system. Also, you can easily add features, such as alarm clock functions (with many alarm setpoints), timestamps, string-formatting utilities to convert date and time to ASCII, etc. Software-implemented clock/calendars are found in a number of familiar appliances such as VCRs, stereos, FAX machines, microwave ovens, etc. If the microprocessor has a low-power standby mode, the software-implemented clock/calendars can be made to maintain correct date and time when the power is removed by also including a battery to power the microprocessor.

Maintaining a clock/calendar is a trivial task for a microprocessor. The first thing you will need is a periodic time source that will interrupt the microprocessor at regular intervals. Such a time source is easy to find. AC power line frequencies (50 or 60 Hz) are generally very accurate over long periods of time. For short-term accuracy, the crystal used to clock the microprocessor is also a good candidate; however, for such an application, the crystal frequency must be divided down. If your application software runs under a real-time multitasking operating system, the OS's *clock tick* is a convenient periodic time source.

If we assumed that the microprocessor was interrupted every one-tenth (0.1) of a second, the software simply needs to maintain integer counters for tenths of a second, seconds, minutes, hours, day, month, and year as follows. The tenths of a second is incremented every interrupt. If the counter overflows from 9 to 0, the seconds counter is incremented. If the seconds counter overflows from 59 to 0, the minutes counter is incremented, etc. Every 24 hours, the days counter is incremented. When the months counter overflows depends on the current month and also, in the case of February, on whether the year is a leap year. The following sections describe how I implemented the software for the clock/calendar module.

6.01 Clock/Calendar Module

The source code for the clock/calendar module is found in the \SOFTWARE\BLOCKS\CLK\SOURCE directory. The source code is found in the files CLK.C (Listing 6.1) and CLK.H (Listing 6.2). All clock/calendar functions and variables related to this module start with Clk, while all #define constants start with CLK_.

6.02 Internals

Figure 6.1 shows a simplified flow diagram of the clock/calendar module. I assume the presence of a real-time kernel but the code can easily be modified to work in a foreground/background environment. Basically, the clock/calendar module consists of a task which executes every second. The task is responsible for updating eight variables that are maintained by the clock/calendar module. You should not directly access these variables from your application. As you might have expected, the variables updated by the clock/calendar module are:

ClkSec: Seconds (0..59)

ClkMin: Minutes (0..59)

ClkHr: Hours (0..23, i.e., military time)

ClkDay: Day (1..31, i.e., day-of-month)

ClkDOW: Day-of-week (0..6, i.e., Sunday, Monday, etc.)

ClkMonth: Month (1..12)

ClkYear: Year (2000..2063)

ClkTS: Timestamp

Figure 6.1   *Clock/Calendar flow diagram.*

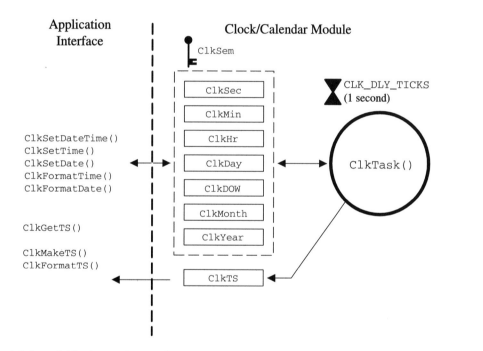

The eighth variable (ClkTS) contains the current date and time in timestamp format (described later).

The date and time counters of the clock/calendar are updated by the task (ClkTask()), which executes every second. The date and time counters are considered shared resources, and thus a mutual exclusion semaphore (ClkSem) must be acquired to access these counters.

ClkTask() calls ClkUpdateTime() to update the hours (ClkHr), minutes (ClkMin), and seconds (ClkSec) counters. ClkUpdateTime() returns TRUE when the clock rolls over from 23:59:59 to 00:00:00 indicating a new day. The Boolean result is used to determine whether the date-updating function, ClkUpdateDate(), is called or not.

At the completion of a day, ClkUpdateDate() is called to update the month (ClkMonth), day (ClkDay), year (ClkYear), and day-of-week (ClkDOW) counters. Updating the date is a little bit more complicated because we need to keep track of the number of days in the current month. The current day-of-week is obtained by calling ClkUpdateDOW(). The day-of-week is a number between 0 and 6, with 0 representing Sunday. The use of a table (ClkMonthTbl[]) greatly simplifies the update of the days in a month and day-of-week counters.

On a lightly loaded system, the clock module should maintain accurate time. As I explained in Chapter 2, specifically in Figure 2.27 on page 96, the clock task could slowly lose track of time if all higher priority tasks (and interrupts) require more processing time than 1 *clock tick*. In other words, on a heavily loaded processor, ClkTask() cannot maintain time accurately the way it is currently implemented. There are two ways to fix this problem. The first and simplest way is to make the clock module task a high priority task. This means that lower priority tasks will not be serviced while the clock task is executing. In general, you should assign the highest priorities to your most critical task and not the clock task because it requires a fair amount of processing time. The processor will maintain the time-of-day correctly as long as the clock task and all high priority tasks can execute in the time between clock ticks.

The second way to fix the problem requires the use of a counting semaphore, as shown in Figure 6.2. The number of *clock ticks* will be "memorized" in the semaphore and thus, the clock task will eventually catch up when the load of the processor is reduced. The clock tick ISR can signal the counting semaphore every clock tick or when a whole second has elapsed. I generally prefer to encapsulate these kind of details, and thus, I wrote a function called ClkSignalClk() that can be called by the clock tick ISR every time a tick occurs. Note that you need to change OSTickISR(), which is found in the file OS_CPU_A.ASM located in the \SOFTWARE\uCOS-II\??\compiler\SOURCE directory of the port you will use with µC/OS-II (see www.uCOS-II.com for details on µC/OS-II ports). To use the counting semaphore, you will need to set CLK_USE_DLY to 0 and modify OSTickISR to call ClkSignalClk(). Setting CLK_USE_DLY to 1 tells the compiler to use OSTimeDlyHMSM().

Figure 6.2   Clock/Calendar flow diagram.

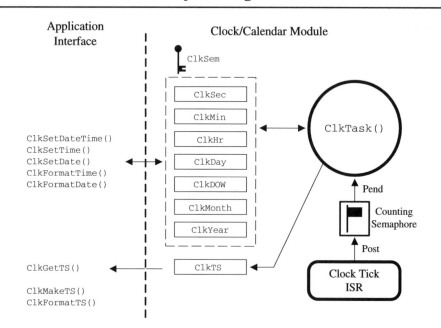

A timestamp (data type TS) packs a date and time into a 32-bit variable. You can use timestamps to mark when certain events have occurred. For example, a timestamp can be used to indicate when a temperature or pressure was exceeded. You can also implement alarm clock type functions using timestamps (described later).

The format of a timestamp is shown in Figure 6.3. Even though I provide you with the format, you should not directly manipulate timestamps in your applications. Instead, you should make use of the functions provided by this module or add functions to this module. This allows for the format to be changed at a later time without affecting your code. You should note that the year uses six bits in the timestamp format and can thus represent only 64 years. The timestamp year is the actual year minus 2000. In other words, a year value of 5 represents 2005.

WARNING
In the previous edition of this book, the timestamp was based on 1990 instead of 2000. If you need to be backwards compatible with the first edition, you can change the value of CLK_TS_BASE_YEAR back to 1990 which is found at the top of CLK.C.

Figure 6.3   Timestamp format.

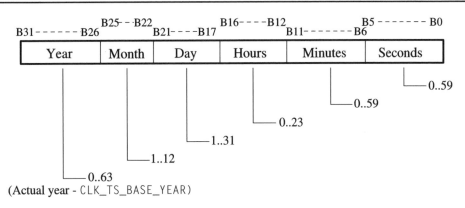

The timestamp format guarantees that later dates and times have larger values. You can thus easily compare timestamps for equality, greater-than, less-than, etc. This feature allows you to design an alarm clock with as many alarm trips as needed.

6.03 Interface Functions

Your application knows about the clock/calendar through the interface functions shown in Figure 6.4.

Figure 6.4   Clock/Calendar module interface functions.

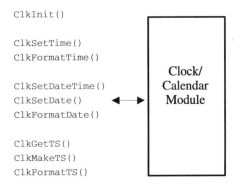

ClkFormatDate()

void ClkFormatDate(INT8U n, char *s);

ClkFormatDate() is also provided for display purposes. This function formats the current date into an ASCII string.

Arguments

n specifies the desired format for the date. ClkFormatDate() currently supports two date formats:

```
n == 1: a condensed date MM-DD-YY
n == 2: full date including:
            day of the week ("Sunday" .. "Saturday"),
            month ("January" .. "December"),
            day of the month (1..31) and
            year (CLK_TS_BASE_YEAR .. CLK_TS_BASE_YEAR + 63).
```

The format is: "DayOfWeek Month Day, Year." For example, 1/1/2000 would be displayed as: "Saturday January 1, 2000." For maximum flexibility, I implemented this function using a switch statement. This allows you to easily add code to support your own date formats. For instance, you could display the date in other languages such as French, Spanish, German, etc.

s is a pointer to the string that will receive the formatted date. You must thus allocate sufficient space for your string. The condensed format (n == 1) requires 9 characters while the other format (n == 2) requires 30 characters (including the NUL character).

Return Value

None

Notes/Warnings

If you are using a preemptive kernel, you should consider making the clock/calendar task priority lower than the application software that will call ClkFormatTime() and ClkFormatDate(). Try to figure out what would happen if you were to format the date and time (these are two separate functions) just before midnight (i.e., 23:59:59)!

Example

```
void Task (void *pdata)
{
    char s[20];


    for (;;) {
        .

        .

        ClkFormatDate(1, s);

        .

        .

    }
}
```

6

ClkFormatTime()

void ClkFormatTime(INT8U n, char *s);

ClkFormatTime() is provided for display purposes. This function formats the current time into an ASCII string.

Arguments

n specifies the desired format for the time. ClkFormatTime() currently supports two time formats:

```
n == 1: 24 hour format, "HH:MM:SS"
n == 2: 12 hour with AM/PM indication, "HH:MM:SS AM"
```

For maximum flexibility, I implemented this function using a switch statement. This allows you to easily add code to support your own formats.

s is a pointer to the string that will receive the formatted time. You must thus allocate sufficient space for your string. The 24-hour format requires nine characters while the 12-hour format requires 12 characters (including the NUL character).

Return Value

None

Notes/Warnings

None

Example

```
void Task (void *pdata)
{
    char s[20];



    for (;;) {
        .
        .
        ClkFormatTime(1, s);
        .
        .
    }
}
```

ClkFormatTS()

void ClkFormatTS(INT8U n, TS ts, char *s);

ClkFormatTS() is provided for display purposes. This function formats a timestamp into an ASCII string.

Arguments

n specifies the desired format for the timestamp. ClkFormatTS() supports only one timestamp format:

    n == 1: "MM-DD-YY HH:MM:SS".
    n == 2: "YYYY-MM-DD HH:MM:SS".

The time is in 24-hour format. For maximum flexibility, I also implemented this function using a switch statement. This allows you to easily add code to support your own timestamp formats.

ts is the timestamp value that you want formatted into an ASCII string.

s is a pointer to the string that will receive the formatted timestamp. You must allocate sufficient space for your string. The timestamp format (n == 1) requires 18 characters (including the NUL character), the timestamp requires 21 characters for format #2 (i.e., n == 2).

Return Value

None

Notes/Warnings

In the previous edition of this book, the timestamp was based on 1990 instead of 2000. If you need to be backwards compatible with the first edition, you can change the value of CLK_TS_BASE_YEAR back to 1990 which is found at the top of CLK.C.

Example

```
void Task (void *pdata)
{
    TS     timestamp;
    char s[20];


    for (;;) {

        .

        .

        timestamp = ClkGetTS();
        ClkFormatTS(1, timestamp, s);
        DispStr(0, 0, s);

        .

        .

    }
}
```

6

ClkGetTS()

```
TS ClkGetTS(void);
```

ClkGetTS() is called by your application to obtain the current date and time in timestamp format. Recall that a timestamp is a 32-bit variable that contains the date and time in a packed format.

Arguments

None

Return Value

The current date and time in timestamp format.

Notes/Warnings

In the previous edition of this book, the timestamp was based on 1990 instead of 2000. If you need to be backwards compatible with the first edition, you can change the value of CLK_TS_BASE_YEAR back to 1990 which is found at the top of CLK.C.

Example

```
void Task (void *pdata)
{
    TS timestamp;


    for (;;) {
        .

        .

        timestamp = ClkGetTS();

        .

        .

    }
}
```

ClkInit()

void ClkInit(void);

ClkInit() is the initialization code for the clock/calendar. ClkInit() must be called before any of the other functions provided in this module. ClkInit() is responsible for the initialization of the clock/calendar variables and the creation of the clock/calendar task.

If you choose to have a clock/calendar chip maintain the correct date and time when power is removed (using a battery), you can use ClkInit() to read the contents of the clock chip and load the corresponding clock/calendar module variables when power is applied to your unit. Note that PCs use this scheme.

Arguments

None

Return Value

None

6

Notes/Warnings

None

Example

```
void main(void)
{
    .
    .
    ClkInit();
    .
    .
    .
}
```

ClkMakeTS()

`TS ClkMakeTS(INT8U month, INT8U day, INT16U year, INT8U hr, INT8U min, INT8U sec);`

`ClkMakeTS()` is called by your application to format a date and time into a timestamp. This function is useful for comparing timestamps. You would use this function to implement an alarm clock feature.

Arguments

month specifies the month of the year and must be a number between 1 and 12.

day corresponds to the day of the month and must be a number between 1 and 31.

year specifies the year. Here I assume you will specify a number between `CLK_TS_BASE_YEAR` (see `CLK.C`) and `CLK_TS_BASE_YEAR+63`. Note that the year is limited to hold 64 years because the year is stored in the timestamp using six bits.

hr specifies the hours and is entered in 24-hour format, i.e., a number between 0 and 23.

min specifies the number of minutes and must be between 0 and 59.

sec specifies the seconds and must also be a number between 0 and 59.

Return Value

The desired date and time in timestamp format.

Notes/Warnings

In the previous edition of this book, the timestamp was based on 1990 instead of 2000. If you need to be backwards compatible with the first edition, you can change the value of `CLK_TS_BASE_YEAR` back to 1990 which is found at the top of `CLK.C`.

Example

```
void Task (void *pdata)
{
    TS alarm;


    alarm = ClkMakeTS(12, 31, 1999, 23, 59, 59);
    for (;;) {
        .

        .
        if (ClkGetTS() > alarm) {
            DispStr(0, 0, "Happy New Year!");
        }
        .

        .
    }
}
```

ClkSetDate()

```
void ClkSetDate(INT8U month, INT8U day, INT16U year);
```

ClkSetDate() is used to set only the calendar portion of the clock/calendar. If you had a clock/calendar chip, you could use this function to also set the date of the chip.

Arguments

month specifies the month of the year and must be a number between 1 and 12.

day corresponds to the day of the month and must be a number between 1 and 31.

year specifies the year. Here I assumed that you will specify a number between CLK_TS_BASE_YEAR and CLK_TS_BASE_YEAR+63.

Return Value

None

Notes/Warnings

None

Example

```
void main(void)
{
    .
    .
    ClkSetDate(1, 1, 2000);
    .
    .
}
```

ClkSetDateTime()

```
void ClkSetDateTime(INT8U month, INT8U day, INT16U year,
                    INT8U hr, INT8U min, INT8U sec);
```

ClkSetDateTime() is used to set the clock/calendar to the desired date and time. If you had a clock/calendar chip, you could use this function to also set the date and time of the chip.

Arguments

month specifies the month of the year and must be a number between 1 and 12.

day corresponds to the day of the month and must be a number between 1 and 31.

year specifies the year. Here I assumed that you will specify a number between CLK_TS_BASE_YEAR and CLK_TS_BASE_YEAR+63.

hr specifies the hours and is entered in 24-hour format, i.e., a number between 0 and 23.

min specifies the number of minutes and must be between 0 and 59.

sec specifies the seconds and must also be a number between 0 and 59.

Return Value

None

Notes/Warnings

None

Example

```
void main(void)
{
    .
    .
    ClkSetDateTime(1, 1, 2000, 23, 59, 59);
    .
    .
}
```

ClkSetTime()

void ClkSetTime(INT8U hr, INT8U min, INT8U sec);

ClkSetTime() is used to set only the clock portion of the clock/calendar. If you had a clock/calendar chip, you could use this function to also set the time of the chip.

Arguments

hr specifies the hours and is entered in 24-hour format, i.e., a number between 0 and 23.

min specifies the number of minutes and must be between 0 and 59.

sec specifies the seconds and must also be a number between 0 and 59.

Return Value

None

Notes/Warnings

None

Example

```
void main(void)
{
    .
    .
    ClkSetTime(23, 59, 59);
    .
    .
}
```

6

6.04 Clock/Calendar Module, Configuration

All you need to do to use the clock/calendar module in your application is to define the value of five #define constants (see file CLK.H and also CFG.H), call ClkInit(), and then initialize the current date and time for the clock/calendar.

CLK_TASK_PRIO defines the priority of ClkTask() in the multitasking environment. The task priority of the clock/calendar module would typically be set relatively low (i.e., a high number under µC/OS-II) because clocks and calendars are generally not considered critical.

CLK_DLY_TICKS defines the number of "clock ticks" needed to obtain one second. I tested the code using an IBM-PC and the tick rate was set to 200 Hz.

CLK_TASK_STK_SIZE defines the size of the stack allocated to the clock/calendar module task. The number of bytes allocated for the stack is given by: CLK_TASK_STK_SIZE times sizeof(OS_STK).

WARNING

In the previous edition of this book, CLK_TASK_STK_SIZE specified the size of the stack for TaskTask() in number of bytes. µC/OS-II assumes the stack is specified in stack width elements.

CLK_DATE_EN is used to allow your application to save ROM space by disabling (when set to 0) the date updating feature of the clock/calendar module.

CLK_TS_EN is used to allow your application to save ROM space by disabling (when set to 0) the timestamp feature of the clock/calendar module. Note that you need to enable the calendar when you enable the timestamp capability.

CLK_USE_DLY is used to indicate that the clock/calendar module will use time delays to delay the clock task every second (when set to 1). The clock/calendar module will be expecting signals from the tick ISR (through ClkSignalClk()) when CLK_USE_DLY is set to 0.

6.05 Bibliography

Viscogliosi, Roberto R.
"C shortcuts and the day of the week"
PC Magazine, May 11, 1993, p.396,401, & 406

Latham, Lance
Standard C Date/Time Library; Programming the World's Calendars and Clocks
R&D Books, Lawrence, KS, 1999
ISBN 0-87930-496-0

Listing 6.1   CLK.C

```
/*
*********************************************************************************************
*                                   Clock/Calendar
*
*                     (c) Copyright 1999, Jean J. Labrosse, Weston, FL
*                                  All Rights Reserved
*
* Filename   : CLK.C
* Programmer : Jean J. Labrosse
*********************************************************************************************
*/


/*
*********************************************************************************************
*                                   INCLUDE FILES
*********************************************************************************************
*/

#define  CLK_GLOBALS                 /* CLK.H is informed to allocate storage for globals      */
#include "includes.h"

/*
*********************************************************************************************
*                                   LOCAL CONSTANTS
*********************************************************************************************
*/

#define  CLK_TS_BASE_YEAR    2000    /* Time stamps start year                                 */

/*
*********************************************************************************************
*                                   LOCAL VARIABLES
*********************************************************************************************
*/

static  OS_EVENT   *ClkSem;          /* Semaphore used to access the time of day clock         */
static  OS_EVENT   *ClkSemSec;       /* Counting semaphore used to keep track of seconds       */

static  OS_STK     ClkTaskStk[CLK_TASK_STK_SIZE];

static  INT8U      ClkTickCtr;       /* Counter used to keep track of system clock ticks       */


/*$PAGE*/
```

6

Listing 6.1 (continued)   CLK.C

```
/*
*********************************************************************************************
*                                   LOCAL TABLES
*********************************************************************************************
*/

#if CLK_DATE_EN
static char *ClkDOWTbl[] = {          /* NAME FOR EACH DAY OF THE WEEK                    */
    "Sunday ",
    "Monday ",
    "Tuesday ",
    "Wednesday ",
    "Thursday ",
    "Friday ",
    "Saturday "
};


static CLK_MONTH ClkMonthTbl[] = {   /* MONTHS TABLE                                     */
    {0,  "",          0},            /* Invalid month                                    */
    {31, "January ",  6},            /* January                                          */
    {28, "February ", 2},            /* February (note leap years are handled by code)   */
    {31, "March ",    2},            /* March                                            */
    {30, "April ",    5},            /* April                                            */
    {31, "May ",      0},            /* May                                              */
    {30, "June ",     3},            /* June                                             */
    {31, "July ",     5},            /* July                                             */
    {31, "August ",   1},            /* August                                           */
    {30, "September ", 4},           /* September                                        */
    {31, "October ",  6},            /* October                                          */
    {30, "November ", 2},            /* November                                         */
    {31, "December ", 4}             /* December                                         */
};
#endif

/*
*********************************************************************************************
*                              LOCAL FUNCTION PROTOTYPES
*********************************************************************************************
*/

       void      ClkTask(void *data);
static BOOLEAN    ClkUpdateTime(void);

#if     CLK_DATE_EN
static BOOLEAN    ClkIsLeapYear(INT16U year);
static void       ClkUpdateDate(void);
static void       ClkUpdateDOW(void);
#endif

/*$PAGE*/
```

Listing 6.1 (continued)   `CLK.C`

```
/*
*********************************************************************************************
*                              FORMAT CURRENT DATE INTO STRING
*
* Description : Formats the current date into an ASCII string.
* Arguments   : n      is the format type:
*                      1   will format the time as "MM-DD-YY"          (needs at least  9 characters)
*                      2   will format the time as "Day Month DD, YYYY" (needs at least 30 characters)
*                      3   will format the time as "YYYY-MM-DD"         (needs at least 11 characters)
*               s      is a pointer to the destination string.  The destination string must be large
*                      enough to hold the formatted date.
*                      contain
* Returns     : None.
* Notes       : - A 'switch' statement has been used to allow you to add your own date formats.  For
*                 example, you could display the date in French, Spanish, German etc. by assigning
*                 numbers for those types of conversions.
*               - This function assumes that strcpy(), strcat() and itoa() are reentrant.
*********************************************************************************************
*/

#if  CLK_DATE_EN
void  ClkFormatDate (INT8U n, char *s)
{
    INT8U   err;
    INT16U  year;
    char    str[5];


    OSSemPend(ClkSem, 0, &err);                 /* Gain exclusive access to time-of-day clock      */
    switch (n) {
        case  1:
             strcpy(s, "MM-DD-YY");             /* Create the template for the selected format     */
             s[0] = ClkMonth / 10 + '0';        /* Convert DATE to ASCII                           */
             s[1] = ClkMonth % 10 + '0';
             s[3] = ClkDay   / 10 + '0';
             s[4] = ClkDay   % 10 + '0';
             year = ClkYear % 100;
             s[6] = year / 10 + '0';
             s[7] = year % 10 + '0';
             break;

        case  2:
             strcpy(s, ClkDOWTbl[ClkDOW]);                  /* Get the day of the week             */
             strcat(s, ClkMonthTbl[ClkMonth].MonthName);    /* Get name of month                   */
             if (ClkDay < 10) {
                 str[0] = ClkDay + '0';
                 str[1] = 0;
             } else {
                 str[0] = ClkDay / 10 + '0';
                 str[1] = ClkDay % 10 + '0';
                 str[2] = 0;
             }
             strcat(s, str);
             strcat(s, ", ");
             itoa(ClkYear, str, 10);
             strcat(s, str);
             break;
```

6

Listing 6.1 (continued)   CLK.C

```
        case  3:
                strcpy(s, "YYYY-MM-DD");           /* Create the template for the selected format      */
                s[0] = year / 1000 + '0';
                year = year % 1000;
                s[1] = year /  100 + '0';
                year = year %  100;
                s[2] = year /   10 + '0';
                s[3] = year %   10 + '0';
                s[5] = ClkMonth / 10 + '0';        /* Convert DATE to ASCII                             */
                s[6] = ClkMonth % 10 + '0';
                s[8] = ClkDay   / 10 + '0';
                s[9] = ClkDay   % 10 + '0';
                break;

        default:
                strcpy(s, "?");
                break;
    }
    OSSemPost(ClkSem);                             /* Release access to clock                           */
}
#endif

/*$PAGE*/
```

Listing 6.1 (continued) CLK.C

```
/*
*********************************************************************************************
*                            FORMAT CURRENT TIME INTO STRING
*
* Description : Formats the current time into an ASCII string.
* Arguments   : n      is the format type:
*                      1   will format the time as "HH:MM:SS"     (24 Hour format)
*                                                                 (needs at least  9 characters)
*                      2   will format the time as "HH:MM:SS AM"  (With AM/PM indication)
*                                                                 (needs at least 13 characters)
*               s      is a pointer to the destination string.  The destination string must be large
*                      enough to hold the formatted time.
*                      contain
* Returns     : None.
* Notes       : - A 'switch' statement has been used to allow you to add your own time formats.
*               - This function assumes that strcpy() is reentrant.
*********************************************************************************************
*/

void  ClkFormatTime (INT8U n, char *s)
{
    INT8U err;
    INT8U hr;


    OSSemPend(ClkSem, 0, &err);                    /* Gain exclusive access to time-of-day clock   */
    switch (n) {
        case  1:
             strcpy(s, "HH:MM:SS");                /* Create the template for the selected format  */
             s[0] = ClkHr  / 10 + '0';             /* Convert TIME to ASCII                         */
             s[1] = ClkHr  % 10 + '0';
             s[3] = ClkMin / 10 + '0';
             s[4] = ClkMin % 10 + '0';
             s[6] = ClkSec / 10 + '0';
             s[7] = ClkSec % 10 + '0';
             break;

        case  2:
             strcpy(s, "HH:MM:SS AM");             /* Create the template for the selected format  */
             s[9] = (ClkHr >= 12) ? 'P' : 'A';     /* Set AM or PM indicator                       */
             if (ClkHr > 12) {                     /* Adjust time to be displayed                  */
                 hr   = ClkHr - 12;
             } else {
                 hr = ClkHr;
             }
             s[0] = hr    / 10 + '0';              /* Convert TIME to ASCII                         */
             s[1] = hr    % 10 + '0';
             s[3] = ClkMin / 10 + '0';
             s[4] = ClkMin % 10 + '0';
             s[6] = ClkSec / 10 + '0';
             s[7] = ClkSec % 10 + '0';
             break;

        default:
             strcpy(s, "?");
             break;
    }
    OSSemPost(ClkSem);                             /* Release access to time-of-day clock          */
}

/*$PAGE*/
```

Listing 6.1 (continued)   CLK.C

```
/*
*********************************************************************************************
*                                  FORMAT TIME-STAMP
*
* Description : This function converts a time-stamp to an ASCII string.
* Arguments   : n          is the desired format number:
*                              1 : "MM-DD-YY HH:MM:SS"       (needs at least 18 characters)
*                              2 : "YYYY-MM-DD HH:MM:SS"     (needs at least 20 characters)
*               ts         is the time-stamp value to format
*               s          is the destination ASCII string
* Returns     : none
* Notes       : - The time stamp is a 32 bit unsigned integer as follows:
*
*       Field: -------Year------ ---Month--- ------Day----- ----Hours----- ---Minutes--- --Seconds--
*       Bit# : 31 30 29 28 27 26 25 24 23 22 21 20 19 18 17 16 15 14 13 12 11 10 9 8 7 6 5 4 3 2 1 0
*
*               - The year is based from CLK_TS_BASE_YEAR.  That is, if bits 31..26 contain 0 it really
*                 means that the year is really CLK_TS_BASE_YEAR.  If bits 31..26 contain 13, the year
*                 is CLK_TS_BASE_YEAR + 13.
*********************************************************************************************
*/

#if CLK_TS_EN && CLK_DATE_EN
void  ClkFormatTS (INT8U n, TS ts, char *s)
{
    INT16U yr;
    INT8U month;
    INT8U day;
    INT8U hr;
    INT8U min;
    INT8U sec;


    yr    = CLK_TS_BASE_YEAR + (ts >> 26);      /* Unpack time-stamp                       */
    month = (ts >> 22) & 0x0F;
    day   = (ts >> 17) & 0x1F;
    hr    = (ts >> 12) & 0x1F;
    min   = (ts >>  6) & 0x3F;
    sec   = (ts & 0x3F);
    switch (n) {
        case 1:
             strcpy(s, "MM-DD-YY HH:MM:SS");    /* Create the template for the selected format    */
             yr    = yr % 100;
             s[ 0] = month / 10 + '0';          /* Convert DATE to ASCII                   */
             s[ 1] = month % 10 + '0';
             s[ 3] = day   / 10 + '0';
             s[ 4] = day   % 10 + '0';
             s[ 6] = yr    / 10 + '0';
             s[ 7] = yr    % 10 + '0';
             s[ 9] = hr    / 10 + '0';          /* Convert TIME to ASCII                   */
             s[10] = hr    % 10 + '0';
             s[12] = min   / 10 + '0';
             s[13] = min   % 10 + '0';
             s[15] = sec   / 10 + '0';
             s[16] = sec   % 10 + '0';
             break;
```

Listing 6.1 (continued)   `CLK.C`

```
        case  2:
               strcpy(s, "YYYY-MM-DD HH:MM:SS");  /* Create the template for the selected format    */
               s[ 0] = yr    / 1000 + '0';        /* Convert DATE to ASCII                          */
               yr    = yr % 1000;
               s[ 1] = yr    /  100 + '0';
               yr    = yr %  100;
               s[ 2] = yr    /   10 + '0';
               s[ 3] = yr    %   10 + '0';
               s[ 5] = month / 10 + '0';
               s[ 6] = month % 10 + '0';
               s[ 8] = day   / 10 + '0';
               s[ 9] = day   % 10 + '0';
               s[11] = hr    / 10 + '0';          /* Convert TIME to ASCII                          */
               s[12] = hr    % 10 + '0';
               s[14] = min   / 10 + '0';
               s[15] = min   % 10 + '0';
               s[17] = sec   / 10 + '0';
               s[18] = sec   % 10 + '0';
               break;

        default:
               strcpy(s, "?");
               break;
    }
}
#endif

/*$PAGE*/
```

6

Listing 6.1 (continued)   CLK.C

```
/*
*********************************************************************************************
*                                    GET TIME-STAMP
*
* Description : This function is used to return a time-stamp to your application.  The format of the
*               time-stamp is shown below:
*
*       Field: -------Year------ ---Month--- ------Day----- ----Hours----- ---Minutes--- --Seconds--
*       Bit# : 31 30 29 28 27 26 25 24 23 22 21 20 19 18 17 16 15 14 13 12 11 10 9 8 7 6 5 4 3 2 1 0
*
* Arguments  : None.
* Returns    : None.
* Notes      : The year is based from CLK_TS_BASE_YEAR.  That is, if bits 31..26 contain 0 it really
*               means that the year is CLK_TS_BASE_YEAR.  If bits 31..26 contain 13, the year is
*               CLK_TS_BASE_YEAR + 13.
*********************************************************************************************
*/

#if CLK_TS_EN && CLK_DATE_EN
TS  ClkGetTS (void)
{
    TS ts;


    OS_ENTER_CRITICAL();
    ts = ClkTS;
    OS_EXIT_CRITICAL();
    return (ts);
}
#endif

/*$PAGE*/
```

Listing 6.1 (continued) `CLK.C`

```
/*
*********************************************************************************************
*                                   TIME MODULE INITIALIZATION
*                               TIME-OF-DAY CLOCK INITIALIZATION
*
* Description : This function initializes the time module.  The time of day clock task will be created
*               by this function.
* Arguments   : None
* Returns     : None.
*********************************************************************************************
*/

void  ClkInit (void)
{
    ClkSem    = OSSemCreate(1);        /* Create time of day clock semaphore                    */
    ClkSemSec = OSSemCreate(0);        /* Create counting semaphore to signal the occurrence of 1 sec. */
    ClkTickCtr =   0;
    ClkSec    =    0;
    ClkMin    =    0;
    ClkHr     =    0;
#if CLK_DATE_EN
    ClkDay    =    1;
    ClkMonth  =    1;
    ClkYear   = 1999;
#endif
#if CLK_TS_EN && CLK_DATE_EN
    ClkTS     = ClkMakeTS(ClkMonth, ClkDay, ClkYear, ClkHr, ClkMin, ClkSec);
#endif
    OSTaskCreate(ClkTask, (void *)0, &ClkTaskStk[CLK_TASK_STK_SIZE], CLK_TASK_PRIO);
}


/*$PAGE*/
```

6

Listing 6.1 (continued) CLK.C

```
/*
*********************************************************************************************
*                                 DETERMINE IF WE HAVE A LEAP YEAR
*
* Description : This function determines whether the 'year' passed as an argument is a leap year.
* Arguments   : year    is the year to check for leap year.
* Returns     : TRUE    if 'year' is a leap year.
*               FALSE   if 'year' is NOT a leap year.
*********************************************************************************************
*/
#if CLK_DATE_EN
static  BOOLEAN  ClkIsLeapYear(INT16U year)
{
    if (!(year % 4) && (year % 100) || !(year % 400)) {
        return TRUE;
    } else {
        return (FALSE);
    }
}
#endif

/*$PAGE*/
```

Listing 6.1 (continued)   CLK.C

```
/*
*********************************************************************************************
*                                    MAKE TIME-STAMP
*
* Description : This function maps a user specified date and time into a 32 bit variable called a
*               time-stamp.
* Arguments   : month    is the desired month   (1..12)
*               day      is the desired day     (1..31)
*               year     is the desired year    (CLK_TS_BASE_YEAR .. CLK_TS_BASE_YEAR+63)
*               hr       is the desired hour    (0..23)
*               min      is the desired minutes (0..59)
*               sec      is the desired seconds (0..59)
* Returns     : A time-stamp based on the arguments passed to the function.
* Notes       : - The time stamp is formatted as follows using a 32 bit unsigned integer:
*
*       Field: -------Year------ ---Month--- ------Day----- ----Hours----- ---Minutes--- --Seconds--
*       Bit# : 31 30 29 28 27 26 25 24 23 22 21 20 19 18 17 16 15 14 13 12 11 10 9 8 7 6 5 4 3 2 1 0
*
*               - The year is based from CLK_TS_BASE_YEAR.  That is, if bits 31..26 contain 0 it really
*                 means that the year is really CLK_TS_BASE_YEAR.  If bits 31..26 contain 13, the year is
*                 CLK_TS_BASE_YEAR + 13.
*********************************************************************************************
*/

#if CLK_TS_EN && CLK_DATE_EN
TS  ClkMakeTS (INT8U month, INT8U day, INT16U yr, INT8U hr, INT8U min, INT8U sec)
{
    TS ts;


    yr -= CLK_TS_BASE_YEAR;
    ts  = ((INT32U)yr << 26) | ((INT32U)month << 22) | ((INT32U)day << 17);
    ts |= ((INT32U)hr << 12) | ((INT32U)min   << 6) |  (INT32U)sec;
    return (ts);
}
#endif

/*$PAGE*/
```

Listing 6.1 *(continued)*   CLK.C

```
/*
*********************************************************************************************
*                                    SET DATE ONLY
*
* Description : Set the date of the time-of-day clock
* Arguments   : month     is the desired month (1..12)
*               day       is the desired day   (1..31)
*               year      is the desired year  (CLK_TS_BASE_YEAR .. CLK_TS_BASE_YEAR+63)
* Returns     : None.
* Notes       : It is assumed that you are specifying a correct date (i.e. there is no range checking
*               done by this function).
*********************************************************************************************
*/

#if  CLK_DATE_EN
void  ClkSetDate (INT8U month, INT8U day, INT16U year)
{
    INT8U err;


    OSSemPend(ClkSem, 0, &err);            /* Gain exclusive access to time-of-day clock       */
    ClkMonth = month;
    ClkDay   = day;
    ClkYear  = year;
    ClkUpdateDOW();                        /* Compute the day of the week (i.e. Sunday ...)    */
    OSSemPost(ClkSem);                     /* Release access to time-of-day clock              */
}
#endif

/*$PAGE*/
```

Listing 6.1 (continued) CLK.C

```
/*
*********************************************************************************************************
*                                       SET DATE AND TIME
*
* Description : Set the date and time of the time-of-day clock
* Arguments   : month     is the desired month    (1..12)
*               day       is the desired day      (1..31)
*               year      is the desired year     (2xxx)
*               hr        is the desired hour     (0..23)
*               min       is the desired minutes  (0..59)
*               sec       is the desired seconds  (0..59)
* Returns     : None.
* Notes       : It is assumed that you are specifying a correct date and time (i.e. there is no range
*               checking done by this function).
*********************************************************************************************************
*/

#if CLK_DATE_EN
void ClkSetDateTime (INT8U month, INT8U day, INT16U year, INT8U hr, INT8U min, INT8U sec)
{
    INT8U err;


    OSSemPend(ClkSem, 0, &err);                      /* Gain exclusive access to time-of-day clock     */
    ClkMonth = month;
    ClkDay   = day;
    ClkYear  = year;
    ClkHr    = hr;
    ClkMin   = min;
    ClkSec   = sec;
    ClkUpdateDOW();                                  /* Compute the day of the week (i.e. Sunday ...)   */
    OSSemPost(ClkSem);                               /* Release access to time-of-day clock             */
}
#endif

/*$PAGE*/
```

6

Listing 6.1 (continued)   CLK.C

```
/*
*********************************************************************************************
*                                    SET TIME ONLY
*
* Description : Set the time-of-day clock
* Arguments   : hr          is the desired hour    (0..23)
*               min         is the desired minutes (0..59)
*               sec         is the desired seconds (0..59)
* Returns     : None.
* Notes       : It is assumed that you are specifying a correct time (i.e. there is no range checking
*               done by this function).
*********************************************************************************************
*/

void  ClkSetTime (INT8U hr, INT8U min, INT8U sec)
{
    OS_ENTER_CRITICAL();                        /* Gain exclusive access to time-of-day clock      */
    ClkHr  = hr;
    ClkMin = min;
    ClkSec = sec;
    OS_EXIT_CRITICAL();                         /* Release access to time-of-day clock             */
}


/*$PAGE*/
/*
*********************************************************************************************
*                     SIGNAL CLOCK MODULE THAT A 'CLOCK TICK' HAS OCCURRED
*
* Description : This function is called by the 'clock tick' ISR on every tick.  This function is thus
*               responsible for counting the number of clock ticks per second.  When a second elapses,
*               this function will signal the time-of-day clock task.
* Arguments   : None.
* Returns     : None.
* Note(s)     : CLK_DLY_TICKS must be set to the number of ticks to produce 1 second.
*               This would typically correspond to OS_TICKS_PER_SEC if you use uC/OS-II.
*********************************************************************************************
*/

void  ClkSignalClk (void)
{
    ClkTickCtr++;                               /* count the number of 'clock ticks' for one second */
    if (ClkTickCtr >= CLK_DLY_TICKS) {
        ClkTickCtr = 0;
        OSSemPost(ClkSemSec);                   /* Signal that one second elapsed                  */
    }
}
```

Listing 6.1 (continued) CLK.C

```c
/*
*********************************************************************************
*                                 TIME-OF-DAY CLOCK TASK
*
* Description : This task is created by ClkInit() and is responsible for updating the time and date.
*               ClkTask() executes every second.
* Arguments   : None.
* Returns     : None.
* Notes       : CLK_DLY_TICKS must be set to produce 1 second delays.
*********************************************************************************
*/

void  ClkTask (void *data)
{
    INT8U err;


    data = data;                        /* Avoid compiler warning (uC/OS requirement)          */
    for (;;) {

#if CLK_USE_DLY
        OSTimeDlyHMSM(0, 0, 1, 0);      /* Delay for one second                                */
#else
        OSSemPend(ClkSemSec, 0, &err);  /* Wait for one second to elapse                       */
#endif

        OSSemPend(ClkSem, 0, &err);     /* Gain exclusive access to time-of-day clock          */
        if (ClkUpdateTime() == TRUE) {  /* Update the TIME (i.e. HH:MM:SS)                     */
#if CLK_DATE_EN
            ClkUpdateDate();            /* And date if a new day (i.e. MM-DD-YY)               */
#endif
        }
#if CLK_TS_EN && CLK_DATE_EN
        ClkTS = ClkMakeTS(ClkMonth, ClkDay, ClkYear, ClkHr, ClkMin, ClkSec);
#endif
        OSSemPost(ClkSem);              /* Release access to time-of-day clock                 */
    }
}

/*$PAGE*/
```

Listing 6.1 (continued) CLK.C

```c
/*
*********************************************************************************************************
*                                          UPDATE THE DATE
*
* Description : This function is called to update the date (i.e. month, day and year)
* Arguments   : None.
* Returns     : None.
* Notes       : This function updates ClkDay, ClkMonth, ClkYear and ClkDOW.
*********************************************************************************************************
*/

#if CLK_DATE_EN
static void ClkUpdateDate (void)
{
    BOOLEAN newmonth;


    newmonth = TRUE;
    if (ClkDay >= ClkMonthTbl[ClkMonth].MonthDays) {  /* Last day of the month?                      */
        if (ClkMonth == 2) {                          /* Is this February?                           */
            if (ClkIsLeapYear(ClkYear) == TRUE) {     /* Yes, Is this a leap year?                   */
                if (ClkDay >= 29) {                   /* Yes, Last day in february?                  */
                    ClkDay = 1;                       /* Yes, Set to 1st day in March                */
                } else {
                    ClkDay++;
                    newmonth = FALSE;
                }
            } else {
                ClkDay = 1;
            }
        } else {
            ClkDay = 1;
        }
    } else {
        ClkDay++;
        newmonth = FALSE;
    }
    if (newmonth == TRUE) {                            /* See if we have completed a month            */
        if (ClkMonth >= 12) {                         /* Yes, Is this december ?                     */
            ClkMonth = 1;                             /* Yes, set month to january...                */
            ClkYear++;                                /*      ...we have a new year!                 */
        } else {
            ClkMonth++;                               /* No,  increment the month                    */
        }
    }
    ClkUpdateDOW();                                   /* Compute the day of the week (i.e. Sunday ...)  */
}
#endif

/*$PAGE*/
```

Listing 6.1 (continued)   CLK.C

```
/*
*********************************************************************************************
*                                 COMPUTE DAY-OF-WEEK
*
* Description : This function computes the day of the week (0 == Sunday) based on the current month,
*               day and year.
* Arguments   : None.
* Returns     : None.
* Notes       : - This function updates ClkDOW.
*               - This function is called by ClkUpdateDate().
*********************************************************************************************
*/
#if CLK_DATE_EN
static void ClkUpdateDOW (void)
{
    INT16U dow;


    dow = ClkDay + ClkMonthTbl[ClkMonth].MonthVal;
    if (ClkMonth < 3) {
        if (ClkIsLeapYear(ClkYear)) {
            dow--;
        }
    }
    dow    += ClkYear + (ClkYear / 4);
    dow    += (ClkYear / 400) - (ClkYear / 100);
    dow    %= 7;
    ClkDOW  = dow;
}
#endif

/*$PAGE*/
```

6

Listing 6.1 (continued)   CLK.C

```
/*
*********************************************************************************************
*                                     UPDATE THE TIME
*
* Description : This function is called to update the time (i.e. hours, minutes and seconds)
* Arguments   : None.
* Returns     : TRUE     if we have completed one day.
*               FALSE    otherwise
* Notes       : This function updates ClkSec, ClkMin and ClkHr.
*********************************************************************************************
*/

static  BOOLEAN  ClkUpdateTime (void)
{
    BOOLEAN newday;


    newday = FALSE;                        /* Assume that we haven't completed one whole day yet   */
    if (ClkSec >= 59) {                    /* See if we have completed one minute yet              */
        ClkSec = 0;                        /* Yes, clear seconds                                   */
        if (ClkMin >= 59) {                /*    See if we have completed one hour yet             */
            ClkMin = 0;                    /*    Yes, clear minutes                                */
            if (ClkHr >= 23) {             /*        See if we have completed one day yet          */
                ClkHr = 0;                 /*        Yes, clear hours ...                          */
                newday    = TRUE;          /*        ... change flag to indicate we have a new day */
            } else {
                ClkHr++;                   /*        No,  increment hours                          */
            }
        } else {
            ClkMin++;                      /*    No,  increment minutes                            */
        }
    } else {
        ClkSec++;                          /* No,  increment seconds                               */
    }
    return (newday);
}
```

Listing 6.2   CLK.H

```
/*
********************************************************************************************
*                                   Clock/Calendar
*
*                       (c) Copyright 1999, Jean J. Labrosse, Weston, FL
*                                   All Rights Reserved
*
* Filename   : CLK.H
* Programmer : Jean J. Labrosse
********************************************************************************************
*/

/*
********************************************************************************************
*                                      CONSTANTS
********************************************************************************************
*/

#ifndef  CFG_H

#define  CLK_DLY_TICKS      OS_TICKS_PER_SEC    /* # of clock ticks to obtain 1
second                */
#define  CLK_TASK_PRIO       50                 /* This defines the priority of
ClkTask()            */
#define  CLK_TASK_STK_SIZE  512                 /* Stack size in BYTEs for
ClkTask()                */

#define  CLK_DATE_EN          1                 /* Enable DATE (when 1)
                   */
#define  CLK_TS_EN            1                 /* Enable TIME-STAMPS (when 1)
                   */
#define  CLK_USE_DLY          1                 /* Task will use OSTimeDly()
instead of pend on sem.  */

#endif

#ifdef   CLK_GLOBALS
#define  CLK_EXT
#else
#define  CLK_EXT  extern
#endif
```

Listing 6.2 (continued) CLK.H

```
/*
*********************************************************************************************
*                                        DATA TYPES
*********************************************************************************************
*/

typedef  INT32U  TS;                    /* Definition of Time Stamp
                    */


#if CLK_DATE_EN
typedef struct clk_month {              /* MONTH RELATED VARIABLES
                    */
    INT8U  MonthDays;                   /* Number of days in each month
                    */
    char   *MonthName;                  /* Name of the month
                    */
    INT8U  MonthVal;                    /* Value used to compute day of the week
                    */
} CLK_MONTH;
#endif

/*
*********************************************************************************************
*                                      GLOBAL VARIABLES
*********************************************************************************************
*/

CLK_EXT  INT8U   ClkHr;
CLK_EXT  INT8U   ClkMin;
CLK_EXT  INT8U   ClkSec;                /* Counters for local TIME
                    */

#if      CLK_DATE_EN
CLK_EXT  INT8U   ClkDay;                /* Counters for local DATE
                    */
CLK_EXT  INT8U   ClkDOW;                /* Day of week (0 is Sunday)
                    */
CLK_EXT  INT8U   ClkMonth;
CLK_EXT  INT16U  ClkYear;
#endif

#if      CLK_TS_EN
CLK_EXT  TS      ClkTS;                 /* Current TIME-STAMP
                    */
#endif
```

Listing 6.2 (continued)   CLK.H

```
/*
*********************************************************************************************
*                                   FUNCTION PROTOTYPES
*********************************************************************************************
*/

void    ClkInit(void);

void    ClkFormatTime(INT8U n, char *s);
void    ClkSetTime(INT8U hr, INT8U min, INT8U sec);

void    ClkSignalClk(void);

#if     CLK_DATE_EN
void    ClkFormatDate(INT8U n, char *s);
void    ClkSetDate(INT8U month, INT8U day, INT16U year);
void    ClkSetDateTime(INT8U month, INT8U day, INT16U year, INT8U hr, INT8U min,
INT8U sec);
#endif

#if     CLK_TS_EN
TS      ClkGetTS(void);
TS      ClkMakeTS(INT8U month, INT8U day, INT16U year, INT8U hr, INT8U min, INT8U
sec);
void    ClkFormatTS(INT8U n, TS ts, char *s);
#endif
```

6

Chapter 7

Timer Manager

Timers are useful in situations where you start an operation, wait a certain amount of time, and then stop the operation. Usually the process looks like this:

1. Start an operation (turn on or turn off an output device).

2. Start the timer.

3. When the timer expires, stop the operation (turn OFF or turn ON the output device).

You can also use timers to detect timeout conditions. For example, you turn on a motor and then start a timer. Here, you are expecting the speed of the motor (i.e., RPM) to increase. If the speed of the motor doesn't exceed a threshold before the timer times out, then you might turn the motor off and notify an operator. In these cases, you start an operation then monitor the process to see if conditions are met before the timer expires:

1. Start an operation.

2. Start the timer.

3. Monitor for desired conditions. If conditions are met, stop the timer.

4. If timer times out, stop the operation and notify operator.

In this chapter, I will describe how I implemented a countdown timer module. The countdown timer module provides your application with as many countdown timers as your application requires (up to 250). Each countdown timer has a resolution of 0.1 second and can be programmed to expire after 99 minutes, 59 seconds and 0.9 seconds. Each countdown timer can be individually started, stopped, set, reset, and checked. Also a user-defined function can be executed when a countdown timer expires (one for each timer).

7.00   Timer Manager Module

The source code for the timer manager module is in the \SOFTWARE\BLOCKS\TMR\SOURCE directory. The source code consist of two files: TMR.C (Listing 7.1) and TMR.H (Listing 7.2). All timer manager functions and variables related to this module start with Tmr while all #define constants start with TMR_.

7.01 Timer Manager Moduler, Internals

Figure 7.1 shows the flow diagram of the timer manager module. Here, I assume the presence of a real-time kernel. This module consists of a single task that executes every tenth of a second. The timer manager task (TmrTask()) is responsible for updating as many countdown timers as your application requires (defined by TMR_MAX_TMR in TMR.H). You can have up to 250 timers.

Figure 7.1   Timer manager module flow diagram.

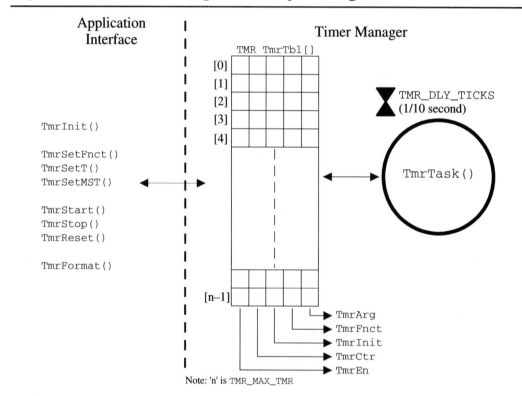

The timer manager is designed around the TMR data structure (TMR.H) which is declared as follows:

```
typedef struct TMR {
    BOOLEAN    TmrEn;
    INT16U     TmrCtr;
    INT16U     TmrInit;
    void       (*TmrFnct)(void *);
    void       *TmrFnctArg;
} TMR;
```

.TmrEn is used to enable and disable the countdown process. Countdown occurs when .TmrEn is set to TRUE by TmrStart(). Countdown is suspended when .TmrEn is set to FALSE by TmrStop().

When the timer is enabled, TmrTask() decrements .TmrCtr towards 0. When .TmrCtr reaches 0, countdown stops. .TmrCtr is loaded when either TmrSetT(), TmrSetMST(), or TmrReset() is called.

The initial value of TmrCtr is stored in .TmrInit. .TmrInit is changed by either TmrSetT() or TmrSetMST().

.TmrFnct is a pointer to a user-defined function that TmrTask() executes whenever its corresponding .TmrCtr reaches 0. The called function is passed .TmrFnctArg (a pointer) as an argument. Both .TmrFnct and .TmrFnctArg are set by TmrCfgFnct() (described later). You must define your timeout function as follows:

```
void UserFnct(void *arg);
```

Note that UserFnct() is passed .TmrFnctArg when it is called. This allows you to design a single function that can be used by more than one timer. The user-defined function will be called by the timer task (TmrTask()) when the timer expires. The execution time of the timer task is thus increased by the execution time of all the functions that will execute when their respective timers expire. You may defer processing of the timeout to another task because the function that executes when the timer expires can signal another task through a semaphore, a mailbox, or even a message queue, as shown:

```
void UserFnct(void *arg)
{
    OSSemPost((OS_EVENT *)arg);
}
```

If you are using µC/OS-II, the argument passed to the user function (in this example) is a pointer to the semaphore.

Some applications do not require the execution of a function upon timeout. In these situations, you will not have to set the pointer because its initial value is NULL. In other words, the timer manager will not execute any function when pointing to NULL.

When the timer manager task executes, it scans all entries in TmrTbl[] for enabled timers. For each timer that has been enabled, TmrTask() decrements TmrTbl[i].TmrCtr towards 0. If the timer reaches 0, the user-defined function (if specified) is executed.

On a lightly-loaded system, the timer manager module should maintain accurate time. As I explained in Chapter 2, specifically Figure 2.27 on page 96, the timer manager task could miss clock ticks if all higher-priority tasks (and interrupts) require more processing time than one *clock tick*. In other words, on a heavily loaded processor, TmrTask() cannot maintain track of time accurately the way it is currently implemented. This is the same problem as with the time-of-day clock described in Chapter 6. Unlike the clock task, however, there is really only one correct way to fix this problem. You really don't want to increase the priority of the timer manager task because its processing time does not depend only on the number of timers it has to manage. Instead, the execution time of the timer manager task depends on the execution time of the functions that will be executed when each timer expires. To fix this problem, you need to use a counting semaphore, as shown in Figure 7.2.

Figure 7.2   Timer manager module flow diagram.

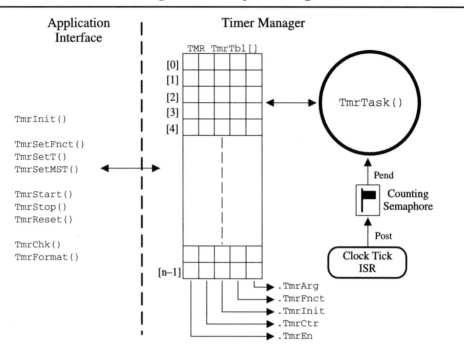

The number of *clock ticks* will be "memorized" in the semaphore, and thus the timer manager task will eventually catch up when the load of the processor is reduced. The clock tick ISR can signal the counting semaphore every clock tick or when 0.1 second has elapsed. I generally prefer to encapsulate the details, so I wrote a function called TmrSignalTmr(), which can be called by the clock tick ISR every time a tick occurs. Note that you need to change OSTickISR(), which is found in the file OS_CPU_A.ASM located in the \SOFTWARE\uCOS-II\??\compiler\SOURCE directory of the port you will use with μC/OS-II (see www.uCOS-II.com for details on μC/OS-II ports). To use the counting semaphore, you will need to set TMR_USE_SEM to 1 and modify OSTickISR() to call TmrSignalTmr().

If you need to manage a large number of timers then you might consider changing the implementation of the module provided in this chapter to a *delta list*. A delta list would maintain a linked list of only the enabled timers. The list would be ordered so that the timer with the least amount of time to timeout is first. TmrTask() would decrement the first entry in the list without scanning the list because the remaining delays are relative to it. For example, if you had five enabled timers with values of 10, 14, 21, 32 then, the list would contain 10, 4, 7, 11, and 7. The total time before the first timer would expire is 10, the second is 10+4, the third is 10+4+7, the fourth is 10+4+7+11, and finally, the fifth timer would be 10+4+7+11+7. The use of a delta list is really only justified when you need many timers. One of the drawbacks of the delta list is that you need one (for a singly-linked list) or two pointers (for a doubly-linked list). You can find a more complete discussion on delta lists in the excellent book by Douglas Comer, *Operating System Design, The XINU approach.*

7.02 Timer Manager Module, Interface Functions

Your application software interfaces with the timer manager through interface functions as shown in Figure 7.3.

Figure 7.3   Timer manager module interface functions.

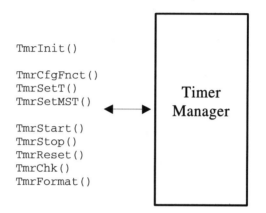

TmrCfgFnct()

```
void TmrCfgFnct(INT8U n, void (*pfnct)(void*), void *arg);
```

Each timer can execute a user-defined function when it expires. In order to use this feature, you must specify the address of the function to execute when the timer expires. This is accomplished by calling TmrCfgFnct().

The execution time of the timer task is augmented by the execution time of all the functions that will execute when their respective timers expire. Some applications do not require the execution of a function upon timeout. In these situations, there is no need to call TmrCfgFnct() because the initial value of the pointer to a function for each timer is NULL. In other words, the timer manager will not execute any function when pfnct is a NULL pointer.

Arguments

n is the timer number to set and must be a number between 0 and TMR_MAX_TMR − 1.

pfnct is a pointer to the function that you would like to execute when the timer expires. You must define this function as follows:

```
void UserFnct(void *arg);
```

Note that UserFnct() is called with the argument you specify in TmrCfgFnct(), that is, arg. This allows you to design a single function that can be used by more than one timer. The user-defined function will be called by the timer task TmrTask() when the timer expires.

Return Value

None

Notes/Warnings

UserFnct() is called with interrupts enabled and you thus need to protect any shared objects.

Example

```
void main (void)
{
    .

    .

    TmrCfgFnct(0, Tmr0TimeoutFnct, (void *)0);

    .

    .

    TmrSetMST(0, 1, 0, 0);      /* Set timer #0 to expire in 1 minute */
    TmrStart(0);                /* Start timer #0                      */

    .

    .
}




void Tmr0TimeoutFnct (void *arg)
{
    DispStr(0, 0, "Timer #0 expired!");
}
```

7

TmrChk ( )

INT16U TmrChk(INT8U n);

TmrChk() allows you to check the progress of the countdown timer. Basically, the function returns the time remaining (in tenths of a second) until the timer expires. The timer expired if the returned value is 0.

Arguments

n is the timer number to start and must be a number between 0 and TMR_MAX_TMR − 1.

Return Value

The time remaining (in tenths of a second) of the desired timer.

Notes/Warnings

This function doesn't tell you whether the timer is running or not.

Example

```
void Task (void)
{
    INT16U time_remaining;


    for (;;) [
        .
        .
        time_remaining = TmrChk(0);    /* Get time left for timer #0 */
        .
        .
    }
}
```

TmrFormat()

```
void TmrFormat(INT8U n, char *s);
```

TmrFormat() is provided for display purposes. This function formats the time remaining of the specified timer into an ASCII string. Timers are always formatted as follows: MM:SS.T where MM is the remaining minutes to timeout, SS is the remaining seconds, and T is the tenths of a second.

Arguments

n is the timer number to format into an ASCII string and must be a number between 0 and TMR_MAX_TMR − 1.

s is a pointer to the string that will receive the formatted timer. Your destination string must allocate at least eight characters (including the NUL character).

Return Value

None

Notes/Warnings

None

7

Example

```
void Task (void)
{
    char s[10];



    for (;;) [
        .

        .

        TmrFormat(0, &s[0]); /* Get time left for timer #0 as "MM:SS.T" */
        .

        .

    }
}
```

TmrInit()

void TmrInit(void);

TmrInit() is the initialization code for the timer manager module. You must call TmrInit() before any other functions provided by this module. TmrInit() is responsible for the initialization of the timer module variables and the creation of the timer manager task.

Arguments

None

Return Value

None

Notes/Warnings

The #define TMR_MAX_TMR (see section 7.03, "Timer Manager Module, Configuration" on page 244) defines the number of timers managed by this module. All timers are disabled and in a non-configured state following initialization.

Example

```
void main (void)
{
    .
    .
    TmrInit();
    .
    .
}
```

TmrReset()

void TmrReset(INT8U n);

You can restart the countdown process to its initial value (established by either TmrSetT() or TmrSetMST()) by calling TmrReset(). This is a convenient function to use if you don't need to reprogram the timer with a new value every time you need to use the timer.

Arguments

n is the timer number to start and must be a number between 0 and TMR_MAX_TMR − 1.

Return Value

None

Notes/Warnings

None

Example

```
void Task (void)
{
    for (;;) [
        .
        .
        TmrReset(0);                   /* Reload timer #0 */
        .
        .
    }
}
```

7

TmrSetMST()

```
void TmrSetMST(INT8U n, INT8U min, INT8U sec, INT8U tenths);
```

This function allows you to set a timer by specifying minutes, seconds, and tenths of a second.

Arguments

n is the timer number to set and must be a number between 0 and TMR_MAX_TMR - 1.

min is the desired number of minutes (0..99).

sec is the desired number of seconds (0..59).

tenths is the desired number of tenths of a second (0..9).

Return Value

None

Notes/Warnings

Note that changing the timer value does not enable the timer. This means that setting the timer value does not initiate countdown. Countdown is initiated by calling TmrStart(). If the timer is enabled, however, TmrSetMST() will reload the timer and countdown will start from the new value.

Example

```
void Task (void)
{
    for (;;) [
        .
        .
        TmrSetMST(0, 0, 15, 0);      /* Reset timer #0 to 15 seconds */
        .
        .
    }
}
```

TmrSetT()

```
void TmrSetT(INT8U n, INT16U tenths);
```

This function allows you to set a timer in tenths of a second.

Arguments

n is the timer number to set and must be a number between 0 and TMR_MAX_TMR -1.

tenths is the desired timeout value of the timer and is specified in tenths of a second. For example, to set a timer to 27.4 seconds, you would specify 274.

Return Value

None

Notes/Warnings

Note that changing the timer value does not enable the timer. This means that setting the timer value does not initiate countdown. Countdown is initiated by calling TmrStart(). If the timer is enabled, however, TmrSetT() will reload the timer and countdown will start from the new value.

Example

```
void Task (void)
{
    for (;;) [
        .
        .
        TmrSetT(0, 150);          /* Reset timer #0 to 15 seconds */
        .
        .
    }
}
```

TmrStart()

void TmrStart(INT8U n);

Countdown of a timer is initiated only when you call TmrStart(). You should set the countdown time prior to calling TmrStart() with either TmrSetT() or TmrSetMST().

Arguments

n is the timer number to start and must be a number between 0 and TMR_MAX_TMR − 1.

Return Value

None

Notes/Warnings

TmrStart() will resume countdown of a timer that has been suspended by TmrStop().

Example

```
void Task (void)
{
    for (;;) [
        .
        .
        TmrSetT(0, 150);        /* Reset timer #0 to 15 seconds */
        TmrStart(0);            /* Start timer #0               */
        .
        .
    }
}
```

TmrStop()

void TmrStop(INT8U n);

Countdown of a timer can be suspended by calling TmrStop(). You can later resume countdown by calling TmrStart().

Arguments

n is the timer number to start and must be a number between 0 and TMR_MAX_TMR - 1.

Return Value

None

Notes/Warnings

TmrStop() doesn't reset the timer value, it simply suspends it.

Example

```
void Task (void)
{
    for (;;) [
        .
        .
        TmrStop(0);             /* Stop (i.e suspend) timer #0 */
        .
        .
    }
}
```

7

7.03 Timer Manager Module, Configuration

Configuration of the timer manager consists of defining the value of four #define constants (see file TMR.H and also, CFG.H).

TMR_TASK_PRIO defines the priority of TmrTask() in the multitasking environment. The task priority of the timer manager module would typically be set relatively low.

TMR_DLY_TICKS defines the number of clock ticks needed to obtain 0.1 second. If you use µC/OS-II, you can set this #define constant to OS_TICKS_PER_SEC / 10.

TMR_TASK_STK_SIZE defines the size of the stack (in bus width units) allocated to the timer manager module task. The number of bytes allocated for the stack is thus given by: TMR_TASK_STK_SIZE times sizeof(OS_STK).

WARNING

In the previous edition of this book, TMR_TASK_STK_SIZE specified the size of the stack for TmrTask() in number of bytes. µC/OS-II assumes the stack is specified in stack width elements.

TMR_MAX_TMR defines the number of timers managed by TmrTask(). If you use this module, you will need to have at least one timer. The timer manager can manage up to 250 timers. The limitation is strictly dictated by the amount of memory available and by the addressing capability of the target microprocessor.

TMR_USE_SEM is used to indicate that the timer manager will be expecting a signal from the tick ISR (through TmrSignalTmr()). When TMR_USE_SEM is set to 0, TmrTask() will use the kernel's time delay service (OSTimeDlyHMSM() for µC/OS-II).

7.04 Bibliography

Comer, Douglas
Operating System Design, The XINU Approach
Englewood Cliffs, New Jersey
Prentice-Hall, Inc., 1984

Listing 7.1   *TMR.C*

```
/*
*********************************************************************************************
*                             Embedded Systems Building Blocks
*                             Complete and Ready-to-Use Modules in C
*
*                                       Timer Manager
*
*                          (c) Copyright 1999, Jean J. Labrosse, Weston, FL
*                                     All Rights Reserved
*
* Filename   : TMR.C
* Programmer : Jean J. Labrosse
*********************************************************************************************
*/


/*
*********************************************************************************************
*                                     INCLUDE FILES
*********************************************************************************************
*/

#define  TMR_GLOBALS
#include "includes.h"

/*
*********************************************************************************************
*                                     LOCAL VARIABLES
*********************************************************************************************
*/

static  OS_EVENT    *TmrSemTenths;
static  OS_STK       TmrTaskStk[TMR_TASK_STK_SIZE];
static  INT8U        TmrTickCtr;

/*
*********************************************************************************************
*                                 LOCAL FUNCTION PROTOTYPES
*********************************************************************************************
*/

static  void        TmrTask(void *data);

/*$PAGE*/
```

7

Listing 7.1 (continued)   TMR.C

```
/*
*********************************************************************************************
*                              CONFIGURE TIMER TIMEOUT FUNCTION
*
* Description : Set the user defined function when the timer expires.
* Arguments   : n         is the timer number 0..TMR_MAX_TMR-1
*               fnct      is a pointer to a function that will be executed when the timer expires
*               arg       is a pointer to an argument that is passed to 'fnct'
* Returns     : None.
*********************************************************************************************
*/

void  TmrCfgFnct (INT8U n, void (*fnct)(void *), void *arg)
{
    TMR *ptmr;


    if (n < TMR_MAX_TMR) {
        ptmr              = &TmrTbl[n];
        OS_ENTER_CRITICAL();
        ptmr->TmrFnct     = fnct;              /* Store pointer to user function into timer     */
        ptmr->TmrFnctArg  = arg;               /* Store user's function arguments pointer       */
        OS_EXIT_CRITICAL();
    }
}

/*$PAGE*/


/*
*********************************************************************************************
*                                      CHECK TIMER
*
* Description : This function checks to see if a timer has expired
* Arguments   : n         is the timer to check
* Returns     : 0         if the timer has expired
*               TmrCtr    the remaining time before the timer expires in 1/10 second
*********************************************************************************************
*/

INT16U  TmrChk (INT8U n)
{
    INT16U val;


    val = 0;
    if (n < TMR_MAX_TMR) {
        OS_ENTER_CRITICAL();
        val = TmrTbl[n].TmrCtr;
        OS_EXIT_CRITICAL();
    }
    return (val);
}

/*$PAGE*/
```

Listing 7.1 (continued)   TMR.C

```
/*
*********************************************************************************************
*                              FORMAT TIMER INTO STRING
*
* Description : Formats a timer into an ASCII string.
* Arguments   : n       is the desired timer
*               s       is a pointer to the destination string.  The destination string must be large
*                       enough to hold the formatted timer value which will have the following format:
*                       "MM:SS.T"
*********************************************************************************************
*/

void  TmrFormat (INT8U n, char *s)
{
    INT8U   min;
    INT8U   sec;
    INT8U   tenths;
    INT16U  val;


    if (n < TMR_MAX_TMR) {
        OS_ENTER_CRITICAL();
        val    = TmrTbl[n].TmrCtr;                       /* Get local copy of timer for conversion    */
        OS_EXIT_CRITICAL();
        min    = (INT8U)(val / 600);
        sec    = (INT8U)((val - min * 600) / 10);
        tenths = (INT8U)(val % 10);                      /* Convert TIMER to ASCII                     */
        s[0]   = min / 10 + '0';
        s[1]   = min % 10 + '0';
        s[2]   = ':';
        s[3]   = sec / 10 + '0';
        s[4]   = sec % 10 + '0';
        s[5]   = '.';
        s[6]   = tenths   + '0';
        s[7]   = NUL;
    }
}

/*$PAGE*/
```

Listing 7.1 (continued)   TMR.C

```
/*
*********************************************************************************************
*                                 TIMER MANAGER INITIALIZATION
*
* Description : This function initializes the timer manager module.
* Arguments   : None
* Returns     : None.
*********************************************************************************************
*/

void  TmrInit (void)
{
    INT8U  err;
    INT8U  i;
    TMR    *ptmr;


    ptmr = &TmrTbl[0];
    for (i = 0; i < TMR_MAX_TMR; i++) {            /* Clear and disable all timers            */
        ptmr->TmrEn   = FALSE;
        ptmr->TmrCtr  = 0;
        ptmr->TmrInit = 0;
        ptmr->TmrFnct = NULL;
        ptmr++;
    }
    TmrTickCtr   = 0;
    TmrSemTenths = OSSemCreate(0);                 /* Create counting semaphore to signal 1/10 second   */
    OSTaskCreate(TmrTask, (void *)0, &TmrTaskStk[TMR_TASK_STK_SIZE], TMR_TASK_PRIO);
}


/*$PAGE*/
```

Listing 7.1 (continued)   TMR.C

```
/*
*********************************************************************************************
*                                     RESET TIMER
*
* Description : This function reloads a timer with its initial value
* Arguments   : n          is the timer to reset
* Returns     : None.
*********************************************************************************************
*/

void  TmrReset (INT8U n)
{
    TMR *ptmr;


    if (n < TMR_MAX_TMR) {
        ptmr        = &TmrTbl[n];
        OS_ENTER_CRITICAL();
        ptmr->TmrCtr = ptmr->TmrInit;        /* Reload the counter                      */
        OS_EXIT_CRITICAL();
    }
}

/*$PAGE*/

/*
*********************************************************************************************
*                   SET TIMER (SPECIFYING MINUTES, SECONDS and TENTHS)
*
* Description : Set the timer with the specified number of minutes, seconds and 1/10 seconds.  The
*               function converts the minutes, seconds and tenths into tenths.
* Arguments   : n          is the timer number 0..TMR_MAX_TMR-1
*               min        is the number of minutes
*               sec        is the number of seconds
*               tenths     is the number of tenths of a second
* Returns     : None.
*********************************************************************************************
*/

void  TmrSetMST (INT8U n, INT8U min, INT8U sec, INT8U tenths)
{
    TMR    *ptmr;
    INT16U  val;


    if (n < TMR_MAX_TMR) {
        ptmr        = &TmrTbl[n];
        val         = (INT16U)min * 600 + (INT16U)sec * 10 + (INT16U)tenths;
        OS_ENTER_CRITICAL();
        ptmr->TmrInit = val;
        ptmr->TmrCtr  = val;
        OS_EXIT_CRITICAL();
    }
}

/*$PAGE*/
```

7

Listing 7.1 (continued)   TMR.C

```
/*
*********************************************************************************************
*                         SET TIMER (SPECIFYING TENTHS OF SECOND)
*
* Description : Set the timer with the specified number of 1/10 seconds.
* Arguments   : n         is the timer number 0..TMR_MAX_TMR-1
*               tenths    is the number of 1/10 second to load into the timer
* Returns     : None.
*********************************************************************************************
*/

void  TmrSetT (INT8U n, INT16U tenths)
{
    TMR *ptmr;


    if (n < TMR_MAX_TMR) {
        ptmr          = &TmrTbl[n];
        OS_ENTER_CRITICAL();
        ptmr->TmrInit = tenths;
        ptmr->TmrCtr  = tenths;
        OS_EXIT_CRITICAL();
    }
}

/*$PAGE*/

/*
*********************************************************************************************
*                SIGNAL TIMER MANAGER MODULE THAT A 'CLOCK TICK' HAS OCCURRED
*
* Description : This function is called by the 'clock tick' ISR on every tick.  This function is thus
*               responsible for counting the number of clock ticks per 1/10 second.  When 1/10 second
*               elapses, this function will signal the timer manager task.
* Arguments   : None.
* Returns     : None.
* Notes       : TMR_DLY_TICKS must be set to produce 1/10 second delays.
*               This can be set to OS_TICKS_PER_SEC / 10 if you use uC/OS-II.
*********************************************************************************************
*/

void  TmrSignalTmr (void)
{
    TmrTickCtr++;
    if (TmrTickCtr >= TMR_DLY_TICKS) {
        TmrTickCtr = 0;
        OSSemPost(TmrSemTenths);
    }
}

/*$PAGE*/
```

Listing 7.1 (continued)   TMR.C

```
/*
*********************************************************************************************
*                                     START TIMER
*
* Description : This function start a timer
* Arguments   : n          is the timer to start
* Returns     : None.
*********************************************************************************************
*/

void  TmrStart (INT8U n)
{
    if (n < TMR_MAX_TMR) {
        OS_ENTER_CRITICAL();
        TmrTbl[n].TmrEn = TRUE;
        OS_EXIT_CRITICAL();
    }
}


/*$PAGE*/

/*
*********************************************************************************************
*                                     STOP TIMER
*
* Description : This function stops a timer
* Arguments   : n          is the timer to stop
* Returns     : None.
*********************************************************************************************
*/

void  TmrStop (INT8U n)
{
    if (n < TMR_MAX_TMR) {
        OS_ENTER_CRITICAL();
        TmrTbl[n].TmrEn = FALSE;
        OS_EXIT_CRITICAL();
    }
}

/*$PAGE*/
```

7

Listing 7.1 (continued) TMR.C

```
/*
*********************************************************************************************
*                                    TIMER MANAGER TASK
*
* Description : This task is created by TmrInit() and is responsible for updating the timers.
*               TmrTask() executes every 1/10 of a second.
* Arguments   : None.
* Returns     : None.
* Note(s)     : 1) The function to execute when a timer times out is executed outside the critical
*                  section.
*********************************************************************************************
*/

static  void  TmrTask (void *data)
{
    TMR    *ptmr;
    INT8U   err;
    INT8U   i;
    void   (*pfnct)(void *);                    /* Function to execute when timer times out   */
    void    *parg;                              /* Arguments to pass to above function        */



    data  = data;                               /* Avoid compiler warning (uC/OS-II req.)     */
    pfnct = (void (*)(void *))0;                 /* Start off with no function to execute      */
    parg  = (void *)0;
    for (;;) {
#if TMR_USE_SEM
        OSSemPend(TmrSemTenths, 0, &err);       /* Wait for 1/10 second signal from TICK ISR  */
#else
        OSTimeDlyHMSM(0, 0, 0, 100);            /* Delay for 1/10 second                      */
#endif

        ptmr = &TmrTbl[0];                      /* Point at beginning of timer table          */
        for (i = 0; i < TMR_MAX_TMR; i++) {
            OS_ENTER_CRITICAL();
            if (ptmr->TmrEn == TRUE) {          /* Decrement timer only if it is enabled      */
                if (ptmr->TmrCtr > 0) {
                    ptmr->TmrCtr--;
                    if (ptmr->TmrCtr == 0) {            /* See if timer expired               */
                        ptmr->TmrEn = FALSE;            /* Yes, stop timer                    */
                        pfnct       = ptmr->TmrFnct;    /* Get pointer to function to execute ... */
                        parg        = ptmr->TmrFnctArg; /* ... and its argument               */
                    }
                }
            }
            OS_EXIT_CRITICAL();
            if (pfnct != (void (*)(void *))0) { /* See if we need to execute function for ... */
                (*pfnct)(parg);                 /* ... timed out timer.                       */
                pfnct = (void (*)(void *))0;
            }
            ptmr++;
        }
    }
}
```

Listing 7.2   *TMR.H*

```
/*
*********************************************************************************************
*                              Embedded Systems Building Blocks
*                              Complete and Ready-to-Use Modules in C
*
*                                        Timer Manager
*
*                           (c) Copyright 1999, Jean J. Labrosse, Weston, FL
*                                        All Rights Reserved
*
* Filename   : TMR.H
* Programmer : Jean J. Labrosse
*********************************************************************************************
*/

/*
*********************************************************************************************
*                                        CONSTANTS
*********************************************************************************************
*/

#ifndef  CFG_H

#define  TMR_DLY_TICKS       (OS_TICKS_PER_SEC / 10)
#define  TMR_TASK_PRIO        45
#define  TMR_TASK_STK_SIZE   512

#define  TMR_MAX_TMR          20

#define  TMR_USE_SEM           0

#endif

#ifdef  TMR_GLOBALS
#define TMR_EXT
#else
#define TMR_EXT   extern
#endif

/*
*********************************************************************************************
*                                        DATA TYPES
*********************************************************************************************
*/

typedef struct tmr {                        /* TIMER DATA STRUCTURE                        */
    BOOLEAN   TmrEn;                        /* Flag indicating whether timer is enabled    */
    INT16U    TmrCtr;                       /* Current value of timer (counting down)      */
    INT16U    TmrInit;                      /* Initial value of timer (i.e. when timer is set) */
    void    (*TmrFnct)(void *);             /* Function to execute when timer times out     */
    void     *TmrFnctArg;                   /* Arguments supplied to user defined function  */
} TMR;
```

7

Listing 7.2 (continued)   TMR.H

```
/*
*********************************************************************************************************
*                                           GLOBAL VARIABLES
*********************************************************************************************************
*/

TMR_EXT  TMR        TmrTbl[TMR_MAX_TMR];             /* Table of timers managed by this module          */

/*
*********************************************************************************************************
*                                          FUNCTION PROTOTYPES
*********************************************************************************************************
*/

void    TmrCfgFnct(INT8U n, void (*fnct)(void *), void *arg);
INT16U  TmrChk(INT8U n);

void    TmrFormat(INT8U n, char *s);

void    TmrInit(void);

void    TmrReset(INT8U n);

void    TmrSetMST(INT8U n, INT8U min, INT8U sec, INT8U tenths);
void    TmrSetT(INT8U n, INT16U tenths);
void    TmrSignalTmr(void);
void    TmrStart(INT8U n);
void    TmrStop(INT8U n);
```

Chapter 8

Discrete I/Os

Discrete inputs and outputs (I/Os) are found in most control and/or monitoring systems. The word *discrete* refers to the fact that the value taken by the input can take only one of two states. For example:

- 1 or 0
- TRUE or FALSE
- ON or OFF
- ENABLED or DISABLED
- PRESENT or ABSENT
- and so on.

Figure 8.1   Discrete inputs.

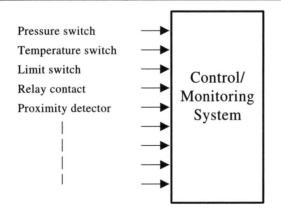

As shown in Figure 8.1, discrete inputs are generally used to monitor the state of manual switches, pressure switches (pressure exceeded or not), temperature switches (temperature exceeded or not), limit switches (device has reach its limit or not), relay contact closures (open or closed), proximity detectors (there or not there), etc. *Discrete inputs* are generally used to determine the state of an input. In some applications, however, you need to know whether a discrete input has changed state or not and, possibly, how many times it did so.

Discrete outputs are used to control lamps, relays, fans, alarms, heaters, valves, etc. (See Figure 8.2.) A discrete output is generally either in one state or the other. A blinking light versus a light that is always ON, however, does a better job of attracting the attention of a user to an error condition.

Figure 8.2   Discrete outputs.

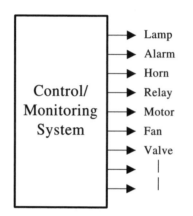

In this chapter, I will provide you with a module that monitors discrete inputs and controls discrete outputs. The module allows you to have as many discrete inputs and outputs as you need (up to 250 each). For each discrete input, you will be able to:

- Determine whether the input is 1 or 0.
- Determine whether a transition from 1 to 0 or from 0 to 1 occurred on the input.
- Determine how many transitions from 1 to 0 or from 0 to 1 occurred on the input.
- Simulate a toggle switch with a momentary closure switch.
- Bypass the hardware for debug purposes.

  For each discrete output, you will be able to:
- Turn the output ON or OFF.
- Blink the output at a user-definable rate (one for each output).
- Bypass your application code to control the output during debugging.

8.00 Discrete Inputs

Reading discrete inputs is a fairly trivial task. You need only provide your microprocessor with as many parallel input lines as you have discrete inputs to read. The microprocessor simply needs to read the input ports, mask off unwanted inputs, and make a decision based on the state of the input.

I generally prefer to put a layer of software between my application code and the hardware so I can change the hardware without affecting the application software. Putting a layer of software also allows you to test your application before you get your hands on the hardware. I like to give a *logical address* to each discrete input, typically from 0 to n. You can thus write a simple function that returns the state of any logical discrete input to your application as shown in the following pseudocode:

```
BOOLEAN DIGet(INT8U n)
{
    Read port where discrete input #n is located;
    Mask off unwanted bits;
    Return the state of the discrete input (either TRUE or FALSE);
}
```

The *mask* is an 8-bit value that selects the desired bit to read. For example, to read the state of bit 4 (bits are numbered 0 to 7 from right to left), the mask would be 0x10. With such a function, your code will be a little bit slower and your code size will increase but the benefits are enormous. Now you can change the hardware as many times as you need and your application code will never know the difference. By encapsulating access to the hardware we can also handle cases where some of the inputs are inverted by the hardware and still return the proper state to the application code. In other words, if an input is considered a logical 0 when it is HIGH, then DIGet() can invert the value of the input read and report a 0 to the application code.

If you have spare address space and a "say" about hardware design, you should consider using one of my favorite chips for discrete inputs: the *74251 8-input data selector/multiplexer,* shown in Figure 8.3. Note that you can have as many discrete inputs as needed by simply adding 74251s.

8

Figure 8.3   *Discrete inputs using 74251.*

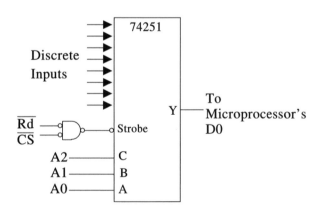

Basically, each discrete input has its own address in the microprocessor address space. Reading a discrete input becomes trivial:

```
BOOLEAN DIGet(INT8U n)
{
    return (Read value from address of port 'n' and mask with 0x01);
}
```

Even with DIGet(), it is still up to your application to determine whether a discrete input has changed state. To determine if an input has changed state, you will need to repeatedly call DIGet() (i.e., poll the input) and compare the previous value with the current one. The input has changed state when both values are different. If you need to know whether the input changed from 0 to 1, you will further need to add code to ensure that the previous state was 0.

What if you had a momentary closure switch connected to a discrete input and needed to simulate a toggle switch? (That is, you press the switch once to turn a device ON and you press the switch again to turn the device OFF.) To accomplish this, you need to change the state of a variable whenever a transition from 0 to 1 is detected.

The discrete I/O module presented in this chapter allows you to configure any discrete input to handle all of the situations described earlier. Each discrete input is considered a *logical channel*. The discrete I/O module allows you to have as many logical channels as you need (up to 250). Figure 8.4 shows a flow diagram of a *discrete input channel*. Note that I used electrical symbols to represent the functions performed by each discrete input channel. Of course, all functions are handled in software.

Figure 8.4   Discrete input channel.

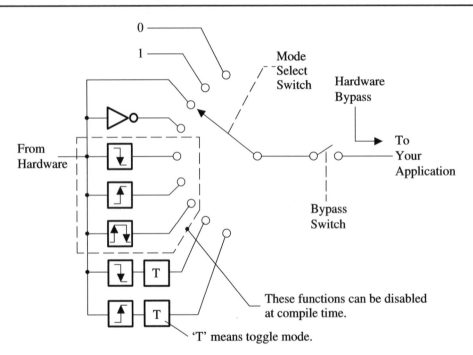

As Figure 8.4 shows, each discrete input channel has the capability to be configured (at run-time) to any of nine modes through the *Mode Select Switch*:

1. Always return a 0.
2. Always return a 1.
3. Return the state of the hardware input.
4. Return the complement of the hardware input.
5. Detect negative-going transitions and return the number of transitions detected.
6. Detect positive-going transitions and return the number of transitions detected.
7. Detect both positive- and negative-going transitions and return the number of transitions detected.
8. Toggle between 0 and 1 when a negative transition is detected.
9. Toggle between 0 and 1 when a positive transition is detected.

To reduce the code size of your application, the edge detection features can be disabled at compile time, as shown in Figure 8.4.

To provide the functionality described earlier, all discrete inputs are read and processed on a continuous basis. In other words, all inputs are polled. Because of this, the maximum rate at which discrete inputs can change state is based on how often inputs are polled. Polling is handled by a task (described later) which executes at a regular interval (you decide at compile time how often the task will execute). Discrete inputs must not change state any faster than half the task execution rate of the discrete I/O module. That is, the task must execute twice as fast as the expected rate of change of discrete inputs.

Your application knows about discrete input channels through interface functions. The interface functions allow you to set the configuration mode of each channel through the *Mode Select Switch*, set the state of the *Bypass Switch* and, if the bypass switch is open, bypass the hardware. Bypassing of the hardware is accomplished by having an interface function deposit a value into the discrete channel. Where your application is concerned, it doesn't know that the value received didn't come from the actual hardware.

8.01 Discrete Outputs

Updating discrete outputs is a straightforward operation but a little trickier than updating discrete inputs. All you need is to provide your microprocessor with enough latched parallel output lines as you have discrete outputs to control. As with discrete inputs, I generally prefer to put a layer of software between my application code and the hardware. This prevents the application code from knowing what kind of hardware is involved and how it is accessed. I can thus port my application code to other environments by simply changing the hardware interface functions. I give a *logical address* to each discrete output, typically from 0 to n. For discrete outputs connected to an 8-bit latched parallel output port, you have two scenarios: either you can read back the contents of the output port (Intel 8255A or Motorola 6821) or else the port is write-only (74273, 74373, etc.). The pseudocode for a port that can be read back would look like this:

```
void DOSet(INT8U n, BOOLEAN val)
{
    Disable interrupts;
    Read the output port;
    if (val == FALSE) {
        AND the port data with complement of 'mask';
    } else {
        OR the port data with mask;
    }
    Write new data to port;
    Enable Interrupts;
}
```

The *mask* is an 8-bit value that selects the desired bit to set or clear. For example, to set or clear bit 6 (bits are numbered 0 to 7 from right to left), the mask would be 0x40. Note that you also need to disable interrupts because updating the discrete output is considered a critical section. Forgetting to disable interrupts is a common mistake. The pseudocode for a port that cannot be read back follows this paragraph. In this case, an image of the output port's content is maintained in memory (i.e., RAM).

```
void DOSet(INT8U n, BOOLEAN val)
{
    Disable interrupts;
    if (val == FALSE) {
        AND the memory image with the complement of the 'mask';
    } else {
        OR the memory image with the mask;
    }
    Write memory image to port;
    Enable Interrupts;
}
```

If you have spare address space and a "say" about hardware design, you should consider using one of my favorite chips for discrete outputs: the *74259 8-bit addressable latch,* as shown in Figure 8.5. Note that you can have as many discrete outputs as needed by simply adding 74259s.

Figure 8.5   Discrete outputs using 74259.

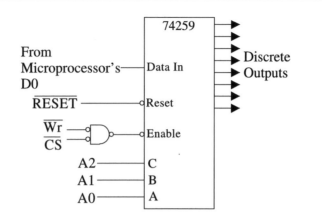

Basically, each discrete output has its own address in the microprocessor address space. Updating a discrete output becomes trivial:

```
void DOSet(INT8U n, BOOLEAN val)
{
    Output value to address of port 'n';
}
```

What if you needed to *blink* one or more discrete outputs? Blinking outputs are quite useful when connected to lights because they can be used to signal alarm conditions to users. To blink an output, you could call DOSet() to change the state of an output at a regular interval from your application code. This obviously complicates your application.

The discrete I/O module presented in this chapter allows you to control discrete outputs and also blink any (or all) of the discrete outputs.

Each discrete output is considered a *logical channel*. The discrete I/O module allows you to have as many logical channels as you need (up to 250). Figure 8.6 shows a flow diagram of a *discrete output channel*. Note that I used electrical symbols to represent the functions performed by each discrete output channel. Of course, all functions are handled in software.

8

Figure 8.6   Discrete output channel.

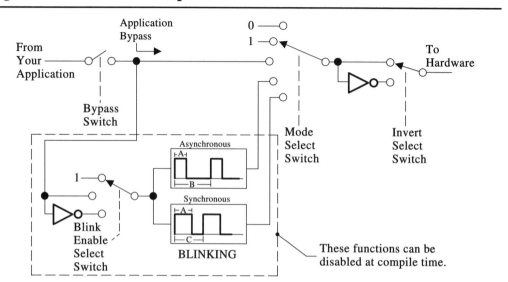

As shown in Figure 8.6, each discrete output channel has the capability to be configured (at run-time) to any of five modes (through the *Mode Select Switch*):

1. Always output a 0.

2. Always output a 1.

3. Directly output what your application desires to put out.

4. Blink the output *asynchronously* (described below).

5. Blink the output *synchronously* (described below).

Your application software can also complement (or invert) the output through the *Invert Select Switch*.

If either of the two blinking modes is selected, your application can determine whether blinking will be enabled through the *Blink Enable Select Switch*. To reduce the code size of your application, the blinking feature can be disabled at compile time, as shown in Figure 8.6.

Your application knows about discrete output channels through interface functions. The interface functions allow you to:

•   Set the configuration mode of each channel through the *Mode Select Switch*.

•   Set the *Blink Enable Select Switch*, which determines how to enable blinking.

•   Determine whether the output will be inverted by setting the *Invert Select Switch*.

•   Set the blinking rate by specifying the values for *A*, *B*, and *C* (see Figure 8.6).

•   Set the state of the *Bypass Switch* and, if the bypass switch is open, bypass your application code. Bypassing of your application is accomplished by having an interface function deposit a value into the discrete channel. Where your application is concerned, it still thinks it is controlling the discrete output channel.

When you choose to blink a discrete output, you need to specify the type of blinking: either *asynchronous* or *synchronous*. In asynchronous mode, you need to specify the duty cycle through two variables: *A* (the *ON* time) and *B* (the total time). Because each discrete output can have different *A* and *B* values, blinking occurs asynchronously. In synchronous mode, you specify the *ON* time (variable *A*) with respect to a common (to all synchronous discrete outputs) total time (variable *C*). The *ON* time and total time are based on how often the discrete I/O module executes. If the discrete I/O modules executes 10 times per second then, an *ON* time of one second requires *A* to be set to 10.

8.02 Discrete I/O Module

The source code for the discrete I/O module is found in the \SOFTWARE\BLOCKS\DIO\SOURCE directory. The source code is found in the files DIO.C (Listing 8.1) and DIO.H (Listing 8.2). As a convention, all functions and variables related to the discrete I/O module start with either DIO (functions or variables common to both discrete inputs and outputs), DI (discrete input functions or variables), or DO (discrete output functions or variables). Similarly, #defines constants will either start with DIO_, DI_, or DO_.

8.03 Discrete I/O Module, Internals

Figure 8.7 shows a flow diagram of the discrete I/O module. (You can also refer to Listings 8.1 and 8.2 for the following description.) The discrete I/O module consists of a single task (DIOTask()) that executes at a regular interval (DIO_TASK_DLY_TICKS). DIOTask() can manage as many discrete inputs and outputs as your application requires (up to 250 each). The discrete I/O manager is initialized by calling DIOInit(). Every DIO_TASK_DLY_TICKS, DIOTask() calls DIRd(), DIUpdate(), DOUpdate(), and DOWr().

8

Figure 8.7   DIO module flow diagram.

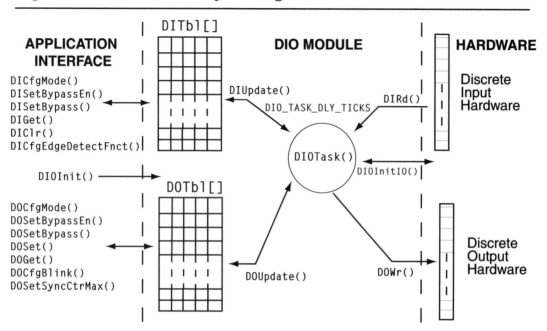

DITbl[] is a table that contains configuration and run-time information for each discrete input channel. An entry in DITbl[] is a structure defined in DIO.H and is called DIO_DI. Discrete inputs are read and mapped to DITbl[i].DIIn by the hardware interface function DIRd(). DIRd() knows about your hardware and thus can be easily changed to adapt to your environment.

Figure 8.8 shows a flow diagram of a discrete input channel. Note that I used electrical symbols to represent functions performed in software for each discrete input channel. .DIIn, .DIModeSel, .DIBypassEn, and .DIVal are structure members of DIO_DI (see DIO.H). DIUpdate() is responsible for updating all the discrete input channels. Discrete input channels that are configured for edge detection are processed by DIIsTrig(). DIIsTrig() keeps track of the previous state (.DIPrev) of the discrete input and is used to determine if an input has changed state.

Figure 8.8   Discrete input channel.

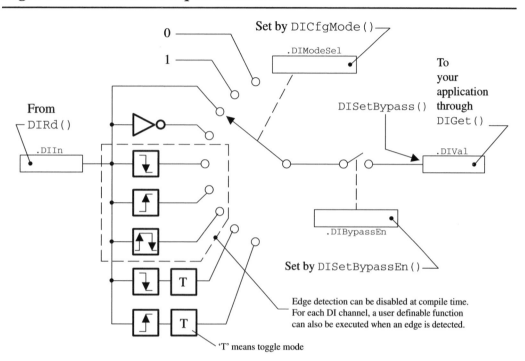

DOTbl[] is a table that contains configuration and run-time information for each discrete output channel. An entry in DOTbl[] is a structure defined in DIO.H and is called DIO_DO. Discrete outputs are mapped from DOTbl[i].DOOut   to your hardware through the interface function DOWr(). DOWr() knows about your hardware and thus can be easily changed to adapt to your environment.

Figure 8.9 shows a flow diagram of a discrete output channel. Note that I used electrical symbols to represent functions performed in software for each discrete output channel. .DOCtrl, .DOBypassEn, .DOBypass, .DOBlinkEnSel, .DOModeSel, .DOInv, and .DOOut are structure members of DIO_DO (see DIO.H). DOUpdate() is responsible for updating all the discrete output channels.

Figure 8.9   Discrete output channel.

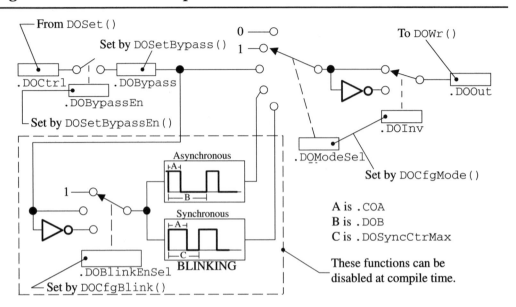

As previously mentioned, there are two blinking modes: synchronous and asynchronous.

Synchronous blinking mode is shown in Figure 8.10. When a discrete output channel is in this mode, its output is HIGH (or LOW depending on the state of .DOInv) when .DOA is less than DOSyncCtr. DOSyncCtr counts from 0 to DOSyncCtrMax (set by DOSetSyncCtrMax()). DOSyncCtr is cleared when it reaches DOSyncCtrMax. This mode is synchronous because all discrete output channels in this mode are referenced to DOSyncCtr.

Figure 8.10   Synchronous blinking mode.

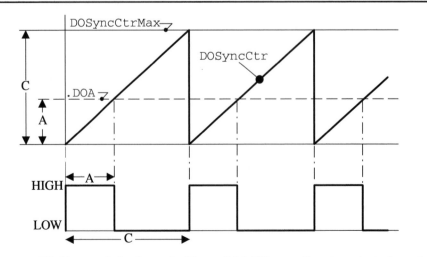

Asynchronous blinking mode is shown in Figure 8.11. When a discrete output channel is in this mode its output is HIGH (or LOW depending on the state of .DOInv) when .DOA is less than .DOBCtr.

`.DOBCtr` counts from 0 to `.DOB` (set by `DOCfgBlink()`). `.DOBCtr` is cleared when it reaches `.DOB`. This mode is asynchronous because all discrete output channels maintain their own `.DOBCtr` and thus can blink at different rates.

Figure 8.11 Asynchronous blinking mode.

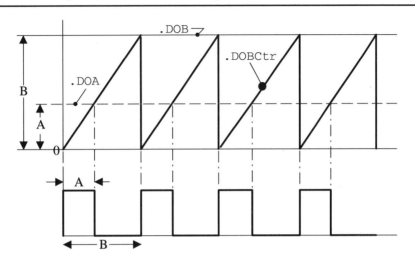

8.04 Discrete I/O Module, Interface Functions

Your application software knows about the discrete I/O module through the interface functions shown in Figure 8.12.

Figure 8.12 Discrete I/O module interface functions.

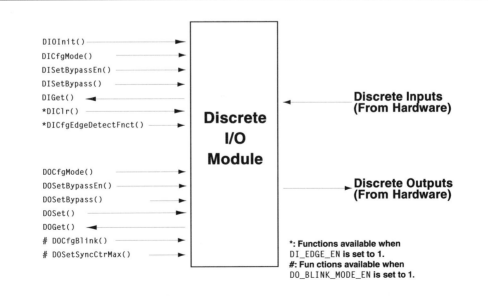

To allow the code size in your application to be reduced, I have added two #defines, which are used to enable/disable code generation for edge detection for discrete inputs (DI_EDGE_EN) and enable/disable code generation for blinking of discrete outputs (DO_BLINK_MODE_EN). Setting these #defines to 1 will enable code generation for the respective code.

DICfgEdgeDetectFnct()

`void DICfgEdgeDetectFnct(INT8U n, void (*fnct)(void *), void *arg);`

When a discrete input channel is configured for edge detection and a transition is detected, a user-definable function can be executed. The function to execute is specified to the discrete input channel by calling `DICfgEdgeDetectFnct()`.

Arguments

n is the discrete input channel you wish to configure. Discrete input channels are numbered from 0 to `DIO_MAX_DI - 1`.

fnct is a pointer to the function that will be executed whenever a transition is detected. Note that passing a `NULL` pointer indicates that no function is to be executed when a transition is detected. All discrete input channels have `NULL` pointers by default. When the function is called, it is passed a pointer to `void` (i.e., the `arg`). This allows different arguments to be passed to a reentrant function. You must declare the function that will be called as follows:

```
void UserFnct (void *arg);
```

Note that `UserFnct()` is called with the argument that you specify in `DICfgEdgeDetect-Fnct()`, that is, `arg`. This allows you to design a single function that can be used by more than one discrete input channel. The user-defined function will be called by the discrete I/O manager task `DIOTask()` when a transition is detected on the input. The execution time of the discrete I/O task is thus augmented by the execution time of all the functions that will execute when a transition is detected in their respective inputs.

Return Value

None

Notes/Warnings

Some applications do not require the execution of a function upon detection of a transition. In these situations, there is no need to call `DICfgEdgeDetectFnct()` because the initial value of the pointer to a function for each discrete input channel is `NULL`. In other words, the discrete I/O task will not execute any function when pointing to `NULL`.

8

Example

The function that executes when a transition is detected can signal another task through a semaphore, a mailbox, or even a message queue. This would allow you to defer processing of input transition detection to either a lower- or higher-priority task.

```
OS_EVENT *DISem;


void Task (void *pdata)
{
    INT8U err;



    DISem = OSSemCreate(0);
    DICfgMode(0, DI_MODE_EDGE_HIGH_GOING);
    DICfgEdgeDetectFnct(0, DIEdgeFnct, (void *)DISem);
    for (;;) {

        .

        .

        OSSemPend(DISem, 0, &err);      /* Wait for DI to transition */

        .

        .

    }
}



void DIEdgeFnct (void *arg)
{
    OSSemPost((OS_EVENT *)arg);       /* DI transitioned          */
}
```

DICfgMode()

void DICfgMode(INT8U n, INT8U mode);

DICfgMode() is used to set the operating mode of a discrete input channel.

Arguments

n is the desired discrete input channel to configure. Discrete input channels are numbered from 0 to DIO_MAX_DI − 1.

mode determines the operating mode of the discrete input channel. The discrete I/O module currently supports nine modes:

1. DI_MODE_LOW allows DIGet() (described later) to always return 0. This function basically simulates grounding an input.

2. DI_MODE_HIGH is similar to DI_MODE_LOW in that it allows DIGet() to always return 1. This function basically simulates tying an input high.

3. DI_MODE_DIRECT allows the discrete input channel to read whatever is present on the hardware input. This is the default mode for a discrete input channel.

4. DI_MODE_INV allows the discrete input channel to read the complement of whatever is present on the hardware input.

5. DI_MODE_EDGE_LOW_GOING allows the discrete input channel to detect and count transitions from 1 to 0 on the hardware input. The frequency of the input signal must be less than the scan rate of the discrete I/O module (determined by DIO_TASK_DLY_TICKS). DIGet() will return the number of 1 to 0 transitions detected. Note that the number of transitions can be cleared by calling DIClr() (described later).

6. DI_MODE_EDGE_HIGH_GOING allows the discrete input channel to detect and count transitions from 0 to 1 on the hardware input. The frequency of the input signal must be less than the scan rate of the discrete I/O module. DIGet() will return the number of 0 to 1 transitions detected. Note that the number of transitions can be cleared by calling DIClr() (described later).

7. DI_MODE_EDGE_BOTH allows the discrete input channel to detect and count either transitions from 0 to 1 or from 1 to 0 on the hardware input. The transition rate of the input signal must be less than the scan rate of the discrete I/O module. DIGet() will return the number of transitions detected. Note that the number of transitions can be cleared by calling DIClr() (described later).

8. DI_MODE_TOGGLE_LOW_GOING allows the state of the discrete input channel to change whenever a transition from a 1 to a 0 is detected. Again, the transition rate of the input signal must be less than the scan rate of the discrete I/O module.

9. DI_MODE_TOGGLE_HIGH_GOING allows the state of the discrete input channel to change whenever a transition from a 0 to a 1 is detected. Again, the transition rate of the input signal must be less than the scan rate of the discrete I/O module.

8

Return Value

None

Notes/Warnings

None

Example

```
void main (void)
{
    .
    .
    DICfgMode(0, DI_MODE_DIRECT);
    .
    .
}
```

DIClr()

void DIClr(INT8U n);

The only way to clear the number of transitions detected when the discrete input channel is configured for edge detection is to call DIClr(). The function has no effect if the channel is not configured for edge detection.

Arguments

n is the discrete input channel you wish to clear. Discrete input channels are numbered from 0 to DIO_MAX_DI - 1.

Return Value

None

Notes/Warnings

None

Example

```
void Task (void *pdata)
{
    DICfgMode(0, DI_MODE_EDGE_HIGH_GOING);
    for (;;) {
        .

        .

        DIClr(0);     /* Clear the number of transitions of channel #0 */

        .

        .

    }
}
```

8

DIGet()

INT16U DIGet(INT8U n);

The current value of the discrete input channel can be obtained by calling DIGet(). If the discrete input channel is configured for edge detection, the returned value will correspond to the number of transitions detected by the channel. If the discrete input channel is not configured for edge detection, the returned value will either be 0 or 1.

Arguments

n is the discrete input channel you wish to read. Discrete input channels are numbered from 0 to DIO_MAX_DI - 1.

Return Value

The current value of the discrete input channel or the number of transitions.

Notes/Warnings

None

Example

```
void Task (void *pdata)
{
    INT16U transitions;


    DICfgMode(0, DI_MODE_EDGE_HIGH_GOING);
    for (;;) {
        .
        .
        transitions = DIGet(0);    /* Get number of transitions on DI #1 */
        .
        .
    }
}
```

DIOInit()

void DIOInit(void);

DIOInit() is the initialization code for the discrete I/O module. DIOInit() must be called before you use any of the other discrete I/O module functions. DIOInit() is responsible for initializing the internal variables used by the module and for the creation of the task that will update the discrete inputs and outputs.

Arguments

None

Return Value

None

Notes/Warnings

None

Example

8

```
void main (void)
{
    .
    .
    DIOInit();
    .
    .
    .
}
```

DISetBypass()

`void DISetBypass(INT8U n, INT16U val);`

Your application software can bypass or override the discrete input channel value by using this function. `DISetBypass()` doesn't do anything unless you have opened the bypass switch by calling `DISetBypassEn()` as described earlier.

Arguments

n is the discrete input channel you wish to bypass. Discrete input channels are numbered from 0 to `DIO_MAX_DI - 1`.

val is the value you want `DIGet()` to return to your application. Because `val` is a INT16U, you can set the number of transitions detected when the discrete input channel is configured for edge detection.

Return Value

None

Notes/Warnings

None

Example

```
void Task (void *pdata)
{
    for (;;) {
        .
        DOSetBypassEn(0, TRUE);   /* Bypass channel #0        */
        .
        DOSetBypass(0, 1);        /* Set value of channel #0 */
        .
        .
    }
}
```

DISetBypassEn()

void DISetBypassEn(INT8U n, BOOLEAN state);

DISetBypassEn() allows your application code to prevent the 'physical' discrete input channel from being updated. This permits your application to set the value returned by DIGet(). The value of the discrete input channel is set by DISetBypass(). DISetBypassEn() and DISetBypass() are very useful for debugging.

Arguments

n is the discrete input channel you wish to bypass. Discrete input channels are numbered from 0 to DIO_MAX_DI – 1.

state is the state of the bypass switch. When TRUE, the bypass switch is open (i.e., the discrete input channel is bypassed). When FALSE, the bypass switch is closed (i.e., the discrete input channel is not bypassed).

Return Value

None

Notes/Warnings

None

8

Example

```
void Task (void *pdata)
{
    for (;;) {
        .
        .
        DISetBypassEn(0, TRUE);  /* Bypass channel */
        .
        .
    }
}
```

DOCfgBlink()

```
void DICfgBlink(INT8U n, INT8U mode, INT8U a, INT8U b);
```

DOCfgBlink() allows you to configure the discrete output blinking mode.

Arguments

n is the discrete output channel you wish to configure for blink mode. Discrete output channels are numbered from 0 to DIO_MAX_DO - 1.

mode sets the state of the *Blink Enable Select Switch* to one of three values:

1. DO_BLINK_EN allows the discrete output to blink continuously.

2. DO_BLINK_EN_NORMAL allows the discrete output to blink only if the input to the discrete output channel is set to 1. Blinking stops when the input to the discrete output channel is set to 0. In this case, the output is forced LOW unless it's inverted.

3. DO_BLINK_EN_INV allows the discrete output to blink only if the input to the discrete output channel is set to 0. Blinking stops when the input to the discrete output channel is set to 1. In this case, the output is forced LOW unless it's inverted.

a specifies the ON time for either synchronous or asynchronous mode (the A value in Figures 8.9, 8.10, and 8.11). The actual ON time is determined by the execution rate of the discrete I/O module. a is given by:

[8.1] $a = $ ON time (sec.) $\times$ Task execution rate (Hz)

b specifies the total period when the discrete output is configured for asynchronous mode (the B value of Figures 8.9 and 8.11). The period is determined by the execution rate of the discrete I/O module. b is given by:

[8.2] $b = $ Period (sec.) $\times$ Task execution rate (Hz)

Return Value

None

Notes/Warnings

None

Example

```
void Task (void *pdata)
{
    DOCfgBlink(0, DO_BLINK_EN, 10, 20);
    for (;;) {
        .
        .
        .
    }
}
```

8

DOCfgMode()

void DOCfgMode(INT8U n, INT8U mode, BOOLEAN inv);

DOCfgMode() is used to set the operating mode of a discrete output channel. Each channel must be individually configured.

Arguments

n is the desired discrete output channel to configure. Discrete output channels are numbered from 0 to DIO_MAX_DO - 1.

mode determines the operating mode of the discrete output channel. The discrete I/O module currently supports five modes:

1. DO_MODE_LOW is the default mode and forces the discrete output LOW.

2. DO_MODE_HIGH is similar to DO_MODE_LOW, except that it forces the discrete output HIGH.

3. DO_MODE_DIRECT allows the discrete output channel to output whatever state you set through DOSet() or DOSetBypass().

4. DO_MODE_BLINK_SYNC allows the discrete output to continuously change from LOW to HIGH and from HIGH to LOW. In this mode, you also need to specify how long the output will be HIGH with respect to a continuously running counter, DOSyncCtr, which is specified through DOSetSyncCtrMax(). If DOSyncCtr is allowed to count from 0 to 100 then, to get a 25 percent duty-cycle, you need to set the HIGH time to 25. This is done by calling DOCfgBlink().

5. DO_MODE_BLINK_ASYNC allows the discrete output to continuously change from LOW to HIGH and from HIGH to LOW. In this mode, you also need to specify how long the output will be HIGH and the total period of the signal. This is done through DOCfgBlink().

inv is used to complement the output. When inv is set to TRUE, the output is complemented as shown in Figure 8.9.

Return Value

None

Notes/Warnings

None

Example

```
void main (void)
{
    .
    .
    DOCfgMode(0, DO_MODE_BLINK_SYNC, FALSE);
    .
    .
}
```

DOGet()

BOOLEAN DOGet(INT8U n);

DOGet() allows your application to get the state of the output that actually goes to the hardware. DOGet() returns either TRUE (the output is set to 1) or FALSE (the output is set to 0).

Arguments

n is the discrete output channel you wish to monitor. Discrete output channels are numbered from 0 to DIO_MAX_DO - 1.

Return Value

None

Notes/Warnings

None

Example

8

```
void Task (void *pdata)
{
    BOOLEAN state;


    for (;;) {
        .
        .
        state = DIGet(0);     /* Get value of channel #0 */
        .
        .
    }
}
```

DOSet ( )

`void DOSet(INT8U n, BOOLEAN state);`

`DOSet()` allows your application to set the state of the discrete output channel. If the discrete output channel is configured for blink mode, the state passed to `DOSet()` is used to enable or disable blinking, as shown in Figure 8.9.

Arguments

n is the discrete output channel you wish to set. Discrete output channels are numbered from 0 to `DIO_MAX_DO - 1`.

state is the desired state of the discrete output and can be either `TRUE` or `FALSE`. Note that the state of the discrete output occurs before any processing is performed on the discrete output channel, as shown in Figure 8.9.

Return Value

None

Notes/Warnings

None

Example

```
void Task (void *pdata)
{
    for (;;) {
        .
        .
        DISetBypass(0, 1);      /* Set value of channel #0's .DIVal */
        .
        .
    }
}
```

DOSetBypass()

void DOSetBypass(INT8U n, BOOLEAN state);

You can bypass what your application code is sending to the discrete output channel by using this function. DOSetBypass() doesn't do anything unless you have opened the bypass switch by calling DOSetBypassEn(), as described earlier.

Arguments

n is the desired discrete output channel to override. Discrete output channels are numbered from 0 to DIO_MAX_DO - 1.

state is the desired state of the discrete output and can be either TRUE or FALSE. Note that the bypass occurs before any processing is performed on the discrete output channel, as shown in Figure 8.9.

Return Value

None

Notes/Warnings

None

8

Example

```
void Task (void *pdata)
{
    for (;;) {
        .
        .
        DISetBypass(0, 1);      /* Set value of channel #0's .DIVal */
        .
        .
    }
}
```

DOSetBypassEn()

```
void DOSetBypassEn(INT8U n, BOOLEAN state);
```

DOSetBypassEn() allows your application code to bypass your application and set the state of the discrete output by calling DOSetBypass(). DOSetBypassEn() and DOSetBypass() are very useful for debugging.

Arguments

n is the desired discrete output channel to bypass. Discrete output channels are numbered from 0 to DIO_MAX_DO − 1.

state is the state of the bypass switch. When TRUE, the bypass switch is open (i.e., the discrete output channel is bypassed). When FALSE, the bypass switch is closed (i.e., the discrete output channel is not bypassed).

Return Value

None

Notes/Warnings

None

Example

```c
void Task (void *pdata)
{
    for (;;) {
        .
        .
        DOSetBypassEn(0, TRUE);  /* Bypass channel */
        .
        .
    }
}
```

DOSetSyncCtrMax()

`void DOSetSyncCtrMax(INT8U val);`

`DOSetSyncCtrMax()` is used to set the period for the synchronous blinking mode. The synchronous blinking mode is useful when you need to have lights blink at the same rate.

Arguments

val specifies the total period when the discrete output is configured for synchronous mode (the C value of Figures 8.9 and 8.10). The period is determined by the execution rate of the discrete I/O module. `val` is given by:

[8.3]   **val** = Period (sec.) × Task execution rate (Hz)

Return Value

None

Notes/Warnings

None

Example

```
void Task (void *pdata)
{
    DOSetSyncCtrMax(100);
    for (;;) {
            .

            .

            .

    }
}
```

8

8.05 Configuration

I added two #defines (DI_EDGE_EN and DO_BLINK_MODE_EN), which are used to enable/disable some of the functions of the discrete I/O module in order to reduce the amount of ROM and RAM. Specifically, DI_EDGE_EN allows you to remove edge detection for all discrete input channels, and DO_BLINK_MODE_EN allows you to remove the blinking capability of discrete output channels.

You could reduce the amount of RAM for each discrete input or output by using bit fields in the DIO_DI and DIO_DO structures. In this case, you would reduce the amount of RAM required at the expense of more code space (manipulation of bit fields requires more code and is slower).

Configuring the discrete I/O module is fairly simple.

1. You need to define the value of seven #defines. The #defines are found in DIO.H and CFG.H.

WARNING

In the previous edition of this book, DIO_TASK_STK_SIZE specified the size of the stack for DIOTask() in number of bytes. µC/OS-II assumes the stack is specified in stack width elements.

---

2. You will need to adapt DIRd(), DIWr(), and DIOInitIO() to your specific environment.

All physical discrete inputs are read by DIRd() and are mapped to their corresponding DIO_DI structures, as shown in Figure 8.13. In the code I provided in Listing 8.1, DIRd() obtains its discrete inputs from an 8-bit parallel port. The least significant bit of the input port corresponds to discrete input channel #0, the next-to-the-least significant bit is channel #1, and so on. Adding more discrete inputs should be a trivial task.

Figure 8.13 Mapping of physical inputs to discrete input channels.

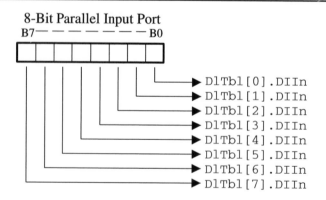

Figure 8.14 shows how discrete output channels are mapped to physical outputs using DOWr(). In the code provided in Listing 8.1, discrete output channels are mapped to an 8-bit parallel port. Discrete output channel #0 is mapped to the least significant bit of the output port (i.e., bit 0), channel #1 is mapped to bit 1, and so on. Adding more discrete outputs should be fairly simple.

Figure 8.14   Mapping of discrete output channels to physical outputs.

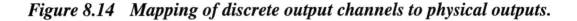

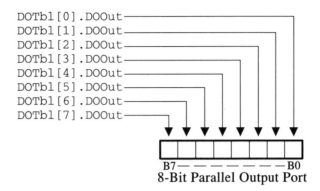

```
DOTbl[0].DOOut
DOTbl[1].DOOut
DOTbl[2].DOOut
DOTbl[3].DOOut
DOTbl[4].DOOut
DOTbl[5].DOOut
DOTbl[6].DOOut
DOTbl[7].DOOut
```

B7 — — — — — — — B0
8-Bit Parallel Output Port

DIOInitIO() is the initialization code which is called by DIOInit() and is used to initialize your physical hardware ports. For example, if you are using Intel's 82C55A Programmable Peripheral Interface (PPI), you would initialize the 82C55A to the desired mode in DIOInitIO().

8.06   How to Use the Discrete I/O Module

To use the discrete I/O module, you will need to call DIOInit() prior to using any of the other functions. You would typically do this in main() as follows:

```
void main(void)
{
    .
    OSInit();                /* Initialize the O.S. (uC/OS-II)        */
    .
    .
    DIOInit();               /* Initialize the discrete I/O module    */
    .
    .
    OSStart();               /* Start multitasking (uC/OS-II)         */
}
```

Once you have initialized the discrete I/O module, you can configure each one of the discrete inputs and outputs by calling DICfgMode(), DICfgEdgeDetect(), DOCfgMode(), and DOCfg-Blink(). You will also need to call DOSetSyncCtrMax() if you are using any of the discrete outputs

8

in synchronous blink mode. You can choose to configure discrete I/O channels immediately after the call to DIOInit() or in your application task, as shown:

```
void AppTask (void *data)
{
    data = data;
    /* Initialize discrete I/O channels here ...*/
    .

    .

    for (;;) {
        /* Application task code ... */
    }
}
```

A traffic light controller would be an ideal application for the discrete I/O module. For the intersection shown in Figure 8.15, you would need eight discrete outputs to control the state of each traffic light (four for North <-> South, four for East <-> West). Each set of four outputs would control:

- 1 green light
- 1 yellow light
- 1 red light
- 1 green light (for left turn arrow)

This traffic light controller caters to pedestrians. Two buttons are needed at each corner so pedestrians can request to cross the intersection. The controller, however, only needs to see two discrete inputs; one to request an East/West crossing and another to request a North/South crossing. Additional lights are required to inform the pedestrian when it is safe to cross the intersection: a *walk* light and a *don't walk* light. The *don't walk* typically blinks when it is no longer safe to cross the intersection. You will need four discrete outputs for pedestrian crossing lights.

Figure 8.16 shows a block diagram of the traffic light controller and the necessary discrete I/Os. The code required to configure the discrete I/Os for the traffic controller follows this paragraph. All discrete outputs are initially configured for direct mode. The mode of the discrete output controlling the *don't walk* light can be changed to blinking mode when it is unsafe to cross the street.

Figure 8.15   Traffic light control using the discrete I/O module.

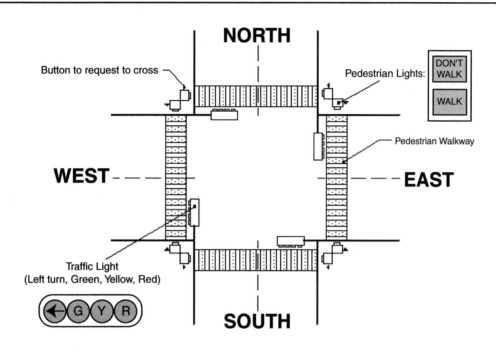

Figure 8.16   Traffic light control block diagram.

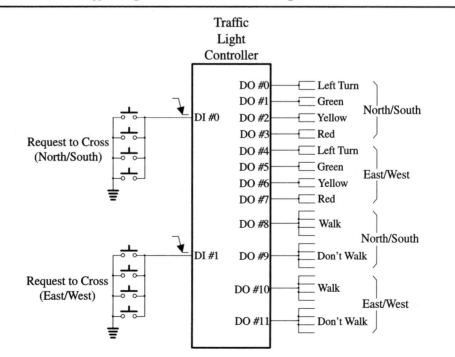

8

```
void TrafficCtrlInitIO(void)
{
    DICfgMode( 0, DI_MODE_EDGE_LOW_GOING); /* Pedestrian buttons        */
    DICfgMode( 1, DI_MODE_EDGE_LOW_GOING);

    DOCfgMode( 0, DO_MODE_DIRECT);              /* Traffic lights            */
    DOCfgMode( 1, DO_MODE_DIRECT);
    DOCfgMode( 2, DO_MODE_DIRECT);
    DOCfgMode( 3, DO_MODE_DIRECT);
    DOCfgMode( 4, DO_MODE_DIRECT);
    DOCfgMode( 5, DO_MODE_DIRECT);
    DOCfgMode( 6, DO_MODE_DIRECT);
    DOCfgMode( 7, DO_MODE_DIRECT);

    DOSet(1, ON);                               /* Turn ON N/S Green light   */
    DOSet(7, ON);                               /* Turn ON E/W Red   light   */

    DOCfgMode( 8, DO_MODE_DIRECT);              /* Pedestrian lights         */
    DOCfgMode( 9, DO_MODE_DIRECT);
    DOCfgMode(10, DO_MODE_DIRECT);
    DOCfgMode(11, DO_MODE_DIRECT);

    DOSet( 9, ON);                              /* Turn ON "DON'T WALK"      */
    DOSet(11, ON);
}
```

Listing 8.1   DIO.C

```
/*
*********************************************************************************************
*                            Embedded Systems Building Blocks
*                          Complete and Ready-to-Use Modules in C
*
*                                  Discrete I/O Module
*
*                     (c) Copyright 1999, Jean J. Labrosse, Weston, FL
*                                  All Rights Reserved
*
* Filename   : DIO.C
* Programmer : Jean J. Labrosse
*********************************************************************************************
*/

/*
*********************************************************************************************
*                                   INCLUDE FILES
*********************************************************************************************
*/

#define  DIO_GLOBALS
#include "includes.h"

/*
*********************************************************************************************
*                                   LOCAL VARIABLES
*********************************************************************************************
*/

#if      DO_BLINK_MODE_EN
static  INT8U       DOSyncCtr;
static  INT8U       DOSyncCtrMax;
#endif

static  OS_STK      DIOTaskStk[DIO_TASK_STK_SIZE];

/*
*********************************************************************************************
*                             LOCAL FUNCTION PROTOTYPES
*********************************************************************************************
*/


static  void       DIIsTrig(DIO_DI *pdi);

static  void       DIOTask(void *data);

static  void       DIUpdate(void);

static  BOOLEAN    DOIsBlinkEn(DIO_DO *pdo);
static  void       DOUpdate(void);

/*$PAGE*/
```

Listing 8.1 (continued) DIO.C

```c
/*
*********************************************************************************************************
*                               CONFIGURE DISCRETE INPUT EDGE DETECTION
*
* Description : This function is used to configure the edge detection capability of the discrete input
*               channel.
* Arguments   : n     is the discrete input channel to configure (0..DIO_MAX_DI-1).
*               fnct  is a pointer to a function that will be executed if the desired edge has been
*                     detected.
*               arg   is a pointer to arguments that are passed to the function called.
* Returns     : None.
*********************************************************************************************************
*/

#if  DI_EDGE_EN
void  DICfgEdgeDetectFnct (INT8U n, void (*fnct)(void *), void *arg)
{
    if (n < DIO_MAX_DI) {
        OS_ENTER_CRITICAL();
        DITbl[n].DITrigFnct    = fnct;
        DITbl[n].DITrigFnctArg = arg;
        OS_EXIT_CRITICAL();
    }
}
#endif
/*$PAGE*/


/*
*********************************************************************************************************
*                               CONFIGURE DISCRETE INPUT MODE
*
* Description : This function is used to configure the mode of a discrete input channel.
* Arguments   : n     is the discrete input channel to configure (0..DIO_MAX_DI-1).
*               mode  is the desired mode and can be:
*                         DI_MODE_LOW               input is forced LOW
*                         DI_MODE_HIGH              input is forced HIGH
*                         DI_MODE_DIRECT            input is based on state of physical sensor (default)
*                         DI_MODE_INV               input is based on the complement of physical sensor
*                         DI_MODE_EDGE_LOW_GOING    a LOW-going  transition is detected
*                         DI_MODE_EDGE_HIGH_GOING   a HIGH-going transition is detected
*                         DI_MODE_EDGE_BOTH         both a LOW-going and a HIGH-going transition are detected
*                         DI_MODE_TOGGLE_LOW_GOING  a LOW-going  transition is detected in toggle mode
*                         DI_MODE_TOGGLE_HIGH_GOING a HIGH-going transition is detected in toggle mode
* Returns     : None.
* Notes       : Edge detection is only available if the configuration constant DI_EDGE_EN is set to 1.
*********************************************************************************************************
*/

void  DICfgMode (INT8U n, INT8U mode)
{
    if (n < DIO_MAX_DI) {
        OS_ENTER_CRITICAL();
        DITbl[n].DIModeSel = mode;
        OS_EXIT_CRITICAL();
    }
}
/*$PAGE*/
```

Listing 8.1 (continued)   `DIO.C`

```
/*
*********************************************************************************************************
*                                   CLEAR A DISCRETE INPUT CHANNEL
*
* Description : This function clears the number of edges detected if the discrete input channel is
*               configured to count edges.
* Arguments   : n     is the discrete input channel (0..DIO_MAX_DI-1) to clear.
* Returns     : none
*********************************************************************************************************
*/

#if  DI_EDGE_EN
void  DIClr (INT8U n)
{
    DIO_DI  *pdi;


    if (n < DIO_MAX_DI) {
        pdi = &DITbl[n];
        OS_ENTER_CRITICAL();
        if (pdi->DIModeSel == DI_MODE_EDGE_LOW_GOING  ||           /* See if edge detection mode selected */
            pdi->DIModeSel == DI_MODE_EDGE_HIGH_GOING ||
            pdi->DIModeSel == DI_MODE_EDGE_BOTH) {
            pdi->DIVal = 0;                                        /* Clear the number of edges detected  */
        }
        OS_EXIT_CRITICAL();
    }
}
#endif

/*$PAGE*/
```

8

Listing 8.1 (continued) `DIO.C`

```
/*
*********************************************************************************************
*                           GET THE STATE OF A DISCRETE INPUT CHANNEL
*
* Description : This function is used to get the current state of a discrete input channel.  If the input
*               mode is set to one of the edge detection modes, the number of edges detected is returned.
* Arguments   : n     is the discrete input channel (0..DIO_MAX_DI-1).
* Returns     : 0     if the discrete input is negated  or, if an edge has not been detected
*               1     if the discrete input is asserted
*               > 0   if edges have been detected
*********************************************************************************************
*/

INT16U  DIGet (INT8U n)
{
    INT16U  val;


    if (n < DIO_MAX_DI) {
        OS_ENTER_CRITICAL();
        val = DITbl[n].DIVal;                            /* Get state of DI channel                 */
        OS_EXIT_CRITICAL();
        return (val);
    } else {
        return (0);                                      /* Return negated for invalid channel      */
    }
}

/*$PAGE*/
```

Listing 8.1 (continued) `DIO.C`

```
/*
***********************************************************************************************
*                                  DETECT EDGE ON INPUT
*
* Description : This function is called to detect an edge (low-going, high-going or both) on the selected
*               discrete input.
* Arguments   : pdi     is a pointer to the discrete input data structure.
* Returns     : none
***********************************************************************************************
*/

#if DI_EDGE_EN
static  void  DIIsTrig (DIO_DI *pdi)
{
    BOOLEAN  trig;


    trig = FALSE;
    switch (pdi->DIModeSel) {
        case DI_MODE_EDGE_LOW_GOING:                     /* Negative going edge                 */
            if (pdi->DIPrev == 1 && pdi->DIIn == 0) {
                trig = TRUE;
            }
            break;

        case DI_MODE_EDGE_HIGH_GOING:                    /* Positive going edge                 */
            if (pdi->DIPrev == 0 && pdi->DIIn == 1) {
                trig = TRUE;
            }
            break;

        case DI_MODE_EDGE_BOTH:                          /* Both positive and negative going    */
            if ((pdi->DIPrev == 1 && pdi->DIIn == 0) ||
                (pdi->DIPrev == 0 && pdi->DIIn == 1)) {
                trig = TRUE;
            }
            break;
    }
    if (trig == TRUE) {                                  /* See if edge detected                */
        if (pdi->DITrigFnct != NULL) {                   /* Yes, see used defined a function    */
            (*pdi->DITrigFnct)(pdi->DITrigFnctArg);      /* Yes, execute the user function      */
        }
        if (pdi->DIVal < 255) {                          /* Increment number of edges counted   */
            pdi->DIVal++;
        }
    }
    pdi->DIPrev = pdi->DIIn;                             /* Memorize previous input state       */
}
#endif

/*$PAGE*/
```

Listing 8.1 (continued)   `DIO.C`

```
/*
********************************************************************************************************
*                                    UPDATE DISCRETE IN CHANNELS
*
* Description : This function processes all of the discrete input channels.
* Arguments   : None.
* Returns     : None.
********************************************************************************************************
*/
```

Listing 8.1 (continued)   `DIO.C`

```
static  void  DIUpdate (void)
{
    INT8U    i;
    DIO_DI   *pdi;


    pdi = &DITbl[0];
    for (i = 0; i < DIO_MAX_DI; i++) {
        if (pdi->DIBypassEn == FALSE) {         /* See if discrete input channel is bypassed   */
            switch (pdi->DIModeSel) {           /* No, process channel                         */
                case DI_MODE_LOW:               /* Input is forced low                         */
                    pdi->DIVal = 0;
                    break;

                case DI_MODE_HIGH:              /* Input is forced high                        */
                    pdi->DIVal = 1;
                    break;

                case DI_MODE_DIRECT:                /* Input is based on state of physical input  */
                    pdi->DIVal = (INT8U)pdi->DIIn;  /* Obtain the state of the sensor             */
                    break;

                case DI_MODE_INV:               /* Input is based on the complement state of input  */
                    pdi->DIVal = (INT8U)(pdi->DIIn ? 0 : 1);
                    break;

#if DI_EDGE_EN
                case DI_MODE_EDGE_LOW_GOING:
                case DI_MODE_EDGE_HIGH_GOING:
                case DI_MODE_EDGE_BOTH:
                    DIIsTrig(pdi);              /* Handle edge triggered mode                  */
                    break;
#endif

                case DI_MODE_TOGGLE_LOW_GOING:
                    if (pdi->DIPrev == 1 && pdi->DIIn == 0) {
                        pdi->DIVal = pdi->DIVal ? 0 : 1;
                    }
                    pdi->DIPrev = pdi->DIIn;
                    break;

                case DI_MODE_TOGGLE_HIGH_GOING:
                    if (pdi->DIPrev == 0 && pdi->DIIn == 1) {
                        pdi->DIVal = pdi->DIVal ? 0 : 1;
                    }
                    pdi->DIPrev = pdi->DIIn;
                    break;
            }
        }
        pdi++;                                 /* Point to next DIO_DO element                */
    }
}

/*$PAGE*/
```

8

Listing 8.1 (continued)   DIO.C

```c
/*
*********************************************************************************************
*                                DISCRETE I/O MANAGER INITIALIZATION
*
* Description : This function initializes the discrete I/O manager module.
* Arguments   : None
* Returns     : None.
*********************************************************************************************
*/

void  DIOInit (void)
{
    INT8U    err;
    INT8U    i;
    DIO_DI  *pdi;
    DIO_DO  *pdo;


    pdi = &DITbl[0];
    for (i = 0; i < DIO_MAX_DI; i++) {
        pdi->DIVal       = 0;
        pdi->DIBypassEn  = FALSE;
        pdi->DIModeSel   = DI_MODE_DIRECT;     /* Set the default mode to direct input          */
#if DI_EDGE_EN
        pdi->DITrigFnct    = (void *)0;        /* No function to execute when transition detected */
        pdi->DITrigFnctArg = (void *)0;
#endif
        pdi++;
    }
    pdo = &DOTbl[0];
    for (i = 0; i < DIO_MAX_DO; i++) {
        pdo->DOOut       = 0;
        pdo->DOBypassEn  = FALSE;
        pdo->DOModeSel   = DO_MODE_DIRECT;     /* Set the default mode to direct output         */
        pdo->DOInv       = FALSE;
#if DO_BLINK_MODE_EN
        pdo->DOBlinkEnSel = DO_BLINK_EN_NORMAL; /* Blinking is enabled by direct user request    */
        pdo->DOA         = 1;
        pdo->DOB         = 2;
        pdo->DOBCtr      = 2;
#endif
        pdo++;
    }
#if DO_BLINK_MODE_EN
    DOSetSyncCtrMax(72);
#endif
    DIOInitIO();
    OSTaskCreate(DIOTask, (void *)0, &DIOTaskStk[DIO_TASK_STK_SIZE], DIO_TASK_PRIO);
}

/*$PAGE*/
```

Listing 8.1 (continued) `DIO.C`

```
/*
*********************************************************************************************
*                               DISCRETE I/O MANAGER TASK
*
* Description : This task is created by DIOInit() and is responsible for updating the discrete inputs and
*               discrete outputs.
*               DIOTask() executes every DIO_TASK_DLY_TICKS.
* Arguments   : None.
* Returns     : None.
*********************************************************************************************
*/

static  void  DIOTask (void *data)
{
    data = data;                                /* Avoid compiler warning (uC/OS requirement)   */
    for (;;) {
        OSTimeDly(DIO_TASK_DLY_TICKS);          /* Delay between execution of DIO manager        */
        DIRd();                                 /* Read physical inputs and map to DI channels   */
        DIUpdate();                             /* Update all DI channels                         */
        DOUpdate();                             /* Update all DO channels                         */
        DOWr();                                 /* Map DO channels to physical outputs            */
    }
}
/*$PAGE*/

/*
*********************************************************************************************
*                            SET THE STATE OF THE BYPASSED SENSOR
*
* Description : This function is used to set the state of the bypassed sensor.  This function is used to
*               simulate the presence of the sensor.  This function is only valid if the bypass 'switch'
*               is open.
* Arguments   : n    is the discrete input channel (0..DIO_MAX_DI-1).
*               val  is the state of the bypassed sensor:
*                         0    indicates a  negated  sensor
*                         1    indicates an asserted sensor
*                         > 0  indicates the number of edges detected in edge mode
* Returns     : None.
*********************************************************************************************
*/

void  DISetBypass (INT8U n, INT16U val)
{
    if (n < DIO_MAX_DI) {
        OS_ENTER_CRITICAL();
        if (DITbl[n].DIBypassEn == TRUE) {      /* See if sensor is bypassed                     */
            DITbl[n].DIVal = val;               /* Yes, then set the new state of the DI channel */
        }
        OS_EXIT_CRITICAL();
    }
}

/*$PAGE*/
```

Listing 8.1 (continued)   DIO.C

```
/*
*********************************************************************************************
*                          SET THE STATE OF THE SENSOR BYPASS SWITCH
*
* Description : This function is used to set the state of the sensor bypass switch.  The sensor is
*               bypassed when the 'switch' is open (i.e. DIBypassEn is set to TRUE).
* Arguments   : n     is the discrete input channel (0..DIO_MAX_DI-1).
*               state is the state of the bypass switch:
*                         FALSE disables sensor bypass (i.e. the bypass 'switch' is closed)
*                         TRUE  enables  sensor bypass (i.e. the bypass 'switch' is open)
* Returns     : None.
*********************************************************************************************
*/

void  DISetBypassEn (INT8U n, BOOLEAN state)
{
    if (n < DIO_MAX_DI) {
        OS_ENTER_CRITICAL();
        DITbl[n].DIBypassEn = state;
        OS_EXIT_CRITICAL();
    }
}

/*$PAGE*/
```

Listing 8.1 (continued)   DIO.C

```
/*
*********************************************************************************************
*                         CONFIGURE THE DISCRETE OUTPUT BLINK MODE
*
* Description : This function is used to configure the blink mode of the discrete output channel.
* Arguments   : n     is the discrete output channel (0..DIO_MAX_DO-1).
*               mode  is the desired blink mode:
*                     DO_BLINK_EN         Blink is always enabled
*                     DO_BLINK_EN_NORMAL  Blink depends on user request's state
*                     DO_BLINK_EN_INV     Blink depends on the complemented user request's state
*               a     is the number of 'ticks' ON  (1..250)
*               b     is the number of 'ticks' for the period (in DO_MODE_BLINK_ASYNC mode) (1..250)
* Returns     : None.
*********************************************************************************************
*/

#if  DO_BLINK_MODE_EN
void  DOCfgBlink (INT8U n, INT8U mode, INT8U a, INT8U b)
{
    DIO_DO  *pdo;


    if (n < DIO_MAX_DO) {
        pdo             = &DOTbl[n];
        a               /= DIO_TASK_DLY_TICKS;       /* Adjust threshold based on how often DIO runs  */
        b               /= DIO_TASK_DLY_TICKS;
        OS_ENTER_CRITICAL();
        pdo->DOBlinkEnSel = mode;
        pdo->DOA        = a;
        pdo->DOB        = b;
        OS_EXIT_CRITICAL();
    }
}
#endif

/*$PAGE*/
```

8

Listing 8.1 (continued)   DIO.C

```
/*
*********************************************************************************************
*                              CONFIGURE DISCRETE OUTPUT MODE
*
* Description : This function is used to configure the mode of a discrete output channel.
* Arguments   : n     is the discrete output channel to configure (0..DIO_MAX_DO-1).
*               mode  is the desired mode and can be:
*                     DO_MODE_LOW              output is forced LOW
*                     DO_MODE_HIGH            output is forced HIGH
*                     DO_MODE_DIRECT          output is based on state of DOBypass
*                     DO_MODE_BLINK_SYNC      output will be blinking synchronously with DOSyncCtr
*                     DO_MODE_BLINK_ASYNC     output will be blinking based on DOA and DOB
*               inv   indicates whether the output will be inverted:
*                     TRUE   forces the output to be inverted
*                     FALSE  does not cause any inversion
* Returns     : None.
*********************************************************************************************
*/

void  DOCfgMode (INT8U n, INT8U mode, BOOLEAN inv)
{
    if (n < DIO_MAX_DO) {
        OS_ENTER_CRITICAL();
        DOTbl[n].DOModeSel = mode;
        DOTbl[n].DOInv     = inv;
        OS_EXIT_CRITICAL();
    }
}


/*$PAGE*/


/*
*********************************************************************************************
*                              GET THE STATE OF THE DISCRETE OUTPUT
*
* Description : This function is used to obtain the state of the discrete output.
* Arguments   : n     is the discrete output channel (0..DIO_MAX_DO-1).
* Returns     : TRUE  if the output is asserted.
*               FALSE if the output is negated.
*********************************************************************************************
*/

BOOLEAN  DOGet (INT8U n)
{
    BOOLEAN  out;


    if (n < DIO_MAX_DO) {
        OS_ENTER_CRITICAL();
        out = DOTbl[n].DOOut;
        OS_EXIT_CRITICAL();
        return (out);
    } else {
        return (FALSE);
    }
}

/*$PAGE*/
```

Listing 8.1 (continued) DIO.C

```
/*
*********************************************************************************************
*                                   SEE IF BLINK IS ENABLED
*
* Description : See if blink mode is enabled.
* Arguments   : pdo    is a pointer to the discrete output data structure.
* Returns     : TRUE   if blinking is enabled
*               FALSE  otherwise
*********************************************************************************************
*/

#if DO_BLINK_MODE_EN
static  BOOLEAN  DOIsBlinkEn (DIO_DO *pdo)
{
    BOOLEAN  en;


    en = FALSE;
    switch (pdo->DOBlinkEnSel) {
        case DO_BLINK_EN:                   /* Blink is always enabled                     */
            en = TRUE;
            break;

        case DO_BLINK_EN_NORMAL:            /* Blink depends on user request's state       */
            en = pdo->DOBypass;
            break;

        case DO_BLINK_EN_INV:               /* Blink depends on the complemented user request's state */
            en = pdo->DOBypass ? FALSE : TRUE;
            break;
    }
    return (en);
}
#endif

/*$PAGE*/
```

Listing 8.1 (continued)   DIO.C

```
/*
*********************************************************************************************************
*                               SET THE STATE OF THE DISCRETE OUTPUT
*
* Description : This function is used to set the state of the discrete output.
* Arguments   : n     is the discrete output channel (0..DIO_MAX_DO-1).
*               state is the desired state of the output:
*                       FALSE indicates a  negated  output
*                       TRUE  indicates an asserted output
* Returns     : None.
* Notes       : The actual output will be complemented if 'DIInv' is set to TRUE.
*********************************************************************************************************
*/

void  DOSet (INT8U n, BOOLEAN state)
{
    if (n < DIO_MAX_DO) {
        OS_ENTER_CRITICAL();
        DOTbl[n].DOCtrl = state;
        OS_EXIT_CRITICAL();
    }
}

/*$PAGE*/

/*
*********************************************************************************************************
*                               SET THE STATE OF THE BYPASSED OUTPUT
*
* Description : This function is used to set the state of the bypassed output.  This function is used to
*               override (or bypass) the application software and allow the output to be controlled
*               directly.  This function is only valid if the bypass switch is open.
* Arguments   : n      is the discrete output channel (0..DIO_MAX_DO-1).
*               state  is the desired state of the output:
*                       FALSE indicates a  negated  output
*                       TRUE  indicates an asserted output
* Returns     : None.
* Notes       : 1) The actual output will be complemented if 'DIInv' is set to TRUE.
*               2) In blink mode, this allows blinking to be enabled or not.
*********************************************************************************************************
*/

void  DOSetBypass (INT8U n, BOOLEAN state)
{
    if (n < DIO_MAX_DO) {
        OS_ENTER_CRITICAL();
        if (DOTbl[n].DOBypassEn == TRUE) {
            DOTbl[n].DOBypass = state;
        }
        OS_EXIT_CRITICAL();
    }
}

/*$PAGE*/
```

Listing 8.1 (continued)   `DIO.C`

```
/*
*****************************************************************************************************
*                              SET THE STATE OF THE OUTPUT BYPASS
*
* Description : This function is used to set the state of the output bypass switch.  The output is
*               bypassed when the 'switch' is open (i.e. DOBypassEn is set to TRUE).
* Arguments   : n    is the discrete output channel (0..DIO_MAX_DO-1).
*               state is the state of the bypass switch:
*                     FALSE disables output bypass (i.e. the switch is closed)
*                     TRUE  enables  output bypass (i.e. the switch is open)
* Returns     : None.
*****************************************************************************************************
*/

void  DOSetBypassEn (INT8U n, BOOLEAN state)
{
    if (n < DIO_MAX_DO) {
        OS_ENTER_CRITICAL();
        DOTbl[n].DOBypassEn = state;
        OS_EXIT_CRITICAL();
    }
}


/*$PAGE*/

/*
*****************************************************************************************************
*                         SET THE MAXIMUM VALUE FOR THE SYNCHRONOUS COUNTER
*
* Description : This function is used to set the maximum value taken by the synchronous counter which is
*               used in the synchronous blink mode.
* Arguments   : val   is the maximum value for the counter (1..255)
* Returns     : None.
*****************************************************************************************************
*/

#if  DO_BLINK_MODE_EN
void  DOSetSyncCtrMax (INT8U val)
{
    OS_ENTER_CRITICAL();
    DOSyncCtrMax = val;
    OS_EXIT_CRITICAL();
}
#endif
/*$PAGE*/
```

8

Listing 8.1 (continued) DIO.C

```
/*
**********************************************************************************************
*                              UPDATE DISCRETE OUT CHANNELS
*
* Description : This function is called to process all of the discrete output channels.
* Arguments   : None.
* Returns     : None.
**********************************************************************************************
*/
```

Listing 8.1 (continued)   DIO.C

```c
static void DOUpdate (void)
{
    INT8U    i;
    BOOLEAN  out;
    DIO_DO   *pdo;

    pdo = &DOTbl[0];
    for (i = 0; i < DIO_MAX_DO; i++) {             /* Process all discrete output channels      */
        if (pdo->DOBypassEn == FALSE) {            /* See if DO channel is enabled              */
            pdo->DOBypass = pdo->DOCtrl;           /* Obtain control state from application      */
        }
        out = FALSE;                               /* Assume that the output will be low unless changed */
        switch (pdo->DOModeSel) {
            case DO_MODE_LOW:                      /* Output will in fact be low                */
                 break;

            case DO_MODE_HIGH:                     /* Output will be high                       */
                 out = TRUE;
                 break;

            case DO_MODE_DIRECT:                   /* Output is based on state of user supplied state */
                 out = pdo->DOBypass;
                 break;
#if DO_BLINK_MODE_EN
            case DO_MODE_BLINK_SYNC:               /* Sync. Blink mode                         */
                 if (DOIsBlinkEn(pdo)) {           /* See if Blink is enabled ...              */
                     if (pdo->DOA >= DOSyncCtr) {  /* ... yes, High when below threshold       */
                         out = TRUE;
                     }
                 }
                 break;

            case DO_MODE_BLINK_ASYNC:              /* Async. Blink mode                        */
                 if (DOIsBlinkEn(pdo)) {           /* See if Blink is enabled ...              */
                     if (pdo->DOA >= pdo->DOBCtr) { /* ... yes, High when below threshold      */
                         out = TRUE;
                     }
                 }
                 if (pdo->DOBCtr < pdo->DOB) {     /* Update the threshold counter             */
                     pdo->DOBCtr++;
                 } else {
                     pdo->DOBCtr = 0;
                 }
                 break;
#endif
        }
        if (pdo->DOInv == TRUE) {                  /* See if output needs to be inverted ...   */
            pdo->DOOut = out ? FALSE : TRUE;       /* ... yes, complement output               */
        } else {
            pdo->DOOut = out;                      /* ... no,  no inversion!                   */
        }
        pdo++;                                     /* Point to next DIO_DO element             */
    }
#if DO_BLINK_MODE_EN
    if (DOSyncCtr < DOSyncCtrMax) {                /* Update the synchronous free running ctr  */
        DOSyncCtr++;
    } else {
        DOSyncCtr = 0;
    }
#endif
}
```

8

Listing 8.1 (continued)   DIO.C

```
#ifndef CFG_C
/*
*********************************************************************************************
*                            INITIALIZE PHYSICAL I/Os
*
* Description : This function is by DIOInit() to initialze the physical I/O used by the DIO driver.
* Arguments   : None.
* Returns     : None.
* Notes       : The physical I/O is assumed to be an 82C55 chip initialized as follows:
*                        Port A = OUT   (Discrete outputs)
*                        Port B = IN    (Discrete inputs)
*                        Port C = OUT   (not used)
*********************************************************************************************
*/

void  DIOInitIO (void)
{
    outp(0x0303, 0x82);                          /* Port A = OUT, Port B = IN, Port C = OUT        */
}


/*
*********************************************************************************************
*                            READ PHYSICAL INPUTS
*
* Description : This function is called to read and map all of the physical inputs used for discrete
*               inputs and map these inputs to their appropriate discrete input data structure.
* Arguments   : None.
* Returns     : None.
*********************************************************************************************
*/

void  DIRd (void)
{
    DIO_DI  *pdi;
    INT8U    i;
    INT8U    in;
    INT8U    msk;


    pdi = &DITbl[0];                             /* Point at beginning of discrete inputs    */
    msk = 0x01;                                  /* Set mask to extract bit 0                */
    in  = inp(0x0301);                           /* Read the physical port (8 bits)          */
    for (i = 0; i < 8; i++) {                     /* Map all 8 bits to first 8 DI channels    */
        pdi->DIIn  = (BOOLEAN)(in & msk) ? 1 : 0;
        msk        <<= 1;
        pdi++;
    }
}
/*$PAGE*/
```

Listing 8.1 (continued)   DIO.C

```
/*
*********************************************************************************************************
*                                     UPDATE PHYSICAL OUTPUTS
*
* Description : This function is called to map all of the discrete output channels to their appropriate
*               physical destinations.
* Arguments   : None.
* Returns     : None.
*********************************************************************************************************
*/

void  DOWr (void)
{
    DIO_DO  *pdo;
    INT8U    i;
    INT8U    out;
    INT8U    msk;


    pdo = &DOTbl[0];                        /* Point at first discrete output channel              */
    msk = 0x01;                             /* First DO will be mapped to bit 0                    */
    out = 0x00;                             /* Local 8 bit port image                              */
    for (i = 0; i < 8; i++) {               /* Map first 8 DO to 8 bit port image                  */
        if (pdo->DOOut == TRUE) {
            out |= msk;
        }
        msk <<= 1;
        pdo++;
    }
    outp(0x0300, out);                      /* Output port image to physical port                  */
}
#endif
```

8

Listing 8.2   DIO.H

```
/*
*********************************************************************************************
*                             Embedded Systems Building Blocks
*                             Complete and Ready-to-Use Modules in C
*
*                             Discrete I/O Module
*
*                             (c) Copyright 1999, Jean J. Labrosse, Weston, FL
*                             All Rights Reserved
*
* Filename   : DIO.H
* Programmer : Jean J. Labrosse
*********************************************************************************************
*/


/*
*********************************************************************************************
*                             CONFIGURATION CONSTANTS
*********************************************************************************************
*/

#ifndef  CFG_H

#define  DIO_TASK_PRIO        40
#define  DIO_TASK_DLY_TICKS    1
#define  DIO_TASK_STK_SIZE    512

#define  DIO_MAX_DI            8        /* Maximum number of Discrete Input  Channels (1..255)   */
#define  DIO_MAX_DO            8        /* Maximum number of Discrete Output Channels (1..255)   */

#define  DI_EDGE_EN            1        /* Enable code generation to support edge trig. (when 1)   */

#define  DO_BLINK_MODE_EN      1        /* Enable code generation to support blink mode (when 1)   */

#endif

#ifdef   DIO_GLOBALS
#define  DIO_EXT
#else
#define  DIO_EXT  extern
#endif
```

Listing 8.2 (continued) `DIO.H`

```
/*
*****************************************************************************************************
*                                         DISCRETE INPUT CONSTANTS
*****************************************************************************************************
*/

                                               /* DI MODE SELECTOR VALUES                              */
#define   DI_MODE_LOW                0         /* Input is forced low                                  */
#define   DI_MODE_HIGH               1         /* Input is forced high                                 */
#define   DI_MODE_DIRECT             2         /* Input is based on state of physical input            */
#define   DI_MODE_INV                3         /* Input is based on the complement of the physical input */
#define   DI_MODE_EDGE_LOW_GOING     4         /* Low  going edge detection of input                   */
#define   DI_MODE_EDGE_HIGH_GOING    5         /* High going edge detection of input                   */
#define   DI_MODE_EDGE_BOTH          6         /* Both low and high going edge detection of input      */
#define   DI_MODE_TOGGLE_LOW_GOING   7         /* Low  going edge detection of input                   */
#define   DI_MODE_TOGGLE_HIGH_GOING  8         /* High going edge detection of input                   */


                                               /* DI EDGE TRIGGERING MODE SELECTOR VALUES              */
#define   DI_EDGE_LOW_GOING          0         /* Negative going edge                                  */
#define   DI_EDGE_HIGH_GOING         1         /* Positive going edge                                  */
#define   DI_EDGE_BOTH               2         /* Both positive and negative going                     */


/*$PAGE*/
```

8

Listing 8.2 *(continued)*   DIO.H

```
/*
*********************************************************************************************
*                                  DISCRETE OUTPUT CONSTANTS
*********************************************************************************************
*/

                                           /* DO MODE SELECTOR VALUES                              */
#define   DO_MODE_LOW              0        /* Output will be low                                   */
#define   DO_MODE_HIGH             1        /* Output will be high                                  */
#define   DO_MODE_DIRECT           2        /* Output is based on state of user supplied state      */
#define   DO_MODE_BLINK_SYNC       3        /* Sync.  Blink mode                                    */
#define   DO_MODE_BLINK_ASYNC      4        /* Async. Blink mode                                    */


                                           /* DO BLINK MODE ENABLE SELECTOR VALUES                 */
#define   DO_BLINK_EN              0        /* Blink is always enabled                              */
#define   DO_BLINK_EN_NORMAL       1        /* Blink depends on user request's state                */
#define   DO_BLINK_EN_INV          2        /* Blink depends on the complemented user request's state */

/*
*********************************************************************************************
*                                       DATA TYPES
*********************************************************************************************
*/

typedef struct dio_di {                    /* DISCRETE INPUT CHANNEL DATA STRUCTURE                */
    BOOLEAN   DIIn;                         /* Current state of sensor input                        */
    INT16U    DIVal;                        /* State of discrete input channel (or # of transitions) */
    BOOLEAN   DIPrev;                       /* Previous state of DIIn for edge detection            */
    BOOLEAN   DIBypassEn;                   /* Bypass enable switch (Bypass when TRUE)              */
    INT8U     DIModeSel;                    /* Discrete input channel mode selector                 */
#if DI_EDGE_EN
    void      (*DITrigFnct)(void *);        /* Function to execute if edge triggered                */
    void      *DITrigFnctArg;               /* arguments passed to function when edge detected      */
#endif
} DIO_DI;


typedef struct dio_do {                    /* DISCRETE OUTPUT CHANNEL DATA STRUCTURE               */
    BOOLEAN   DOOut;                        /* Current state of discrete output channel             */
    BOOLEAN   DOCtrl;                       /* Discrete output control request                      */
    BOOLEAN   DOBypass;                     /* Discrete output control bypass state                 */
    BOOLEAN   DOBypassEn;                   /* Bypass enable switch (Bypass when TRUE)              */
    INT8U     DOModeSel;                    /* Discrete output channel mode selector                */
    INT8U     DOBlinkEnSel;                 /* Blink enable mode selector                           */
    BOOLEAN   DOInv;                        /* Discrete output inverter selector (Invert when TRUE) */
#if DO_BLINK_MODE_EN
    INT8U     DOA;                          /* Blink mode ON time                                   */
    INT8U     DOB;                          /* Asynchronous blink mode period                       */
    INT8U     DOBCtr;                       /* Asynchronous blink mode period counter               */
#endif
} DIO_DO;
/*$PAGE*/
```

Listing 8.2 (continued)   DIO.H

```
/*
*********************************************************************************************
*                                        GLOBAL VARIABLES
*********************************************************************************************
*/

DIO_EXT  DIO_DI      DITbl[DIO_MAX_DI];
DIO_EXT  DIO_DO      DOTbl[DIO_MAX_DO];

/*
*********************************************************************************************
*                                        FUNCTION PROTOTYPES
*********************************************************************************************
*/

void     DIOInit(void);

void     DICfgMode(INT8U n, INT8U mode);
INT16U   DIGet(INT8U n);
void     DISetBypassEn(INT8U n, BOOLEAN state);
void     DISetBypass(INT8U n, INT16U val);

#if      DI_EDGE_EN
void     DIClr(INT8U n);
void     DICfgEdgeDetectFnct(INT8U n, void (*fnct)(void *), void *arg);
#endif


void     DOCfgMode(INT8U n, INT8U mode, BOOLEAN inv);
BOOLEAN  DOGet(INT8U n);
void     DOSet(INT8U n, BOOLEAN state);
void     DOSetBypass(INT8U n, BOOLEAN state);
void     DOSetBypassEn(INT8U n, BOOLEAN state);

#if      DO_BLINK_MODE_EN
void     DOCfgBlink(INT8U n, INT8U mode, INT8U a, INT8U b);
void     DOSetSyncCtrMax(INT8U val);
#endif

/*
*********************************************************************************************
*                                        FUNCTION PROTOTYPES
*                                        HARDWARE SPECIFIC
*********************************************************************************************
*/

void     DIOInitIO(void);
void     DIRd(void);
void     DOWr(void);
```

8

## Chapter 9

Fixed-Point Math

Most low-end microprocessors (typical of embedded processors) do not provide hardware-assisted floating-point math. Microprocessor manufacturers unfortunately seem to feel that floating-point math is not very important in embedded systems. This has not been my experience. Fortunately, ANSI C compilers allow you to use floating-point math but at a cost; floating-point libraries require extra ROM and RAM but most importantly, they require more processing time than integer math. For example, floating-point addition could take hundreds of microseconds on a low-end, 8-bit microprocessor, whereas it typically takes only a few microseconds to perform a 16-bit integer addition. Multiplications and especially divisions are even worse. As an embedded system programmer, you are often confronted with the task of writing the fastest and smallest possible code for real-time operations. This chapter will show you how to perform basic arithmetic operations on fractional numbers by using only integers. In other words, this chapter will answer the questions: "Without using floating-point arithmetic, how would you add 12.34 and 987.654, multiply 3.1416 by 5.4, or divide 0.00456 by 98.7?"

Throughout this chapter, I will be using 16-bit integers, but most of the concepts presented here apply to any integer size. This chapter will show you how to use the concept of fixed-point math to get the most out of integer arithmetic. Chapter 10 will make use of the information presented in this chapter.

9.00 Fixed-Point Numbers

Fixed-point is an alternative form for expressing numerical values. Fixed-point math is integer math, but because it allows fractions, it is much more versatile and often can substitute for slower and more cumbersome floating-point operations. The idea of fixed-point math is to trick the computer into thinking you are talking about an integer when in fact you, the programmer, know that you are dealing with a number that has a fractional component.

Figure 9.1.a shows a 16-bit integer. The computer thinks only in bits. In integer arithmetic, the bit positions are said to represent 2 to progressively higher powers starting from the right. The bit string 0000000000010000, therefore, represents the number 16.

Figure 9.1.a Signed and unsigned 16-bit integers.

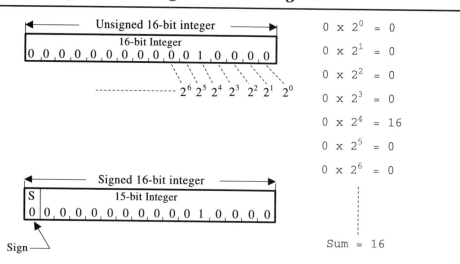

A practitioner of fixed-point math would observe that there is an implied decimal point (called a *radix point*) to the right of the rightmost bit position and would ask, "Why must it fall there? Why can't I put the radix point somewhere else?" In other words, why must the rightmost bit represent 2^0?

Figure 9.1.b shows the same 16-bit string. In this case, the programmer decides to place the radix point between the 5th and 6th bit positions, which make the rightmost bit 2^{-5}. The string 0000000000010000 is now not 16, but 0.5. Another way to look at this is to say that the integer 16 has been *scaled* by 2^{-5} (multiplied by 2^{-5}, or .03125):

$$16 \times 2^{-5} = 0.5$$

Figure 9.1.b Signed and unsigned fixed-point numbers with radix point between 5th and 6th bits.

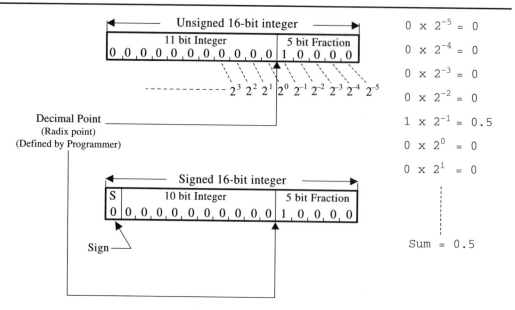

The computer, then, thinks it is working with the integer 16, but the programmer independently maintains a record of how the 16 should be *scaled*.

By manipulating the position of the radix point, a programmer can scale integers into fractional values. The location of the radix point defines a convention for how the program will interpret a 16-bit string. As the radix point moves to the left (increasing the fractional portion of the string) the fraction becomes more precise and the overall range of the number diminishes (because there are fewer whole-number places).

The unsigned integer of Figure 9.1.b can be used to represent numbers having a range of 0.0 to 2047.96875, while the signed integer can represent numbers between –1024.0 to 1023.96875 (assuming twos complement). Both signed and unsigned numbers have a resolution of 1/32nd (0.03125). You can used fixed-point to represent distances, surfaces, volumes, temperatures, pressures, etc. Depending on the application, you can fix the position of the radix point elsewhere to suit the range of numbers you have to deal with.

Figure 9.2 shows how you can represent temperatures from –459.67 °F (0° Kelvin, absolute 0) to +2048 °F by using an 11-bit integer and a 4-bit fraction. An integer value of 11528 represents a temperature of 720.5 °F (11528×2^{-4}). Using this format, temperatures can be represented with a 1/16th °F resolution. The temperature scale is an ideal use for fixed-point math because the range is well defined, so the programmer can easily set the location of the radix point in advance.

Figure 9.2   *Representing temperatures from –459.67 °F to 2047 °F.*

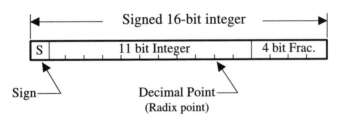

When your program performs arithmetic operations (add, subtract, multiply, or divide) on fixed-point numbers, it actually manipulates integers. (Microprocessors do not provide mechanisms to represent fixed-point numbers.) This means that the programmer must personally keep track of the position of the radix points. To represent fixed-point numbers, I will use the following notation:

        Fixed-point number = *<mantissa>***S***<exponent>*

where **S** means that the mantissa needs to be scaled by $2^{exponent}$ to determine the value of the fixed-point number. The exponent is sometimes called the *scale factor*. The mantissa is always an integer number. I use this notation to differentiate the fixed-point notation from the floating-point notation *<mantissa>***E***<exponent>*. Following are some examples of the use of this notation.

5S–3	represents 0.6250 or, 5×2^{-3} or, $5 \div 8$
31S–8	represents 0.1211 or, 31×2^{-8} or, $31 \div 256$
–123S–16	represents –0.001877 or, -123×2^{-16} or, $-123 \div 65536$

The mantissa is shown in bold to emphasize that the fixed-point number is actually represented using an integer whereas the exponent is maintained mentally by the programmer.

Scaling is done to allow almost any number to be represented using a 16-bit integer. The position of the radix point is determined from the largest number that you need to represent. Equation [9.1] shows how to obtain the mantissa and the exponent (scale factor) for any positive value x between 0.0 and 65535.0.

POSITIVE NUMBERS (0.0 < x ≤ 65535.0):

[9.1]
$$factor = -INT\left(\frac{\log\left(\frac{65535}{x}\right)}{\log(2)}\right)$$

$$mantissa = INT(2^{-factor} \times x + 0.5)$$

where `INT()` means that you take the integer portion of the result. In other words, the result is truncated. `log()` is the logarithm of the number in parentheses (either `logn()` or `log10()`). When x is 0.0 both the mantissa and the factor are 0. To represent the number 1.2345 using the fixed-point number notation, you would substitute 1.2345 in Equation [9.1] as follows:

$$-15 = -INT\left(\frac{\log\left(\frac{65535}{1.2345}\right)}{\log(2)}\right)$$

$$40452 = INT(2^{15} \times 1.2345 + 0.5)$$

Thus, the number 1.2345 is written as **40452S–15**.

Equation [9.2] shows how to obtain the mantissa and the exponent for a positive value of x that is greater than 65535.0.

POSITIVE NUMBERS(x > 65535.0):

[9.2]
$$factor = INT\left(\frac{\log\left(\frac{x}{65535}\right)}{\log(2)}\right) + 1$$

$$mantissa = \frac{x}{2^{factor}}$$

Again, `INT()` means that we take the integer portion of the result. `log()` is the logarithm of the number in parentheses. For example, the number 107573 is represented as:

$$1 = INT\left(\frac{\log\left(\frac{107573}{65535}\right)}{\log(2)}\right) + 1$$

$$53786 = \frac{107573}{2^1}$$

Thus, the number 107573 is written as **53786S1**. Note that in this case, we lose resolution because we would actually need 17 bits to represent 107573 but we only have 16-bits.

Equation [9.3] shows how to obtain the mantissa and the exponent for any signed value *x* between −32767.0 to +32767.0 (inclusively).

SIGNED NUMBERS (−32767.0 ≤ *x* ≤ +32767, except 0.0):

[9.3]
$$factor = -INT\left(\frac{\log\left(\frac{32767}{|x|}\right)}{\log(2)}\right)$$

$$mantissa = 2^{-factor} \times x$$

where `INT()` means that we take the integer portion of the result. In other words, the result is truncated. |*x*| means the absolute value of the number to scale. `log()` is the logarithm of the number in parentheses. When *x* is 0.0, both the mantissa and the factor are 0.

Equation [9.4] shows how to obtain the mantissa and the exponent for a signed integer that is less than −32767.0 and greater than +32767.0.

SIGNED NUMBERS (−32767.0 > *x* > +32767):

[9.4]
$$factor = INT\left(\frac{\log\left(\frac{|x|}{32767}\right)}{\log(2)}\right) + 1$$

$$mantissa = \frac{x}{2^{factor}}$$

Again, `INT()` means that we take the integer portion of the result. |*x*| is the absolute value of the number to scale, and `log()` is the logarithm of the number in parentheses.

9.01 Fixed-Point Addition and Subtraction

To add or subtract two fixed-point numbers, the exponent of both numbers must be the same. For example, you could not add the signed fixed-point number **20480S−15** (0.6250) with **31745S−18** (0.1211) because they do not represent the same order of magnitude. In order to add these numbers, you would first convert the smaller number (**31745S−18**) to the order of magnitude of the larger number. You would do this by adding 3 to the exponent (which is the same as multiplying by 2^3, or 8) and then dividing the mantissa by 8. The number would be **3968S−15** (i.e., 3968/32768). The result of the addition is thus

24448S–15 (0.746094). Pretty simple, right? Actually, things gets a little trickier when you add two numbers and the result exceeds unity. For example:

0.99 + 0.99 = 1.98 or

32440S–15 + **32440**S–15 = **64880**S–15

What actually happens here is that the addition overflows because the maximum value for a signed 16-bit fixed-point number can only be 32767! In this case, you can avoid the overflow by scaling both numbers to S–14 instead of S–15 as shown following this paragraph. You will thus need to be careful when you add or subtract two fixed-point numbers.

0.99 + 0.99 = 1.98 or

16220S–14 + **16220**S–14 = **32440**S–14

9.02 *Fixed-Point Multiplication*

To multiply fixed-point numbers, you simply multiply the mantissa of the two numbers and add the exponents. For example, we can multiply the two signed 16-bit fixed-point numbers:

0.6250 × 0.1211 = 0.075688 or

20480S–15 × **31745**S–18 = **650137600**S–33

One thing to note here is that when you multiply two signed 16-bit numbers, the result is a 30-bit number. Because of this, your C compiler needs to support signed longs (32-bit numbers). In the previous example, you must divide the number by **32768**S–15 (i.e., this is a division by 1.0 and does not change the result) to obtain a signed 16-bit result. A division by **32768**S–15 simply involves shifting the mantissa right 15 places. In this case, the result would be **19840**S–18 (or 0.075684).

For unsigned fixed-point numbers, the multiplication yields a 32-bit result. For example, 0.6250 × 0.1211 looks like this:

40960S–16 × **63491**S–19 = **2600591360**S–35

A division of 65536 would make the previous result fit back into an unsigned 16-bit integer: **39681**S–19 (or 0.075686). Note that the result is more accurate than its signed version because more bits were used in the unsigned multiplication.

9.03 *Fixed-Point Division*

Divisions are always trickier (and slower) than multiplications. For example, instead of dividing a number by 10, you should consider multiplying the number by 0.1 (or **26214**S–18, signed). If you have to perform a division, however, you simply divide the mantissas and subtract the exponents as:

0.2345 ÷ –10.987 = –0.021343 or

30736S–17 ÷ –**22501**S–11 = –**1**S–6 (–0.015625)

Note how the result is totally incorrect. This is because the division produced a result of –1 and a remainder of 8235. C compilers don't know what to do with remainders. To avoid this problem, you simply need to scale the dividend by **32768**S–15 and remember that the final result has been multiplied by 32768:

$$(\textbf{30736}S\text{--}17 \times \textbf{32768}S\text{--}15) \div \textbf{--22501}S\text{--}11 = \textbf{--44760}S\text{--}21 \text{ (or --0.021343)}$$

Note that the mantissa of the result doesn't fit in a 16-bit signed number. Because of this, the result needs to be adjusted as follows:

$$\textbf{--44760}S\text{--}21 \div \textbf{2}S\text{--}1 = \textbf{--22380}S\text{--}20 \text{ (or --0.021343)}$$

The overflow problem will occur whenever the mantissa of the numerator is greater than the mantissa of the denominator. Your code will have to check for this situation.

9.04 Fixed-Point Comparison

Comparing two fixed-point numbers presents a problem similar to the problem of adding and subtracting: the exponent of both numbers must be the same. For example, comparing **20480**S–15 with **31745**S–18 requires that you adjust the smaller of the two numbers to match the scale of the larger. **31745**S–18 would thus become **3968**S–15 (i.e., 3968/32768). Once both numbers represent the same order of magnitude, comparing the two numbers is simply a matter of comparing the mantissas.

9.05 Using Fixed-Point Arithmetic, Example #1

9

Suppose you needed to compute the circumference of a circle that can vary in diameter from 1.22 to 20.8 inches. The circumference of a circle is given by:

[9.5] $Circumference = \pi \times Diameter$

Because diameters are positive quantities, we will use unsigned fixed-point numbers. π can be represented as **51472**S–14 (actually 3.141602). As shown in Figure 9.3, we need a 5-bit integer to represent the diameter of the circle; the other 11 bits of an unsigned 16-bit integer number are used to hold the fraction. In other words, the diameter will be scaled by 2^{11}. Numbers for the diameter will be represented as **<mantissa>**S–11.

Figure 9.3   Fixed-point representation for circle diameter.

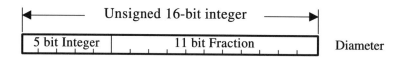

The circumference of the circle is computed in C as follows:

```
INT16U Circumference(INT16U diameter)
{
    INT16U x;


    x = (INT16S)((51472L * (INT32U)diameter) >> 16);
    return (x);
}
```

Multiplying two 16-bit unsigned integers will yield a 32-bit result, so you must adjust the resultant mantissa by dividing by 65536 (i.e., shifting right 16 places). The exponent of the result is determined as follows. π has the exponent of S–14 and the diameter has an exponent of S–11. However, the right shift is the same as dividing by **65536S–16** and thus, the exponent of the result is ((–14) + (–11) – (–16)) = S–9 (S–14 × S–11 ÷ S–16).

Our minimum circumference is obtained by substituting a 1.22 (**2498S–11**) inch diameter circle in the previous code. The multiplication yields **128577056S–25**. After the shift, the result is **1961S–9** (3.830078) which is within about 0.07 percent of the correct result of 3.832743. Our maximum circumference is obtained by substituting a 20.8 (**42598S–11**) inch diameter circle in the previous code. The multiplication yields **2192604256S–25**. After the shift, the result is **33456S–9** (65.343750) which is within about 0.002 percent of the correct result of 65.345127.

9.06 Using Fixed-Point Arithmetic, Example #2

Computing the volume of a cylinder involves more multiplications. The formula for the volume of a cylinder is:

[9.6]
$$Volume = \frac{\pi \times (Diameter)^2 \times Length}{4}$$

Suppose the cylinder length varies from 9 to 24 inches, and the diameter varies from 1 to 12 inches. To compute the volume of a cylinder, I will again use unsigned integer math because all arguments are strictly positive. π can be represented as **51472S–14** (actually 3.141602). To represent the length of the cylinder, we need 5 bits for the integer portion (up to 31 inches). The other 11 bits of an unsigned 16-bit integer number are used to hold the fraction; in other words, the length will be scaled by 2^{11}. Similarly, the diameter will require 4 bits for the integer portion and 12 bits for the fraction. This is shown in Figure 9.4. Numbers representing the length will be represented as **<mantissa>S**–11while numbers for the diameter will be represented as **<mantissa>S**–12.

Figure 9.4   Fixed-point representation for cylinder length and diameter.

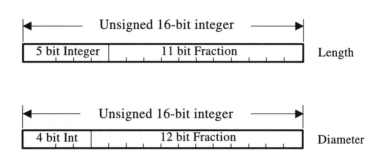

The volume of the cylinder is computed in C as follows:

```
INT16U Volume(INT16U length, INT16U diameter)
{
    INT32U x;
    INT32U dia;



    dia  = (INT32U)diameter;
    x    = (51472L * dia) >> 16;        /* S- 10 Result    */
    x    = (x * dia) >> 16;             /* S- 6 Result     */
    x    = (x * (INT32U)length) >> 16;  /* S- 1 Result     */
    return ((INT16U)x);                 /* S- 3 Result     */
}
```

Each multiplication is carried out separately because you must convert the resulting 32-bit mantissa to a 16-bit mantissa. The exponent of the result is S–10 (S–14 × S–12 ÷ S– 16). The diameter is multiplied by the intermediate result and again, the new result is adjusted. The exponent of this new result is S–6 (S–10 × S–12 ÷ S–16). Finally, the length is multiplied by the surface of the circle to obtain the volume. The exponent of the result is S–1 (S–6 × S–11 ÷ S–16), however, you can avoid dividing by 4 simply by changing the scale of the result. Thus, the final exponent is S–3.

Our minimum volume is obtained by substituting a 9-inch long (**18432**S–11) 1-inch diameter cylinder (**4096**S–12).

1st Multiplication	**51472**S–14 × **4096**S–12 is **210829312**S–26 or **3217**S–10 after the shift.
2nd Multiplication	**3217**S–10 × **4096**S–12 is **13176832**S–22 or **201**S–6 after the shift.
3rd Multiplication	**201**S–6 × **18432**S–11 is **3704832**S–17 or **56**S–1 after the shift.

The returned value is actually scaled S–3 and thus, the final result is **56**S–3 (or 7.00). The real volume should be 7.06858, which results in an error of 0.98 percent. Performing the same operations using our maximum values (12-inch diameter (**49152**S–12) and a 24-inch length (**49152**S–11)) will yield the following results:

1st Multiplication	**51472**S–14 × **49152**S–12 is **2529951744**S–26 or **38604**S–10 after the shift.
2nd Multiplication	**38604**S–10 × **49152**S–12 is **1897463808**S–22 or **28953**S–6 after the shift.
3rd Multiplication	**28953**S–6 × **49152**S–11 is **1423097856**S–17 or **21714**S–1 after the shift.

The returned value is then **21714**S–3 (2714.25). The actual volume is 2714.336 yielding an error of only 0.003 percent. One thing to note is that the second multiplication produced a number that is less than half of the full scale. In other words, 28953 is less than half the full range of an unsigned 16-bit number (0 to 65535). By shifting left by 15 places instead of 16 places, you could actually obtain better accuracy from that point on, as shown:

1st Multiplication	**51472**S–14 × **49152**S–12 is **2529951744**S–26 or **38604**S–10 after the shift.
2nd Multiplication	**38604**S–10 × **49152**S–12 is **1897463808**S–22 or **57906**S–7 after a shift of only 15 places.
3rd Multiplication	**57906**S–7 × **49152**S–11 is **2846195712**S–18 or **43429**S–4 after the shift.

The returned value is this case is **43429**S–4, which is 2714.3125, but the computation was performed with better accuracy throughout. This improvement in accuracy would help when computing smaller volumes. The final code would be:

```
INT16U Volume(INT16U length, INT16U diameter)
{
    INT32U x;
    INT32U dia;


    dia   = (INT32U)diameter;
    x     = (51472L * dia) >> 16;        /* S- 10 Result     */
    x     = (x * dia) >> 15;             /* S- 7 Result      */
    x     = (x * (INT32U)length) >> 16;  /* S- 2 Result      */
    return ((INT16U)x);                  /* S- 4 Result      */
}
```

9.07 Using Fixed-Point Arithmetic, Example #3

You can use fixed-point arithmetic to convert °C (degrees Celcius) to °F (degrees Fahrenheit). The equation for converting °F to °C is:

$$[9.7] \qquad °C \ = \ \frac{(°F - 32) \times 5}{9}$$

In order to determine how to implement the conversion equation using fixed-point arithmetic, you need to know the range of temperatures that you will be dealing with. Suppose that you are interested in temperatures from –40 °F to 250 °F. The range chosen forces you to use signed integer arithmetic. Also, you need 8 bits to represent temperatures up to 250 °F, and thus, 7 bits will be used to represent fractional degrees. The bias of 32 °F is represented as **4096**S–7, while the constant multiplier 5/9 can be represented as **18204**S–15. The code to perform the conversion is:

```
INT16S FtoC(INT16S f)
{
    return (((INT32S)(f -  4096) * 18204) >> 15);        /* Result is S- 7 */
}
```

The temperature in °C is scaled S– 7 (i.e., S–7 $\times$ S–15 $\div$ S–15). Performing the conversion from °C to °F is just as simple. The equation is:

$$[9.8] \qquad °F \ = \ \frac{°C \times 9}{5} + 32$$

Again, the 32 °F constant is **4096**S–7, while the constant multiplier 9/5 is **29491**S–14. The conversion code is:

```
INT16S CtoF(INT16S c)
{
    INT16S x;


    x = ((INT32S)c * 29491) >> 14;
    return (x + 4096);              /* Result is S- 7 */
}
```

Note that to obtain an S–7 result, I had to divide the result of the multiplication by 16384 instead of 32768.

9

9.08 Conclusion

To use fixed-point arithmetic, you need to know the range of values that the variables can take. Fixed-point arithmetic operations will generally execute quickly because most microprocessors are good at performing integer operations. This performance is at the expense of accuracy and complexity. To improve the accuracy you have to use more bits. Using fixed-point arithmetic produces large errors when using small numbers (i.e., numbers at the bottom of the scale) and decent results using large numbers. For large numbers, the improvement in accuracy is a result of using more bits. Fixed-point works very well when the dynamic range of the numbers is small.

9.09 Bibliography

Crowell, Charles
"Floating-Point Arithmetic with the TMS32010"
Houston, TX
Texas Instruments Inc., 1986

Institute of Electrical and Electronics Engineers, Inc.
ANSI/IEEE Std 754–1985, IEEE Standard for Binary Floating-Point Arithmetic
345 East 47th Street
New York, NY 10017

Knuth, Donald E.
The Art of Computer Programming, Vol. 2, Seminumerical Algorithms
Reading, Massachusetts
Addison-Wesley Publishing Company
ISBN 0-201-03822-6

Morgan, Don
Numerical Methods, Real-Time and Embedded Systems Programming
San Mateo, CA
M&T Publishing, Inc.
ISBN 1-55851-232-2

Prosise, Jeff
"Questions & Answer"
Microsoft Systems Journal
March 1993, p85,86

Simar, Ray Jr.
"Floating-Point Arithmetic with the TMS32010"
Houston, TX
Texas Instruments Inc., 1986

Chapter 10

Analog I/Os

Natural parameters such as temperature, pressure, displacement, altitude, humidity, flow, etc., are *analog*. In other words, the value taken by these parameters can change continuously instead of in discrete steps. To be manipulated by a computer, these analog parameters must be converted to digital. This is called *analog-to-digital conversion*.

Certain analog parameters can also be controlled. For example, the speed of an automobile is adjusted by changing the position of the *throttle*. The exact position of the throttle depends on many factors, such as wind resistance, whether you are going uphill or downhill, etc. You can control the flow of liquids or gases by adjusting the opening of a valve. (Flow, in this case, is not necessarily proportional to the opening of the valve, but this is a different issue.) The position of the heads in some hard disk drives is controlled by *voice coil* type *actuators*. An actuator is a device that converts electrical or pneumatic signals into linear motion. To be controlled by a computer, analog parameters must be converted from their digital form to analog. This is called *digital-to-analog conversion*.

This chapter discusses software issues relating to analog-to-digital conversions and digital-to-analog conversions. I will also describe how I implemented an analog I/O module. The analog I/O module offers the following features:

- Reads and scales from 1 to 250 analog inputs.
- Updates and scales from 1 to 250 analog outputs.
- Each analog I/O channel can define its own scaling function.
- Your application obtains *Engineering Units* from analog input channels instead of ADC counts.
- Your application provides *Engineering Units* to analog output channels instead of DAC counts.

    This chapter assumes you understand the concept of fixed-point math, described in Chapter 9.

10.00 Analog Inputs

A typical analog-to-digital system generally consists of the following circuit elements:

- transducer
- amplifier
- filter
- multiplexer
- analog-to-digital converter (ADC)

The interconnection of these components is shown in Figure 10.1. The inputs to the system are the *physical parameters* to measure (pressure, temperature, flow, position, etc.).

Figure 10.1   Analog-to-digital conversion.

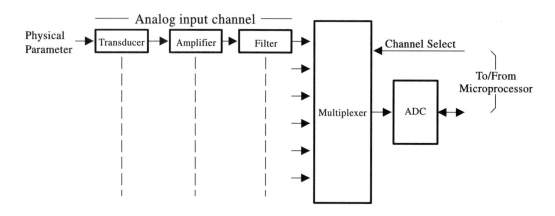

The physical parameter is first converted into an electrical signal by a *transducer*. Transducers are available to convert temperature, pressure, humidity, position, etc., to electrical signals. An *amplifier* is generally used to increase the amplitude of the transducer output to a more usable level for further processing (typically between 1 and 10 volts); the output of a transducer may produce a signal in the microvolt to millivolt range. The amplifier is frequently followed by a *low pass filter*, which is used to reduce unwanted high-frequency electrical noise. The process described previously is usually called *input conditioning* and each conditioned input is also referred to as an *analog input channel*. Analog input channels are *multiplexed* into an *analog-to-digital converter* (ADC) because ADCs are often expensive devices. The ADC converts each analog input signal to digital form. The microprocessor is responsible for selecting which analog input it wants to convert and also for initiating the conversion process for the selected channel. The block diagram of Figure 10.1 can be augmented by adding a *sample-and-hold* stage between the multiplexer and the ADC which would be used to ensure that the level of the signal is constant while a conversion is taking place.

The process of converting analog signals to digital is a complex topic and is covered in great details in many books (see "Bibliography" on page 374). In this book, I will concentrate mostly on some of the software aspects. Analog-to-digital conversion basically consists of transforming a continuous analog signal into a set of digital codes. This is called *quantizing*. Figure 10.2 shows how a 0-to-10 volt signal is quantized into a 3-bit code.

Figure 10.2   Quantizing an analog signal.

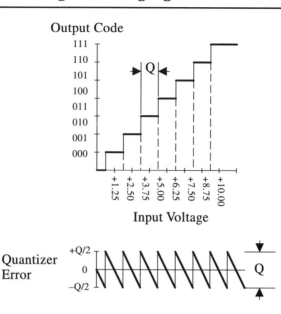

There are several important points to note about Figure 10.2. First, the *resolution* of the quantizer is defined by the number of bits it uses. An 8-bit quantizer will divide the input level into 256 steps. A 12-bit quantizer will divide the input level into 4,096 steps. Thus, a 12-bit quantizer has a higher resolution than an 8-bit quantizer. The number of steps for the quantizer is 2^n where n corresponds to the number of bits used. Quantizers (or ADCs) are commercially available from 4 to 24 bits. The required resolution is dictated by the application. There are literally hundreds of ADCs to choose from, and generally cost increases with resolution.

An important point to make is that the maximum value of the digital code of an ADC, namely all 1s (ones), does not correspond with the analog Full Scale (FS) but rather, one Least Significant Bit (LSB) less than full scale or:

[10.1] $Maximum\_value\_of\_digital\_code = FS \times (1 - 2^{-n})$

For example, a 12-bit ADC with a 0 to +10V analog range has a maximum digital code of 0x0FFF (4095) and a maximum analog value of +10V $\times$ (1 – 2^{-12}) or +9.99756V. In other words, the maximum analog value of the converter never quite reaches the point defined as full scale. At any part of the input range of the ADC, there is a small range of analog values within which the same code is produced. This small range in values is known as the *quantization size*, or *quantum*, Q. The quantum in Figure 10.2 is 1.25V and is found by dividing the full scale analog range by the number of steps of the quantizer. Q is thus given by the following equation:

[10.2] $Q = \dfrac{FSV}{2^n}$

Q is the smallest analog difference that can be distinguished by the quantizer.
FSV is the full scale voltage range.
n corresponds to the number of bits used by the quantizer (i.e., ADC).

As shown in Figure 10.2 (Quantizer Error), a sawtooth error function is obtained if the ADC input is moved through its range of analog values and the difference between output and input is taken. For example, any voltage between 1.875V and 3.125V will produce the binary code 010.

All ADCs require a small but significant amount of time to quantize an analog signal. The time it takes to make the conversion depends on several factors: the converter resolution, the conversion technique, and the technology used to manufacture the ADC. The *conversion speed* (how fast an analog voltage is converted to digital) required for a particular application depends on how fast the signal to be converted is changing and on the desired accuracy. The *conversion time* (inverse of conversion speed) is frequently called *aperture time*. If the analog signal to measure varies by more than the resolution of the quantizer during the conversion time, then a *sample-and-hold* circuit should be used. ADCs are available with conversion speeds ranging from about three conversions per second to well over 100 million conversions per second.

10.01 Reading an ADC

The method used to read the ADC depends on how fast the ADC converts an analog voltage to a binary code. In most cases, however, the ADC must be explicitly triggered to perform a conversion. In other words, you must issue a command to the ADC to start the conversion process. Very fast ADCs, those that can convert an analog signal in less than 1 μS, generally have dedicated hardware to handle the fast conversion rate and will typically buffer the *samples*. When the buffer is full, the analog samples are processed *offline*. This is basically how a digital storage oscilloscope works. At the other end of the spectrum, ADCs used in voltmeters are generally slow (about 200 mS) but accurate (4 1/2 digits, or 0.005 percent).

The actual method used to read an ADC depends on many factors: the conversion time of the ADC, how often you need the analog value converted, how many channels you have to read, etc. The next three sections describe some possible methods of reading an ADC.

10.01.01 Reading an ADC, Method #1

The scheme shown in Figure 10.3 assumes that the ADC conversion time is relatively slow (greater than about 5 mS). Here a driver (a function) reads an analog input channel and returns the result of the conversion to your application. Your application calls the driver in Figure 10.3 and passes it the desired channel to read. The driver starts by selecting (through the multiplexer) the desired analog channel (①) to read. Before starting the conversion, you may want to wait a few microseconds to allow for the signal to propagate through the multiplexer and stabilize. If you don't wait for the multiplexer's output to stabilize, your readings may be unstable. Next, the ADC is triggered to start the conversion (②). The driver then delays to allow for the conversion to complete (③). Note that the delay time must be longer than the conversion time of the ADC. After the delay, the driver assumes that the conversion is complete and reads the ADC (④). The binary result is then returned to your application (⑤). The pseudocode is:

```
ReadAnalogInputChannel(Channel#)
{
    Select the desired analog input channel;
    Wait for MUX output to stabilize;
    Start ADC conversion;
    Delay 'x' mS to allow for conversion to complete;
    Read ADC and return result to the caller;
}
```

Figure 10.3   Reading an ADC (Method #1).

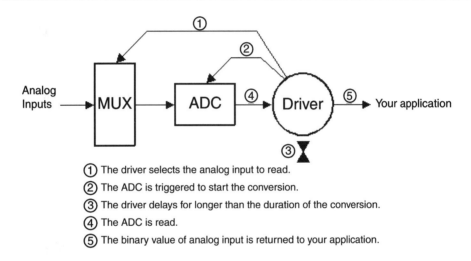

① The driver selects the analog input to read.
② The ADC is triggered to start the conversion.
③ The driver delays for longer than the duration of the conversion.
④ The ADC is read.
⑤ The binary value of analog input is returned to your application.

10

This method is simple and can be used with slow-changing analog signals. For example, you can use this method when measuring the temperature of a room (which doesn't change very quickly).

10.01.02 Reading an ADC, Method #2

You can actually use a signal provided by most ADCs (i.e., the *End Of Conversion* (EOC) signal) to tell your driver when the ADC has completed its conversion. The code and your hardware in this case will be a little more complicated, but this method is more efficient.

Figure 10.4   Reading an ADC (Method #2).

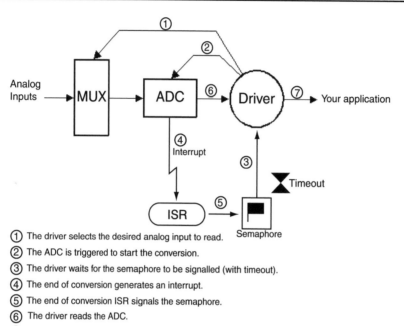

① The driver selects the desired analog input to read.
② The ADC is triggered to start the conversion.
③ The driver waits for the semaphore to be signalled (with timeout).
④ The end of conversion generates an interrupt.
⑤ The end of conversion ISR signals the semaphore.
⑥ The driver reads the ADC.
⑦ The binary value of the analog input is returned to your application.

Again, your application calls the driver by passing it the analog input channel to read. The driver shown in Figure 10.4 starts by selecting (through the multiplexer) the desired analog channel (①). At this point, you should again wait a few microseconds to allow for the signal to propagate through the multiplexer and stabilize. The ADC is then triggered to start the conversion (②). The driver then waits for a semaphore (③) with a timeout. A timeout is used to detect a hardware malfunction. In other words, you don't want the driver to wait forever if the ADC fails (i.e., never finishes the conversion). When the analog conversion completes, the ADC generates an interrupt (④). The ADC conversion-complete ISR signals the semaphore (⑤), which notifies the driver that the ADC has completed its conversion. When the driver gets to execute, it reads the ADC (⑥) and returns the binary result to your application (⑦).

The pseudocode for both the driver and the ISR follows.

You would use this method if the conversion time of the ADC is greater than the execution time of the ISR and the call to wait for the semaphore. For example, your ADC takes 1 mS to perform a conversion, and the total execution time of the ISR and the call to wait for the semaphore requires only about 50 µS. If the execution time of the ISR and the call to wait for the semaphore is greater than the conversion time of the ADC, you might as well wait in a software loop (polling the ADC's EOC line) until the ADC completes its conversion. This method will be discussed next.

```
ReadAnalogInputChannel (Channel#)
{
    Select the desired analog input channel;
    Wait for MUX output to stabilize;
    Start ADC conversion;
    Wait for signal from ADC ISR (with timeout);
    if (Timed out) {
        Signal error;
    } else {
        Read ADC and return result to the caller;
    }
}
Conversion complete ISR
{
    Signal conversion complete semaphore;
}
```

10.01.03 Reading an ADC, Method #3

The third method can be used if the conversion time of the ADC is less than the time needed to process the interrupt and wait for the semaphore, as described in the previous method. For example, depending on the microprocessor, an ADC with a conversion time less than 25 µS cannot afford the overhead of an interrupt and a semaphore which could take over 50 µS. In other words, the execution time to handle the interrupt overhead and the time to signal and wait for the semaphore can take more than 25 µS. This is true of most 8-bit and some 16-bit microprocessors.

Your application calls the driver shown in Figure 10.5 by passing it the desired analog input channel to read. The driver starts by selecting (through the multiplexer) the channel to read (①). Again, before starting the conversion, you may want to wait a few microseconds to allow for the signal to propagate through the multiplexer and stabilize. The ADC is then triggered to start the conversion (②). The driver then waits (③) in a software loop for the ADC to complete its conversion. While waiting in the loop, the driver monitors the status (the EOC) or the *BUSY* signal of the ADC. You need to ensure that you have a way to prevent an infinite loop if your hardware becomes defective. An infinite loop is avoided by using a software counter which is decremented every time through the polling loop (see the pseudocode following this paragraph). The initial value of the counter is determined from the execution time of each iteration of the polling loop. For example, if you have an ADC that should perform a conversion in 50 µS and each iteration through the polling loop takes 5 µS, you will need to load the counter with a value of at least 10. You want to use the loop counter as an indication of a hardware malfunction and not to indicate when the ADC is done converting. Based on experience, you should load the loop counter so that a timeout occurs when the polling time exceeds the ADC conversion by about 25 to 50 percent. In other words, you would load the counter with a value between 13 and 15 in my example. When the ADC finally signals an end of conversion, the driver reads the ADC (④) and returns the binary result to your application (⑤).

10

Figure 10.5   Reading an ADC (Method #3)

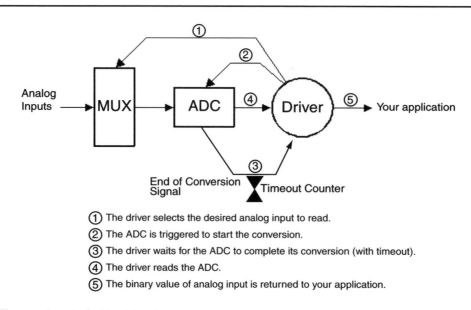

① The driver selects the desired analog input to read.
② The ADC is triggered to start the conversion.
③ The driver waits for the ADC to complete its conversion (with timeout).
④ The driver reads the ADC.
⑤ The binary value of analog input is returned to your application.

The pseudocode for the driver is:

```
ReadAnalogInputChannel (Channel#)
{
    Select the desired analog input channel (i.e. MUX);
    Wait for MUX output to stabilize;
    Start ADC conversion;
    Load timeout counter;
    while (ADC Busy && Counter-- > 0)   /* Polling Loop                    */
        ;
    if (Counter == 0) {                 /* Check for hardware malfunction */
        Signal error;
    } else {
        Read ADC and return result to the caller;
    }
}
```

Actually, I prefer this method because:

- You can get fairly inexpensive fast ADCs (~25 µS conversion time).
- You don't have the added complexity of an ISR.
- Your signal has less time to change during a conversion.
- This method imposes very little overhead on your CPU.
- The polling loop can be interrupted to service interrupts.

10.01.04 Reading an ADC, Miscellaneous

The nice thing about reading analog input channels through drivers is that the implementation details are hidden from your application. You can use any of the three drivers shown without changing your application code.

By always returning the same number of bits to your application, you can make your application insensitive to the actual number of bits of the ADC. In other words, if the ADC driver always returned a signed 16-bit number irrespective of the actual number of bits for the ADC, your application would not have to be adjusted every time you changed the word size of your ADC. This is actually quite easy to accomplish, as shown in Figure 10.6. All you need to do is to shift left the binary value of the ADC until the most significant bit of the ADC value is in bit position number 14 of the result. I use a 16-bit signed result because the computations required to scale the result of the ADC need to be signed. This will be described in the next section. If you deal with higher resolution ADCs, you may want to write your drivers and application code to assume signed 32-bit values.

Figure 10.6   ADC driver always returning a signed 16-bit result.

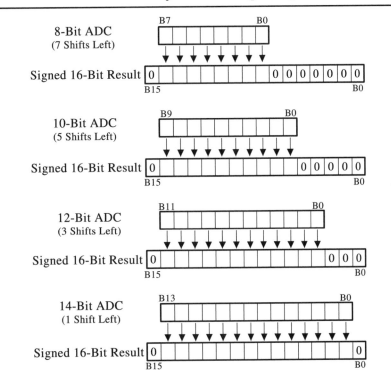

For example, an 8-bit ADC can measure a voltage between 0 and 0.996094 (255/256) of the full scale voltage (see Equation [10.1]). This is the same as (255 << 7) / 32768, or 0.996094. Similarly, a 12-bit ADC can measure a voltage between 0 and 4095/4096 or 0.999756, which is the same as (4095 << 3) / 32768 (i.e., 0.999756). You can thus hide the details about how many bits each ADC has with respect to your application without losing any accuracy.

10.02 Temperature Measurement Example

As we have seen, an ADC produces a binary code based on a full scale voltage. If you are measuring a temperature, for example, this information means very little to you. What you really want to know is the temperature of what you are measuring. The circuit in Figure 10.7 shows a commonly used temperature sensor *Integrated Circuit* (IC), the National Semiconductor LM34A.

Figure 10.7 Temperature measurement using an LM34A.

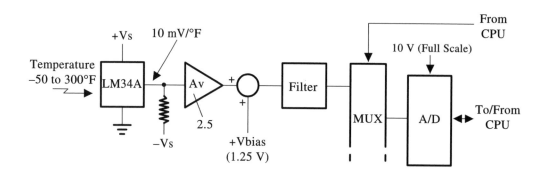

The LM34A produces a voltage that is directly proportional to the temperature surrounding it, specifically, 10 mV/°F. Note that you can also obtain the temperature in degrees Celsius by using an LM35A. The amplifier is designed to have a gain of 2.5, and thus –50 to 300 °F will produce a voltage of –1.25 to 7.50 volts. By using a 10-bit ADC, you can obtain a resolution of about 0.342 °F (350 °F/1024). Note that the ADC can only convert positive voltages, and thus a bias of 1.25 volts is introduced following the amplification stage to ensure that a positive voltage is present at the input of the ADC for the complete temperature range. With this bias, –50 °F will appear as 0 V, 0 °F will be 1.25 V and 300 °F will be 8.75 V. The value obtained at the ADC is given by:

[10.3]

$$ADC_{counts} = \frac{\left(Temperature_{(°F)} \times 0.01_{V/(°F)} \times 2.5_{A_V} + 1.25_{V_{bias}}\right) \times 1023_{counts}}{10_{V_{FullScale}}}$$

counts is an industry standard convention that means the *binary* value of the ADC.

$0.01_{V/(°F)}$ corresponds to the transducer transfer function — 10 mV/°F — specified by National Semiconductor.

2.5 is the gain of the amplifier stage and is established by the hardware designer.

1.25 is the bias voltage to ensure that the ADC always reads a positive voltage.

1023 is the maximum binary value taken by a 10-bit converter.

$10_{V_{FullScale}}$ is the full scale voltage.

For example, a temperature of 100 °F would have a value of 383 counts (actually, 383.625). Note that the ADC can produce only integer values, and thus the actual value of 383.625 is truncated to 383. To obtain the temperature read at the sensor, you need to rearrange Equation [10.3] so that temperature is given as a function of ADC counts, as shown in Equation [10.4]. This process is often called converting ADC counts to *engineering units* (E.U.):

[10.4] $$Temperature_{(°F)} = \frac{\dfrac{ADC_{counts} \times 10_{V_{FullScale}}}{1023_{counts}} - V_{bias}}{0.01_{V/(°F)} \times 2.5_{A_V}}$$

The general form for this equation is:

[10.5] $$E.U. = \frac{\dfrac{ADC_{counts} \times FSV}{(2^n - 1)} - V_{bias}}{Transducer_{V/(EU)} \times A_V}$$

E.U. is the engineering unit of the transducer (°F, PSI, Feet, etc.).

V_{bias} is the bias voltage added to the output of the amplifier stage to allow the ADC to read negative values.

FSV is the full scale voltage of the ADC.

Transducer $_{(V/EU)}$ corresponds to the number of volts produced by the transducer per engineering unit.

A_V is the gain of the amplifier stage.

n is the resolution of the ADC (in number of bits).

You can also write Equation [10.5] as follows:

[10.6] $$E.U. = \frac{\left(ADC_{counts} - Bias_{counts}\right) \times FSV}{Transducer_{V/(EU)} \times A_V \times (2^n - 1)}$$

In this case, *Bias*$_{counts}$ corresponds to the ADC counts of the bias voltage as is given by the following equation:

[10.7] $$Bias_{counts} = \frac{V_{bias} \times (2^n - 1)}{FSV}$$

Note that most of the terms in Equation [10.6] are known when the system is designed, and thus, to save processing time, they should not be evaluated at run time. In other words, you could rewrite the equation as follows:

[10.8] $$E.U. = \left(ADC_{counts} - ConvOffset_{counts}\right) \times ConvGain_{(EU)/(count)}$$

where:

[10.9] $$ConvGain_{(EU)/(count)} = \frac{FSV}{Transducer_{V/(EU)} \times A_V \times (2^n - 1)}$$

Note that the units of the conversion gain (*ConvGain*) are E.U. per ADC count.

10

[10.10]
$$ConvOffset_{counts} = -\left(\frac{V_{bias} \times (2^n - 1)}{FSV}\right)$$

In the temperature measurement example, the conversion gain would be 0.391007 and the conversion offset would be 127.875. You can apply fixed-point arithmetic and scale factors (see Chapter 9) to the temperature measurement example. The temperature of the LM34A sensor is given by:

[10.11]
$$Temperature_{(°F)} = \left(ADC_{counts} + ConvOffset_{counts}\right) \times ConvGain_{(°F)/(count)}$$

Remember that you have a 10-bit ADC, and thus the range of counts is from 0 to 1023. You can scale this number by multiplying the ADC counts by 32 (shifting left five places). To perform the subtraction with the bias, you need to scale the bias (i.e., conversion offset) by the same value, or 127.875 × 32 = **4092S–5**. The gain (0.391007) can be scaled by multiplying by 65536, and thus the conversion gain is **25625S–16**. The temperature is thus given by:

[10.12]   Temperature(°F) S–21 = ((ADC counts << 5)S–5 – **4092S–5**) × **25625S–16**

or

[10.13]   Temperature(°F) S–6 = (((ADC counts << 5)S–5 – **4092S–5**) × **25625S–16**) >> 15

From Equation [10.3], 150 °F would produce 511 ADC counts. Substituting 511 counts in Equation [10.12] produces the following:

        Temperature (°F) S–21 = (**16352S–5** – **4092S–5**) × **25625S–16**, or
        Temperature (°F) S–21 = **314162500S–21** (i.e., 149.80)

or using Equation [10.13]:

        Temperature (°F) S–6   = **9587S–6** (i.e., 149.80)

The C code to convert the ADC counts to temperature is:

```
INT16S RdTemp(INT16S raw)
{
    INT16S cnts;
    INT16S temp;


    cnts    = (raw << 5) -   4092;
    temp    = (INT16S)(((INT32S)cnts * (INT32S)25625) >> 15L);
    return (temp);              /* Result is scaled S- 6 */
}
```

Note that `raw` corresponds to the ADC counts (10 bits). The total counts (`cnts`) number is computed separately because a good compiler should perform this operation using 16-bit arithmetic instead of 32-bit (which would be faster). Counts and gain are then converted to INT32S because the multiplication needs 30-bit precision. The result is divided by 32768 so that it fits back into a 16-bit signed variable. Finally, the temperature is returned in °F scaled S–6. You could obtain the temperature to the nearest degree by first adding 32 (0.5) and then dividing the result by 64. In other words, by rounding the result.

The electronic components used to provide the amplication and the bias voltage are generally inaccurate. Oddly enough, extra components can be added to allow the amplification stage and bias voltage to be precisely adjusted (that is, *calibrated*). Adding such components, however, adds recurring cost to your system. Component inaccuracies easily can be compensated in software by modifying Equation [10.8] as:

[10.14]

$$EU = \left(ADC_{counts} + ConvOffset_{counts} + CalOffset_{counts} \right) \times ConvGain_{(EU)/(count)} \times CalGain$$

The calibration gain (*CalGain*) and calibration offset (*CalOffset*) would be entered by a calibration technician using a keyboard/display or through a communications port. Both calibration parameters could then be stored in a non-volatile memory device such as battery backed-up RAM, EEPROM, or even a floppy disk. The adjustment range of the calibration parameters is based on the accuracy of the electronic components used. A 10 percent adjustment range should be sufficient for most situations. For the calibration gain, all we need is an adjustment range between 0.90 (**14745**S–14) and 1.10 (**18022**S–14). In our example, all we need is an adjustment range between –100 (**–3200**S–5) and +100 (**3200**S–5) for the calibration offset when using a 10-bit ADC. The new C code to convert raw ADC counts to a temperature is:

```
INT16S RdTemp(INT16S raw)
{
    INT16S cnts;
    INT16S temp;


    cnts    = (raw << 5) -  4092 + CalOffset;
    temp    = (INT16S)(((INT32S)cnts  * (INT32S)25625) >> 15L);
    temp    = (INT16S)(((INT32S)temp * (INT32S)CalGain) >> 14L);
    return (temp);            /* Result is scaled S- 6 */
}
```

For example, if the actual gain of the amplification stage of our temperature measurement example was 2.45 instead of 2.50 then, `CalGain` would be set to 1.020408 (**16718**S–14). Similarly, if the bias voltage was 1.27V instead of 1.25V then, you would have to subtract 0.02V, or 65 counts (see Equation [10.10]). In other words, `CalOffset` would be set to **–65**S–5.

10.03 Analog Outputs

A typical digital to analog system generally consist of the following circuit elements:

- digital to analog converter (DAC)
- filter
- amplifier
- transducer

Digital-to-analog converters (DACs) are generally inexpensive devices, and thus each analog output channel can have its own DAC, as shown in Figure 10.8. The DAC converts a binary value provided by a microprocessor to either a current or a voltage (depending on the DAC). The voltage or current is *filtered* to smooth out the step changes. An *amplifier* stage is sometimes used to increase the amplitude or power drive capability of the analog output channel in order to properly interface with the transducer. The *transducer* is used to convert the electrical signal to a physical quantity. For example, transducers are available to convert electrical signals to pressures (known as *current-to-pressure transducers*, or I to P). These pressures can be — and often are — used to control other physical devices.

Figure 10.8   Digital-to-analog conversion.

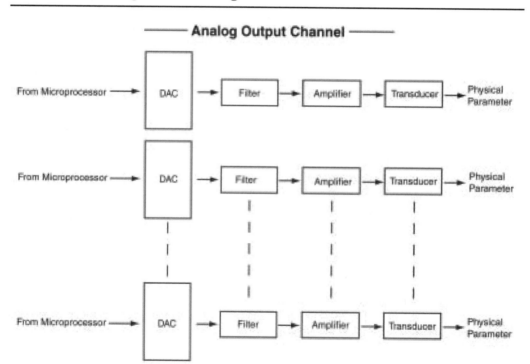

DACs are commercially available with resolutions from 4 to 16 bits. The resolution to choose from is application specific. There are literally hundreds of DACs to choose from. Generally, the cost of DACs increases with resolution and conversion speed. DACs are much faster than ADCs. Conversion time (also called *settling time*) is always less than a few microseconds and can be as fast as 5 nS (nano-second). Very fast DACs are used in video applications, and because of their higher cost and lower resolution (8 bits), very fast DACs are seldom used in industrial applications.

A digital-to-analog conversion is handled exclusively in hardware. From a software standpoint, updating a DAC is as simple as writing the binary value to one or more (if more than 8 bits) I/O port locations or memory locations (when DACs are memory mapped).

10.04 Temperature Display Example

Suppose you wanted to display the temperature read by our LM34A (see Section 10.02) on a meter, as shown in Figure 10.9.

An 8-bit DAC is deemed sufficient considering the accuracy of these types of meters. The DAC is followed by a circuit that converts the voltage output of the DAC to a current (a *V–>I Converter*). The Full Scale Voltage (FSV) of the DAC is set to 2.5 volts. The current converter is designed to produce about 42 µA/V, and the meter requires 100 µA for full scale. Your task is to write a function that takes the temperature (–50 °F to +300 °F) as an input and produces the proper output current (0 to 100 µA) to drive the meter.

Figure 10.9 Temperature display.

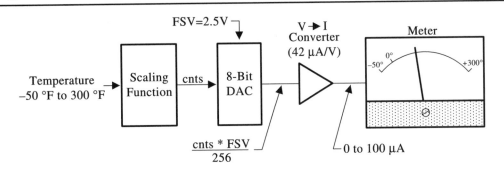

The relationship between the temperature and the meter current is shown in Figure 10.10.

Figure 10.10 Temperature to DAC counts scaling.

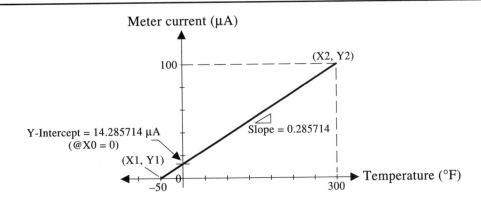

The graph can also be represented by the following linear equation:

[10.15] $y = m \times x + b$

where m is the *slope* and b is the *Y-intercept* (the value on the y-axis when x is 0). The slope gives us the current per degree of temperature and is given by:

[10.16] $$m = \frac{\left(Y_2 - Y_1\right)}{\left(X_2 - X_1\right)}$$

In this case, the slope is 100µA / 350 °F , or 0.285714 µA/°F. The Y-intercept (i.e., Y_0) is given by:

[10.17] $$Y_0 = m \times \left(X_0 - X_1\right) + Y_1$$

By substituting the values of m, Y_1, X_1, and X_0 (i.e., 0) in Equation [10.17], you obtain a Y-intercept of 14.285714 µA. The meter current thus is given by:

[10.18] $Meter_{µA} = 0.285714_{(µA)/(°F)} \times Temperature_{°F} + 14.285714_{µA}$

The meter current is also given by:

[10.19] $$Meter_{µA} = \frac{DAC_{counts} \times FSV}{256} \times 42_{(µA)/V}$$

Combining Equations [10.18] and [10.19], I obtain:

[10.20]

$$0.285714_{(µA)/(°F)} \times Temperature_{°F} + 14.285714_{µA} = \frac{DAC_{counts} \times 2.5}{256} \times 42_{(µA)/V}$$

Solving for DAC_{counts}, I obtain:

[10.21] $$DAC_{counts} = INT\left(\frac{0.285714 \times 256}{2.5 \times 42_{(µA)/V}} \times Temperature_{°F} \times \frac{14.285714 \times 256}{2.5 \times 42_{(µA)/V}}\right)$$

Note that *INT()* means that only the integer portion of the result is retained. As you can see, Equation [10.21] is also a linear equation, where m is 0.696598 and b is 34.829931. DAC_{counts} thus are given by:

[10.22] $$DAC_{counts} = INT\left(0.696598_{(counts)/(°F)} \times Temperature_{°F} + 34.829931_{counts}\right)$$

Substituting –50 °F in Equation [10.22], I obtain 0 counts (as I should). Similarly, substituting 300 °F in Equation [10.22], I obtain 243 counts, which should produce 100 µA.

As with analog inputs, the electronic components used in circuits such as the voltage-to-current converter are generally inaccurate. You can compensate for component inaccuracies in software by modifying Equation [10.22] as:

[10.23]

$$DAC_{counts} = INT\left(0.696598_{(counts)/(°F)} \times Temperature_{°F} \times CalGain + 34.829931_{counts} + CALOffset\right)$$

The effect of the calibration gain and offset is shown in Figure 10.11, which has been exaggerated for sake of discussion. The actual curve that you get from an incorrect gain and offset needs to be adjusted, as shown in Figure 10.11.

Figure 10.11 Calibration gain and offset adjustments (exaggerated).

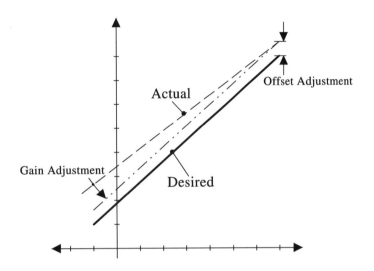

The adjustment range of the calibration parameters is based on the accuracy of the electronic components. Based on experience, a 10 percent adjustment range should be sufficient for most situations. For the calibration gain, you only need an adjustment range between 0.90 and 1.10. For the calibration offset, you need an adjustment range between –25 and +25 for an 8-bit ADC. What would happen if the voltage-to-current converter was actually putting out 40 µA/V instead of 42 (a 5 percent error)? In this case, the slope in Equation [10.23] (see Equation [10.21], substituting 40 instead of 42) would need to be adjusted to 0.731428 and the intercept would need to be 36.571428. This can be accomplished by setting `CalGain` and `CalOffset` to 1.05 and 1.741497 respectively.

The general form for Equation [10.23] is:

[10.24]

$$DAC_{counts} = INT\left(ConvGain_{(counts)/(EU)} \times CalGain \times Input_{EU} + ConvOffset_{counts} + CalOffset_{counts}\right)$$

10

10.05 Analog I/O Module

In this chapter, I provide you with a complete analog I/O module that will allow you to read and scale up to 250 analog inputs and scale and update up to 250 analog output channels. Each analog input channel is scanned at a regular interval and the scan rate for each channel can be programmed individually. This allows you to determine whether some analog inputs are scanned more often than others. Similarly, each analog output channel is updated at a regular interval and the update rate for each channel can also be programmed individually. This allows you to establish which analog outputs are to be updated more often.

The source code for the analog I/O module is found in the \SOFTWARE\BLOCKS\AIO\SOURCE directory. The source code is found in the files AIO.C (Listing 10.1) and AIO.H (Listing 10.2). As a convention, all functions and variables related to the analog I/O module start with either AIO (functions and variables common to both analog inputs and outputs), AI (analog input functions and variables) or AO (analog output functions and variables). Similarly, #defines constants will either start with AIO_, AI_, or AO_.

10.06 Internals

The analog I/O module makes extensive use of floating-point arithmetic (additions, multiplications, and divisions). The reason I chose to use floating-point instead of integer arithmetic is that it is very difficult to make a general purpose analog I/O module using integer arithmetic. The analog I/O module can become CPU-intensive unless you have hardware-assisted floating-point (i.e., a math coprocessor). The analog I/O module, however, can be easily modified to make use of integer arithmetic if you have a dedicated application.

Figure 10.12 shows a block diagram of the analog I/O module. You should also refer to Listings 10.1 and 10.2 for the following description. As shown, the analog I/O module consists of a single task (AIOTask()) that executes at a regular interval (set by AIO_TASK_DLY). AIOTask() can manage as many analog inputs and outputs as your application requires (up to 250 each). The analog I/O module must be initialized by calling AIOInit(). AIOInit() initializes all analog input channels, all analog output channels, the hardware (ADCs and DACs), a semaphore used to ensure exclusive access to the internal data structures used by the analog I/O module, and finally, AIOInit() creates AIOTask().

AITbl[] is a table that contains configuration and run-time information for each analog input channel. An entry in AITbl[] is a structure defined in AIO.H and is called AIO. AIUpdate() is charged with reading all of the analog input channels on a regular basis. AIUpdate() calls AIRd() and passes it a logical channel number (0..AIO_MAX_AI - 1). AIRd() is responsible for selecting the proper analog input through one or more multiplexers (based on the logical channel number), starting and waiting for the proper ADC to convert (if more than one is used), and for returning raw counts to AIUpdate(). AIRd() is the only function that knows about your hardware, and thus AIRd() can easily be adapted to your environment.

AOTbl[] is a table that contains configuration and run-time information for each analog output channel. An entry in AOTbl[] also uses the AIO structure. AOUpdate() is responsible for updating all of the analog output channels on a regular basis. AOUpdate() calls AOWr() and passes it a logical channel number (0..AIO_MAX_AO - 1) and the raw DAC counts. AOWr() is responsible for outputting the raw counts to the proper DAC based on the logical channel. AOWr() is the only function that knows about your hardware, and thus AOWr() can easily be adapted to your environment.

Figure 10.12 `AIO` module flow diagram.

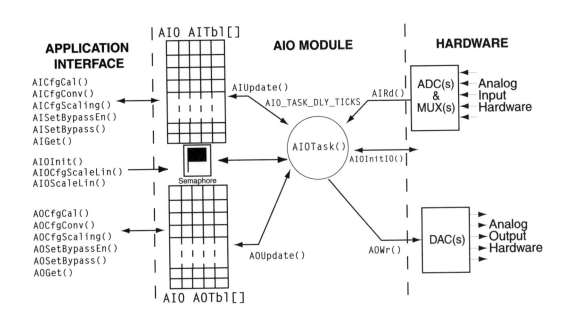

Figure 10.13 shows a flow diagram of a single analog input channel. Note that I used electrical symbols to represent functions performed in software. `.AIO???` are all members of the `AIO` structure. `AIUpdate()` updates each channel as described in the following paragraphs.

Figure 10.13 Analog input channel flow diagram.

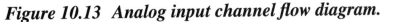

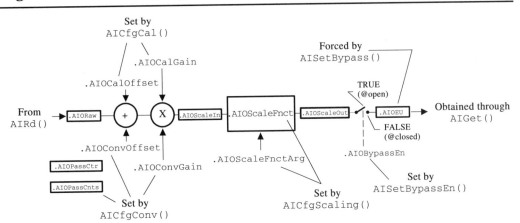

The raw counts obtained from `AIRd()` are placed in the channel's `.AIORaw` variable. The raw counts are then added to `.AIOCalOffset` and `.AIOConvOffset`. The result of this operation is then multiplied by `.AIOCalGain` and `.AIOConvGain`. These mathematical operations are basically used to implement Equation [10.14]:

[10.25]   .AIOScaleIN =(.AIORaw + .AIOConvOffset + .AIOCalOffset) ✕
                          .AIOConvGain ✕ .AIOCalGain

.AIScaleFnct is a pointer to a function that is executed when the channel is updated. The function allows you to apply further processing when reading an analog input. For example, a *Resistance Temperature Detector* (RTD) is a device that requires special processing. The temperature at the RTD is proportional to the resistance of the RTD (but is nonlinear). A scaling function can thus be written to convert .AIOScaleIn (the resistance of the RTD) to a temperature in degrees Fahrenheit (placed in .AIOScaleOut). There are many types of RTDs, and thus you need to be able to specify the actual type used. This is where .AIOScaleFnctArg comes in. .AIOScaleFnctArg is a pointer to any arguments that your scaling function requires. In the case of an RTD, this argument can specify the type of RTD used. The scaling function that you write must be declared as:

```
void AIOScale???(AIO *paio);
```

When called, your scaling function will receive a pointer to the AIO channel to scale (or linearize). The input to your function is available in paio->AIOScaleIn, and your function must place the result in paio->AIOScaleOut. Any arguments to the scaling function are found through paio->AIOScaleFnctArg. If you do not have any linearization function, the value of .AIOScaleIn is simply copied to .AIOScaleOut by AIUpdate().

.AIOBypassEn is a software switch that is used to prevent the analog input from being updated. This feature allows your application code to "bypass" the channel and force a value into .AIOEU. When another part of your application code tries to read the analog input channel, it will actually be getting the forced value instead of what the sensor is measuring. I have found this feature to be invaluable.

.AIOEU is the value that your application code will obtain when it needs the latest value read by the analog input channel (by calling AIGet()). .AIOEU contains engineering units. This means that if the analog input channel monitors a pressure, your application code will obtain a value in either PSI, KPa, InHgg, etc.

.AIOPassCnts allows your application code to specify how often the analog input channel is to be updated. In fact, .AIOPassCnts specifies how many analog input scans are needed before the channel is updated. In other words, if analog inputs are read every 50 mS and you specify a pass count of 20, then the analog input channel will be read every 1000 mS (i.e., 1 second).

Figure 10.14 shows a flow diagram of a single analog output channel. Note that I used electrical symbols to represent functions performed in software. As with analog input channels, .AIO??? are all members of the AIO structure. AOUpdate() updates each channel as described in the following paragraphs.

Figure 10.14 Analog output channel flow diagram.

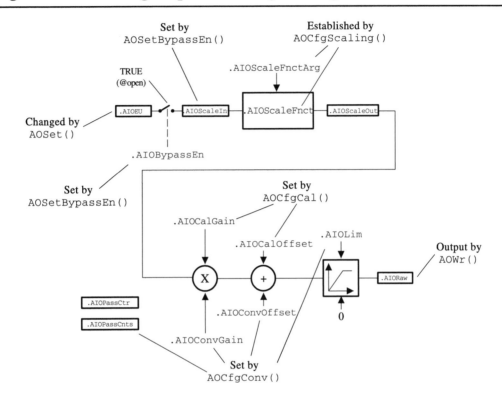

Your application deposits the value for the analog output channel by calling AOSet(). This value is passed in engineering units. This means that if the analog output channel controls a meter that displays the RPM of a rotating device, you call AOSet() by specifying an RPM and the analog output channels takes care of figuring out how much voltage or current is needed to display the RPM.

.AIOBypassEn is a software switch used to override the value that your application code is trying to put out on the analog output channel. Another function provided by the analog I/O module is used to load .AIOScaleIn. This feature is very useful for debugging purposes because it allows you to test your output independently of the application code.

.AIScaleFnct is a pointer to a function that is executed when the analog output channel is updated. The function allows you to apply further processing prior to updating an analog output. For example, a 0 to 100 mA output may be controlling a valve. If the flow through the valve is proportional to the output — but nonlinear, the function can make the valve action look linear with respect to your application. If your software needs to support different types of valves, you can specify which valve is being used through .AIOScaleFnctArg. .AIOScaleFnctArg is a pointer to any arguments that your scaling function requires. The scaling function that you write must be declared as follows:

```
void AIOScale???(AIO *paio);
```

When called, your scaling function will receive a pointer to the AIO channel to scale (or linearize). The input to your function is available in paio->AIOScaleIn, and your function must place the result in paio->AIOScaleOut. Any arguments to your function are found through `paio->AIOScaleFnctArg`. If you do not have any linearization function, the value of `.AIOScaleIn` is simply copied to `.AIOScaleOut` by `AOUpdate()`.

`.AIOScaleOut` is then multiplied by `.AIOCalGain` and `.AIOConvGain`. The result of the multiplication is the added to `.AIOCalOffset` and `.AIOConvOffset`. The result of this operation is deposited in `.AIORaw` so that it can be sent to the proper DAC by `AOWr()`.

[10.26] $.AIORaw = .AIOScaleOut \times .AIOConvGain \times .AIOCalGain + .AIOConvOffset + .AIOCalOffset$

`.AIOLim` is used to ensure that `.AIORaw` does not exceed the maximum counts allowed by the DAC. For example, an 8-bit DAC has a range of 0 to 255 counts. An output of 256 counts to a DAC would appear to the DAC as 0 (the lower eight bits of 100000000_2). `.AIOLim` contains the maximum count that can be sent to the DAC (255 for an 8-bit DAC).

`.AIOPassCnts` allows your application code to specify how often the analog output channel is to be updated. In fact, `.AIOPassCnts` specifies how many analog output scans are needed before the channel is updated. In other words, if analog outputs are updated every 50 mS and you specify a pass count of 5, the analog output channel will only be updated every 250 mS.

10.07 Interface Functions

Your application software knows about the analog I/O module through the interface functions shown in Figure 10.15.

Figure 10.15 Analog I/O module interface functions.

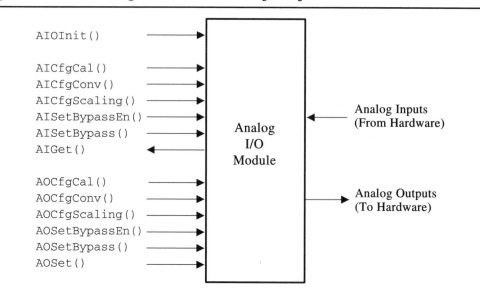

AICfgCal()

INT8U AICfgCal(INT8U n, FP32 gain, FP32 offset);

AICfgCal() is used to set the calibration gain and offset of an analog input channel. The analog I/O module implements Equation [10.14], and this function is used to set the value of CalGain and CalOffset.

Arguments

n is the desired analog input channel to configure. Analog input channels are numbered from 0 to AIO_MAX_AI – 1.

gain is a multiplying factor that is used to compensate for component inaccuracies and doesn't have any units. The gain would be entered by a calibration technician and stored in some form of non-volatile memory device such as an EEPROM or battery-backed-up RAM.

offset is a value that is added to the raw counts of the ADC to compensate for offset type errors caused by component inaccuracies. The offset would also be entered by a calibration technician and stored in some form of non-volatile memory device such as an EEPROM or battery-backed-up RAM.

Return Value

AICfgCal() returns 0 upon success and 1 if the analog input channel you specified is not within 0 and AIO_MAX_AI – 1.

Notes/Warnings

None

Example

```
void main (void)
{
    .
    .
    /* Calibration gain and offset obtained by technician */
    AICfgCal(0, (FP32)1.09, (FP32)10.0);
    .
    .
}
```

10

AICfgConv()

```
INT8U AICfgConv(INT8U n, FP32 gain, FP32 offset, INT8U pass);
```

AICfgConv() is used to set the conversion gain, offset, and the value of the pass counter for an analog input channel. The analog I/O module implements Equation [10.14], and this function is used to set the value of ConvGain and ConvOffset.

Arguments

n is the desired analog input channel to configure. Analog input channels are numbered from 0 to AIO_MAX_AI - 1.

gain is the conversion gain of the ADC channel in engineering units per count (E.U./count). gain is given by Equation [10.9] which is repeated in Equation [10.27] for your convenience:

[10.27]
$$gain_{(EU)/(count)} = \frac{FSV}{Transducer_{V/(EU)} \times A_V \times (2^{bits} - 1)}$$

FSV is the Full Scale Voltage of the ADC and typically is the reference voltage used with the ADC.

Transducer(V/EU) corresponds to the number of volts produced by the transducer per engineering unit. For example, the LM34A produces 0.01 volt per degree Fahrenheit.

A_V is the gain of the amplifier stage of an analog input channel (see Figure 10.1).

bits is the number of bits of the ADC.

offset is used to bias the ADC counts. offset is given by Equation [10.10] which is repeated in Equation [10.28] for your convenience.

[10.28]
$$offset_{counts} = -\frac{V_{bias} \times (2^{bits} - 1)}{FSV}$$

Vbias is the bias voltage added to the output of the amplifier stage to allow the ADC to read negative values (see Figure 10.7 on page 336 for an example on how to use the bias).

pass is used to specify a *pass count*. The pass count specifies to the module how often the analog channel will be read. The analog I/O module reads all analog input channels on a regular basis every so many clock ticks. This is called *scanning*. pass specifies how many scans are needed to read the analog input channel. For example, suppose the analog I/O module's scan rate is 10 Hz and you specify a *pass count* of 5 for analog input channel #0. Analog input channel #0 will be read every half second. I included a pass count because some analog input channels may not need to be read as often as others. For example, if you wanted the program to read the temperature of a room, you could tell it to read the temperature every 250 scans (or every 25 seconds, as in my example).

Return Value

AICfgConv() returns 0 upon success and 1 if the analog input channel you specified is not within 0 and AIO_MAX_AI - 1.

Notes/Warnings

None

Example

```
void main (void)
{
    .

    .

    /* Conversion gain and offset obtained by hardware engineer */
    AICfgConv(0, (FP32)1.987, (FP32)123.0, 1);

    .

    .
}
```

10

AICfgScaling()

INT8U AICfgScaling(INT8U n, void (fnct)(AIO *paio), void *arg);

AICfgScaling() is used to specify a scaling function to be executed when the analog input channel is read. The scaling function allows you to apply further processing when reading an analog input. There is no need to call AICfgScaling() if the analog input channel does not need a scaling function. In fact, if you don't define a scaling function the member .AIOScalingIn will simply be copied to .AIOScalingOut by AIUpdate() (see code).

Arguments

n is the desired analog input channel to configure. Analog input channels are numbered from 0 to AIO_MAX_AI - 1.

fnct is a pointer to the scaling function that will be executed when the analog input channel is read. You must write fnct to expect an argument. Specifically, fnct must be written to receive a pointer to the analog I/O data structure called AIO as shown in the code fragment following this paragraph. You specify a NULL pointer to prevent a previously configured channel from using a scaling function:

```
void fnct (AIO *paio);
```

arg is a pointer to any arguments or parameters needed for the scaling function. This argument can be used to specify specific options about the scaling being performed.

Return Value

AICfgScaling() returns 0 upon success and 1 if the analog input channel you specified is not within 0 and AIO_MAX_AI - 1.

Notes/Warnings

The scaling function is assumed to take its input from paio->AIOScaleIn and produce its result in paio->AIOScaleOut.

Example

```
INT8U ThermoType = THERMO_TYPE_J;



void main (void)
{
    .

    .

    AICfgScaling(0, ThermoLin, (void *)&ThermoType);

    .

    .

}




void ThermoLin (AIO *paio)
{
    /* Function to linearize a thermocouple */
    paio->AIOScaleIn is assumed to contain the number of millivolts for
        the thermocouple.
    paio->AIOScaleOut is where the temperature of the thermocouple
        is assumed to be saved to.
    paio->AIOScaleFnctArg could have also indicated the type of
        thermocouple used as well as whether the temperature is in
        degrees F or C.
}
```

10

AIGet()

INT8U AIGet(INT8U n, FP32 *pval);

The current value of the analog input channel can be obtained by calling AIGet(). The value obtained is in engineering units or, physical units. For example, if the analog input channel is measuring a temperature from a thermocouple then the value returned is the number of degrees at the thermocouple.

Arguments

n is the desired analog input channel. Analog input channels are numbered from 0 to AIO_MAX_AI - 1.

pval is a pointer to where the value of the analog input channel will be stored.

Return Value

AIGet() returns 0 upon success and 1 if the analog input channel you specified is not within 0 and AIO_MAX_AI - 1.

Notes/Warnings

The value returned is the last 'scanned' value. In other words, an ADC conversion is not performed when you call this function — AIOTask() is responsible for 'scanning' the analog input on a continuous basis.

Example

```
void Task (void *pdata)
{
    INT8U err;
    FP32  eu;


    for (;;) {
        .
        .
        err = AIGet(0, &eu);   /* Get current value of analog input #0 */
        .
    }
}
```

AIOInit()

void AIOInit(void);

AIOInit() is the initialization code for the analog I/O module. AIOInit() must be called before you use any of the other analog I/O module functions. AIOInit() is responsible for initializing the internal variables used by the module and for creating the task that will update the analog inputs and outputs.

Arguments

None

Return Value

None

Notes/Warnings

You are expected to provide the value of the following compile-time configuration constants (see Section 10.08, "Analog I/O Module, Configuration"):

```
AIO_TASK_STK_SIZE
AIO_TASK_PRIO
AIO_MAX_AI
AIO_MAX_AO
```

Example

```
void main (void)
{
    .
    .
    AIOInit();
    .
    .
}
```

10

AISetBypass()

INT8U AISetBypass(INT8U n, FP32 val);

Your application software can bypass or override the analog input channel value by using this function. `AISetBypass()` doesn't do anything unless you *open* the bypass *switch* by calling `AISetBypassEn()`.

Arguments

n is the desired analog input channel to override. Analog input channels are numbered from 0 to `AIO_MAX_AI - 1`.

val is the value you want `AIGet()` to return to your application. The value you pass to to `AISetBypass()` is in engineering units.

Return Value

`AISetBypass()` returns 0 upon success and 1 if the analog input channel you specified is not within 0 and `AIO_MAX_AI - 1`.

Notes/Warnings

`AISetBypass()` forces the value of `.AIOEU` in Figure 10.13 when `.AIOBypassEn` is set to TRUE.

Example

```
void Task (void *pdata)
{
    FP32 val;



    for (;;) {
        .

        .

        val = Get value from keyboard;
        AISetBypass(0, (FP32)val);

        .

    }
}
```

AISetBypassEn()

INT8U AISetBypassEn(INT8U n, BOOLEAN state);

AISetBypassEn() allows your application code to prevent the analog input channel from being updated. This permits another part of your application to set the value returned by AIGet(). In other words, you can "fool" the application code that monitors the analog input channel into thinking that the value is coming from a sensor, when in fact, the value returned by the analog input channel can come from another source. The value of the analog input channel is set by AISetBypass(). AISetBypassEn() and AISetBypass() are very useful functions for debugging.

Arguments

n is the desired analog input channel to bypass. Analog input channels are numbered from 0 to AIO_MAX_AI − 1.

state is the state of the bypass switch. When TRUE, the bypass switch is open (i.e., the analog input channel is bypassed). When FALSE, the bypass switch is closed (i.e., the analog input channel is not bypassed).

Return Value

AISetBypassEn() returns 0 upon success and 1 if the analog input channel you specified is not within 0 and AIO_MAX_AI − 1.

Notes/Warnings

AISetBypassEn() forces the value of .AIOBypassEn in Figure 10.13.

10

Example

```
void main (void)
{
    .
    .
    AISetBypassEn(0, TRUE);
    .
    .
}
```

AOCfgCal()

INT8U AOCfgCal(INT8U n, FP32 gain, FP32 offset);

AOCfgCal() is used to set the calibration gain and offset of an analog output channel. An analog output channel basically implements a generalization of Equation [10.23], as shown in Equation [10.29]:

[10.29] $\quad$ $DAC_{counts} = INT$ (.AIOConvGain$_{(counts/EU)}$ X .AIOCalGain X .AIOScaleOut$_{(EU)}$ + .AIOConvOffset$_{(counts)}$ + .AIOCalOffset$_{(counts)}$)

You can specify a calibration gain (.AIOCalGain) and offset (.AIOCalOffset) to compensate for component inaccuracies.

Arguments

n is the desired analog output channel. Analog output channels are numbered from 0 to AIO_MAX_AO − 1.

gain is a multiplying factor that is used to compensate for component inaccuracies and doesn't have any units. gain sets the value of .AIOCalGain in Figure 10.13. The gain would be entered by a calibration technician and stored in some form of non-volatile memory device such as an EEPROM or battery-backed-up RAM.

offset is a value that is added to the raw counts before outputing to a DAC to compensate for offset-type errors caused by component inaccuracies. offset sets the value of .AIOCalOffset in Figure 10.13. The offset would also be entered by a calibration technician and stored in some form of non-volatile memory device such as an EEPROM or battery-backed-up RAM.

Return Value

AOCfgCal() returns 0 upon success and 1 if the analog output channel you specified is not within 0 and AIO_MAX_AO − 1.

Notes/Warnings

None

Example

```
void main (void)
{
    .

    .
    AOCfgCal(0, (FP32)1.05, (FP32)10.6);
    .

    .
}
```

AOCfgConv()

INT8U AOCfgConv(INT8U n, FP32 gain, FP32 offset, INT16S lim, INT8U pass);

AOCfgConv() is used to set the conversion gain, conversion offset, and the value of the pass counter for an analog output channel. An analog output channel basically implements a generalization of Equation [10.20], as shown in Equation [10.29] (see page 358). AOCfgConv() is used to set the value of .AIOConvGain and .AIOConvOffset.

Arguments

n is the desired analog output channel to configure. Analog output channels are numbered from 0 to AIO_MAX_AO − 1.

gain is the conversion gain for the analog output channel in counts per engineering unit (counts/E.U.). gain sets the .AIOConvGain field of Figure 10.14.

offset is used to bias the DAC counts and sets the .AIOConvOffset field of Figure 10.14.

lim is used to specify the maximum count that can be sent to the DAC. This argument ensures that the DAC will never be written with a count larger than lim. For example, an 8-bit DAC has a maximum count of 255 (2^n − 1). lim sets the .AIOLim field of Figure 10.14.

pass is used to specify a *pass count*. The pass count is used to specify to the module how often the analog channel will be updated. The analog I/O module updates all analog output channel on a regular basis every so many clock ticks. This is called *scanning*. pass specifies how many scans are needed to update the specific analog output channel. For example, suppose the analog I/O module scan rate is 10 Hz and you specify a *pass count* of 2 for analog output channel #4. In this case, analog output channel #4 will be updated five times per second. I included a pass count because some analog output channels may not need to be updated as often as others. pass sets the .AIOPassCnts field of Figure 10.14.

Return Value

AOCfgConv() returns 0 upon success and 1 if the analog output channel you specified is not within 0 and AIO_MAX_AO − 1.

Notes/Warnings

None

Example

```
void main (void)
{
    .
    .
    AOCfgConv(0, (FP32)1.05, (FP32)10.6, 0x0FFF, 1);
    .
    .
}
```

10

AOCfgScaling()

`INT8U AOCfgScaling(INT8U n, void (*fnct)(AIO *paio), void *arg);`

`AOCfgScaling()` is used to specify a scaling function to be executed when the analog output channel is updated. The scaling function allows you to apply further processing before updating an analog output. You don't need to call this function if your analog output channel doesn't need a scaling function. In this case, the `.AIOScaleIn` field will simply be copied to the `.AIOScalingOut field` by `AOUpdate()` (see code).

Arguments

n is the desired analog output channel. Analog output channels are numbered from 0 to `AIO_MAX_AO` − 1.

fnct is a pointer to the scaling function that will be executed when the analog output channel is updated. `fnct` sets the value of `.AIOScaleFnct` in Figure 10.14. `fnct` must be written to receive a pointer to the analog I/O data structure called `AIO` as follows:

```
void fnct (AIO *paio);
```

arg is a pointer to any arguments or parameters needed for the scaling function. `arg` sets the value of `.AIOScaleFnctArg` in Figure 10.14. This argument can be used to specify specific options about the scaling being performed.

Return Value

`AOCfgScaling()` returns 0 upon success and 1 if the analog output channel you specified is not within 0 and `AIO_MAX_AO` − 1.

Notes/Warnings

The scaling function is assumed to take its input from `paio->AIOScaleIn` and produce its result in `paio->AIOScaleOut`.

Example

```
void main (void)
{
    .

    .

    AOCfgScaling(0, ActLin, (void *)0);

    .

    .
}



void ActLin (AIO *paio)
{
    /* Linearize actuator function */
    paio->AIOScaleIn is the input value to the scaling function.
    paio->AIOScaleOut is where the scaling function will place the result.
    paio->AIOScaleFnctArg in this case is not used but could be made
        to tell ActLin() the type of actuator to linearize.
}
```

10

AOSet()

INT8U AOSet(INT8U n, FP32 val);

This function is used by your application software to set the value of the analog output channel. The value you set the channel to is specified in engineering units. In other words, if your analog output channel has been configured to control the position of a valve in percent then, you would pass the desired percentage of position you desire (a number between 0.0 and 100.0).

Arguments

n is the desired analog output channel. Analog output channels are numbered from 0 to AIO_MAX_AO – 1.

val is the desired value for the analog output channel and is specified in engineering units.

Return Value

AOSet() returns 0 upon success and 1 if the analog output channel you specified is not within 0 and AIO_MAX_AO – 1.

Notes/Warnings

None

Example

```
void Task (void *pdata)
{
    FP32 valve;


    for (;;) {
        .

        .

        valve = Get desired value position from user;
        AOSet(0, (FP32)valve);

        .

    }
}
```

AOSetBypass()

`INT8U AOSetBypass(INT8U n, FP32 val);`

Your application software can bypass or override the analog output channel value by using this function. `AOSetBypass()` doesn't do anything unless you open the bypass switch by calling `AOSetBypassEn()`, as described previously. As with `AOSet()`, the value you set the channel to is specified in engineering units.

Arguments

n is the desired analog output channel. Analog output channels are numbered from 0 to `AIO_MAX_AO` – 1.

val is the value that you want to force into the analog output channel (in engineering units).

Return Value

`AOSetBypass()` returns 0 upon success and 1 if the analog output channel you specified is not within 0 and `AIO_MAX_AO` – 1.

Notes/Warnings

None

Example

```
void Task (void *pdata)
{
    FP32 val;


    for (;;) {

        .

        .

        val = Get value from keyboard;
        AOSetBypass(0, (FP32)val);

        .

    }
}
```

10

AOSetBypassEn()

INT8U AOSetBypassEn(INT8U n, BOOLEAN state);

AOSetBypassEn() allows you to prevent your application from changing the value of an analog output channel. This allows you to gain control of the analog output channel from elsewhere in your application code. This is a quite useful feature because it allows you to test your analog output channels one by one. In other words, you can set an analog output to any desired value even though your application software is trying to control the output. The value of the analog output channel is set by AOSetBypass(). AOSetBypassEn() and AOSetBypass() are very useful for debugging.

Arguments

n is the desired analog output channel. Analog output channels are numbered from 0 to AIO_MAX_AO − 1.

state is the state of the bypass *switch*. When TRUE, the bypass switch is opened (i.e., the analog output channel is bypassed). When FALSE, the bypass switch is closed (i.e., the analog output channel is not bypassed).

Return Value

AOSetBypassEn() returns 0 upon success and 1 if the analog output channel you specified is not within 0 and AIO_MAX_AO − 1.

Notes/Warnings

None

Example

```
void main (void)
{
    .
    .
    AOSetBypassEn(0, TRUE);
    .
    .
}
```

10.08 Analog I/O Module, Configuration

Configuration of the analog I/O module is quite simple.

1. You need to define the value of five #defines. The #defines are found in AIO.H (or CFG.H).

   AIO_TASK_PRIO is used to set the priority of the analog I/O module task.

   AIO_TASK_DLY is used to establish how often the analog I/O module will be executed. AIO_TASK_DLY determines the number of milliseconds to delay between execution of the analog I/O task.

WARNING

In the previous edition of this book, you needed to specify AIO_TASK_DLY_TICKS which specified the number of ticks between execution of AIOTask(). Because µC/OS-II provides a more convenient function (i.e., OSTimeDlyHMSM()) to specify the task execution period in hours, minutes, seconds and milliseconds, AIO_TASK_DLY_TICKS is no longer used and AIO_TASK_DLY now specifies the scan period in milliseconds instead of ticks.

   AIO_TASK_STK_SIZE specifies the size of the stack (in bus width units) allocated to the analog I/O task. The number of bytes allocated for the stack is thus given by: AIO_TASK_STK_SIZE times sizeof(OS_STK).

WARNING

In the previous edition of this book, AIO_TASK_STK_SIZE specified the size of the stack for AIOTask() in number of bytes. µC/OS-II assumes the stack is specified in stack width elements.

10

   AIO_MAX_AI determines the number of analog input channels that will be handled by the analog I/O task.

   AIO_MAX_AO determines the number of analog output channels handled by the analog I/O task.

2. You will need to define how analog inputs are read (i.e., how to read your ADC(s). ADCs must all be handled through AIRd(). The function prototype for AIRd() is:

```
INT16S AIRd (INT8U ch);
```

AIRd() is called by AIUpdate() (see code) and is passed the logical channel number (0 to AIO_MAX_AI − 1). You must translate this logical channel into code that selects the proper multiplexer for the desired channel, start the ADC, wait for the conversion to complete, read the ADC, and finally, return the ADC's counts.

3. You will need to provide the code for the function that writes to all DACs (i.e., AOWr()). The function prototype for AOWr() is:

```
void AOWr (INT8U ch, INT16S raw);
```

AOWr() is called by AOUpdate() (see code) and is passed the logical channel number (0 to AIO_MAX_AO − 1). You must translate this logical channel into code that selects the proper DAC for the desired channel. AOWr() is also passed the counts to send to the DAC. Your code must thus write the counts to the proper DAC.

4. You will need to provide the hardware initialization function (AIOInitIO()), which is called by AIOInit().The function prototype for AIOInit() is:

```
void AIOInit (void);
```

10.09 How to Use the Analog I/O Module

Let's assume that you need to read the analog inputs and control the analog outputs shown in Figure 10.16.

Figure 10.16 Using the analog I/O module.

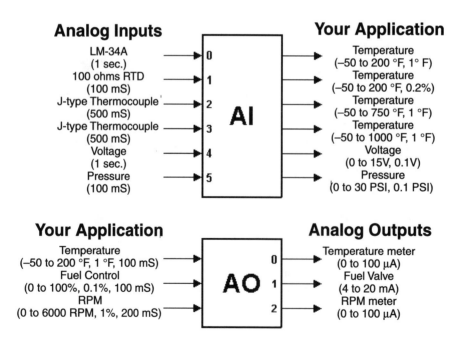

The analog I/O module has to read six analog inputs, and thus you will configure AIO_MAX_AI to 6. Similarly, to update three analog outputs, you need to set AIO_MAX_AO to 3. We can set AIO_TASK_DLY to 100 (i.e., milliseconds) because all analog I/Os need to be read or updated in multiples of 100 mS.

Obviously, you need to allocate sufficient stack space (i.e., AIO_TASK_STK_SIZE) for AIOTask() as well as determine what priority (i.e., AIO_TASK_PRIO) you want to give to that task.

To initialize the analog I/O module, you need to call AIOInit() prior to using any of the analog I/O module functions. You would typically do this in main():

```
void main (void)
{
    .
    OSInit();                       /* Initialize the O.S.  (mC/OS-II)      */
    .
    .
    AIOInit();                      /* Initialize the analog I/O module     */
    .
    .
    OSStart();                      /* Start multitasking   (mC/OS-II)      */
}
```

You would initialize each one of the analog I/O channels from an application task, as shown in the code fragment following this paragraph. It is important that you do this at the task level because some of the analog I/O module services assume that the operating system is running in order to access the mutual exclusion semaphore (AIOSem).

```
void AppTask (void *data)
{
    data = data;
    /* Initialize analog I/O channels here ...*/
    .
    .
    for (;;) {
        /* Application task code ... */
    }
}
```

Let's assume the hardware designer came up with the circuit shown in Figure 10.17 to read the analog inputs. As you can see, each input has signal conditioning circuitry which feeds into a multiplexer. The multiplexer selects one of the analog inputs to be converted by a 12-bit analog-to-digital converter (ADC).

10

Figure 10.17 Analog inputs.

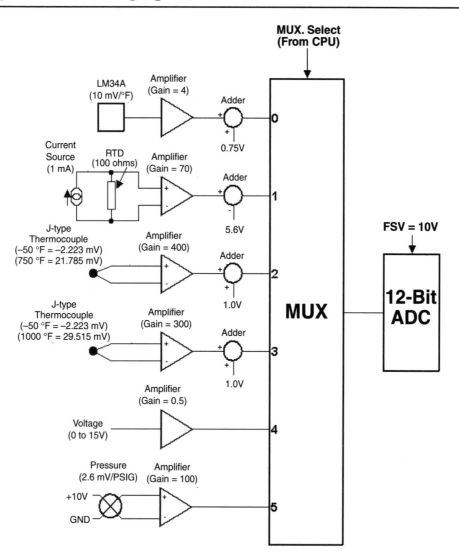

10.09.01 How to Use the Analog I/O Module, AI #0

Analog input channel #0 is an LM–34A temperature sensor used to read temperatures from –50 to 200 F°. Using Equation [10.9], the conversion gain is:

[10.30]
$$ConvGain_{(EU)/(count)} = \frac{FSV}{Transducer_{V/(EU)} \times A_V \times (2^n - 1)}$$

$$ConvGain_{(°F)/(count)} = \frac{10}{0.01_{V/(°F)} \times 4 \times (2^{12} - 1)}$$

$$ConvGain_{(°F)/(count)} = 0.061050$$

From Equation [10.10], the conversion offset is:

[10.31]
$$ConvOffset_{counts} = -\left(\frac{V_{bias} \times (2^n - 1)}{FSV}\right)$$

$$ConvOffset_{counts} = -\left(\frac{0.75 \times (2^{12} - 1)}{10}\right)$$

$$ConvOffset_{counts} = -307.125$$

The temperature at the LM34A is given by Equation [10.11] and is:

[10.32]
$$Temperature_{°F} = \left(ADC_{counts} + ConvOffset_{counts}\right) \times ConvGain_{(EU)/(count)}$$

$$Temperature_{°F} = \left(ADC_{counts} - 307.125\right) \times 0.061050$$

Because the LM-34A only needs to be read once per second, the pass counter for the channel will be set to 10 (i.e., 10×100 mS scan period).

10.09.02 How to Use the Analog I/O Module, AI #1

Analog input channel #1 is a 100-ohm Resistance Temperature Device (RTD). The RTD has about 80 ohms of resistance when the temperature at the RTD is –50 °F and 139 ohms when the temperature at the RTD is 200 °F. Unfortunately, the temperature at the RTD is not a linear function of resistance, and thus you will have to write a linearization function (beyond the scope of this chapter). The current source is used to develop a voltage across the RTD so that the resistance of the RTD can be measured. The circuit produces 1 mV per ohm (which is before the amplifier). By using Equations [10.9], [10.10], and [10.11], the resistance of the RTD is given by:

[10.33]
$$ConvGain_{(ohms)/(count)} = 0.034886$$

$$ConvOffset_{counts} = -2293.2$$

$$Resistance_{ohms} = \left(ADC_{counts} - 2293.2 \right) \times 0.034886$$

The pass counter for analog input channel #1 will be set to 1 in order to read the RTD every 100 mS.

10.09.03 How to Use the Analog I/O Module, AI #2

Analog input channel #2 is a J-Type thermocouple (another temperature measurement device). If you want to get the official reference on thermocouples, you should get the *NIST Monograph 175* (see "Bibliography" on page 374). A thermocouple produces a small voltage (called the *Seebeck voltage*) that varies as a function of temperature. The temperature at the thermocouple is not a linear function of the voltage produced. To further complicate things, the temperature at the thermocouple is also a function of a reference temperature called the *Cold Junction*. Determining the temperature at the thermocouple is beyond the scope of this book. Let's say for now that all you need to do is to measure the voltage (actually milli-volts) produced by the thermocouple. It is thus up to you to write a linearization function (also called *thermocouple compensation* function). A J-Type thermocouple produces –2.223 mV at –50 °F and 21.785 mV at 750 °F. This voltage is amplified by 400 so that it can be read by the ADC. A bias voltage is introduced to ensure that the ADC only sees positive voltages. From Equations [10.9], [10.10], and [10.11], the number of milli-volts at the thermocouple is given by:

[10.34] $ConvGain_{(mV)/(count)} = 0.006105$

$ConvOffset_{counts} = -409.5$

$$Thermocouple_{mV} = \left(ADC_{counts} - 409.5 \right) \times 0.006105$$

All you have to do is linearize the thermocouple based on the number of milli-volts read from the thermocouple. The pass counter for analog input channel #2 will be set to 5 in order to read the thermocouple every 500 mS.

10.09.04 How to Use the Analog I/O Module, AI #3

Analog input channel #3 is also a J-Type thermocouple. A J-Type thermocouple produces –2.223 mV at –50 °F and 29.515 mV at 1000 °F. This voltage is amplified by 300 so that it can be read by the ADC. The bias voltage is also introduced to ensure that the ADC only sees positive voltages. From Equations [10.9], [10.10], and [10.11], the number of milli-volts at the thermocouple is given by:

[10.35] $ConvGain_{(mV)/(count)} = 0.008140$

$ConvOffset_{counts} = -409.5$

$$Thermocouple_{mV} = \left(ADC_{counts} - 409.5 \right) \times 0.008140$$

Again, all you have to do is linearize the thermocouple based on the number of milli-volts read from the thermocouple. The pass counter for analog input channel #3 will also be set to 5 in order to read the thermocouple every 500 mS.

10.09.05 How to Use the Analog I/O Module, AI *#4*

Analog input channel #4 reads a voltage directly (maybe a battery). Because the voltage to read exceeds the FSV of the ADC, the hardware designer decided to simply divide the voltage in half. From Equations [10.9], [10.10], and [10.11], the voltage at the input is given by:

[10.36]
$$ConvGain_{(V)/(count)} = 0.004884$$

$$ConvOffset_{counts} = -0$$

$$Voltage_V = \left(ADC_{counts} - 0\right) \times 0.004884$$

The pass counter for analog input channel #4 will also be set to 10 in order to read the thermocouple every second.

10.09.06 How to Use the Analog I/O Module, AI *#5*

Analog input channel #5 reads a pressure from a pressure transducer which produces 2.6 mV/PSIG (pounds per square inch gauge). From Equations [10.9], [10.10], and [10.11], the pressure read by the transducer is given by:

[10.37]
$$ConvGain_{(PSIG)/(count)} = 0.009392$$

$$ConvOffset_{counts} = -0$$

$$Pressure_{PSIG} = \left(ADC_{counts} - 0\right) \times 0.009392$$

The pass counter for analog input channel #5 will be set to 1 in order to read the pressure every 100 mS.

Let's assume that the hardware designer came up with the circuit shown in Figure 10.18 to update the analog outputs.

10

Figure 10.18 Analog outputs.

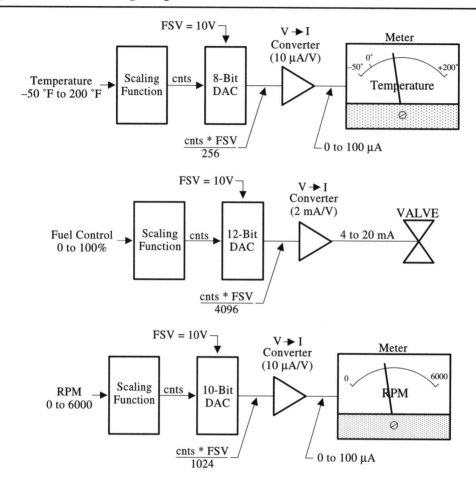

10.09.07 How to Use the Analog I/O Module, AO #0

Analog output channel #0 is used to display temperatures from –50 °F to 200 °F on a 0 to 100 µA meter. A display of –50 µF is obtained with 0 DAC counts (0 µA) while 200 °F is obtained with 255 DAC counts (99.609 µA). The DAC counts are given by:

[10.38]
$$ConvGain_{(counts)/(°F)} = 1.02$$

$$ConvOffset_{counts} = 51$$

$$DAC_{counts} = 1.02 \times Temperature_{°F} + 51$$

The pass counter for analog output channel #0 will be set to 1 in order to update the meter every 100 mS.

10.09.08 How to Use the Analog I/O Module, AO #1

Analog output channel #1 is used to control the opening of a valve. The valve is closed when the control current is 4 mA and wide open when the control current is 20 mA. The counts vs. output current is given by:

[10.39]
$$DAC_{counts} = \frac{2^n \times Out_{mA}}{FSV \times 2_{(mA)/V}}$$

A 12-bit DAC is used because a 10-bit DAC would not have the required resolution. Using a 10-bit DAC, 4 mA would require 205 counts (Equation [10.36]), while 20 mA would require 1023 counts, a range of 818 counts, or 0.122 percent. Note that 11-bit DACs are not commercially available. A 12-bit DAC requires 819.2 counts for a 4 mA output and 4095 counts for 20 mA (actually 19.995 mA). The DAC counts required to control the DAC are given by:

[10.40]
$$ConvGain_{(counts)/\%} = \frac{4095 - 819.2}{100\% - 0\%} = 32.758$$

$$ConvOffset_{counts} = 819.2$$

$$DAC_{counts} = 32.758 \times Input_{\%} + 819.2$$

The pass counter for analog output channel #1 will be set to 1 in order to update the valve every 100 mS.

10.09.09 How to Use the Analog I/O Module, AO #2

Analog output channel #2 is used to display the RPM of a rotating device on a 0 to 100 μA meter. A display of 0 RPM is obtained with 0 DAC counts (0 μA), while 6000 RPM is obtained with 1023 DAC counts (99.902 μA). The DAC counts are given by:

[10.41]
$$ConvGain_{(counts)/(RPM)} = 0.1705$$

$$ConvOffset_{counts} = 0$$

$$DAC_{counts} = 0.1705 \times RPM + 0$$

The pass counter for analog output channel #2 will be set to 2 in order to update the meter every 200 mS.

The code to initialize the analog I/O channels is:

10

```
void AppInitAIO (void)
{
    AICfgConv(0, 0.061050,  307.125, 10);      /* Analog Inputs       */
    AICfgConv(1, 0.034886, 2293.2,    1);
    AICfgConv(2, 0.006105,  409.5,    5);
    AICfgConv(3, 0.008140,  409.5,    5);
    AICfgConv(4, 0.004884,    0.0,   10);
    AICfgConv(5, 0.009392,    0.0,    1);

    AICfgScaling(1, /* Pointer to RTD code */, /* Pointer to args */);
    AICfgScaling(2, /* Pointer to TC  code */, /* Pointer to args */);
    AICfgScaling(3, /* Pointer to TC  code */, /* Pointer to args */);

    AOCfgConv(0,  1.02,       51.0,  255, 1);  /* Analog Outputs      */
    AOCfgConv(1, 32.758,     819.2, 4095, 2);
    AOCfgConv(2,  0.1705,      0.0, 1023, 2);
}
```

You can now obtain the value read by any analog input channels by using AIGet() and set any analog output channel by calling AOSet().

10.10 Bibliography

Burns, G.W., Scroger, M.G., Strouse, G.F., Croarkin, M.C. and Guthrie, W.F.
Temperature-Electromotive Force Reference Functions and Tables for the Letter-Designated Thermocouple Types Based on the ITS-90 (NIST Monograph 175)
United States Department of Commerce
National Institute of Standards and Technology (NIST)
Gaithersburg, MD 20899
(301) 975-3058

Morgan, Don
Numerical Methods, Real-Time and Embedded Systems Programming
San Mateo, CA
M&T Publishing, Inc.
ISBN 1-55851-232-2

U.S. Software
14215 NW Science Park Dr.
Portland, OR 97229
(503) 641-8446

Zuch, Eugene L.
Data Acquisition and Conversion Handbook
Mansfield, MA
Datel/Intersil, 1979

Listing 10.1 **AIO.C**

```
/*
********************************************************************************************************
*                                       Analog I/O Module
*
*                          (c) Copyright 1999, Jean J. Labrosse, Weston, FL
*                                       All Rights Reserved
*
* Filename   : AIO.C
* Programmer : Jean J. Labrosse
********************************************************************************************************
*/

/*
********************************************************************************************************
*                                       INCLUDE FILES
********************************************************************************************************
*/

#define   AIO_GLOBALS
#include "includes.h"

/*
********************************************************************************************************
*                                       LOCAL VARIABLES
********************************************************************************************************
*/

static OS_STK      AIOTaskStk[AIO_TASK_STK_SIZE];
static OS_EVENT    *AIOSem;

/*
********************************************************************************************************
*                                       LOCAL FUNCTION PROTOTYPES
********************************************************************************************************
*/

       void       AIOTask(void *data);

static void       AIInit(void);
static void       AIUpdate(void);

static void       AOInit(void);
static void       AOUpdate(void);

/*$PAGE*/
```

Listing 10.1 (continued) `AIO.C`

```
/*
*********************************************************************************************
*                        CONFIGURE THE CALIBRATION PARAMETERS OF AN ANALOG INPUT CHANNEL
*
* Description : This function is used to configure an analog input channel.
* Arguments   : n          is the analog input channel to configure:
*               gain       is the calibration gain
*               offset     is the calibration offset
* Returns     : 0          if successfull.
*               1          if you specified an invalid analog input channel number.
*********************************************************************************************
*/

INT8U  AICfgCal (INT8U n, FP32 gain, FP32 offset)
{
    INT8U err;
    AIO   *paio;


    if (n < AIO_MAX_AI) {
        paio               = &AITbl[n];                /* Point to Analog Input structure       */
        OSSemPend(AIOSem, 0, &err);                    /* Obtain exclusive access to AI channel  */
        paio->AIOCalGain   = gain;                     /* Store new cal. gain and offset into struct */
        paio->AIOCalOffset = offset;
        paio->AIOGain      = paio->AIOCalGain   * paio->AIOConvGain;    /* Compute overall gain   */
        paio->AIOOffset    = paio->AIOCalOffset + paio->AIOConvOffset;  /* Compute overall offset */
        OSSemPost(AIOSem);                                             /* Release AI channel      */
        return (0);
    } else {
        return (1);
    }
}

/*$PAGE*/
```

Listing 10.1 (continued) `AIO.C`

```
/*
*********************************************************************************************
*                     CONFIGURE THE CONVERSION PARAMETERS OF AN ANALOG INPUT CHANNEL
*
* Description : This function is used to configure an analog input channel.
* Arguments   : n        is the analog channel to configure (0..AIO_MAX_AI-1).
*               gain     is the conversion gain
*               offset   is the conversion offset
*               pass     is the value for the pass counts
* Returns     : 0        if successfull.
*               1        if you specified an invalid analog input channel number.
*********************************************************************************************
*/

INT8U  AICfgConv (INT8U n, FP32 gain, FP32 offset, INT8U pass)
{
    INT8U err;
    AIO  *paio;


    if (n < AIO_MAX_AI) {
        paio               = &AITbl[n];                    /* Point to Analog Input structure            */
        OSSemPend(AIOSem, 0, &err);                        /* Obtain exclusive access to AI channel      */
        paio->AIOConvGain  = gain;                         /* Store new conv. gain and offset into struct */
        paio->AIOConvOffset = offset;
        paio->AIOGain      = paio->AIOCalGain  * paio->AIOConvGain;    /* Compute overall gain       */
        paio->AIOOffset    = paio->AIOCalOffset + paio->AIOConvOffset; /* Compute overall offset     */
        paio->AIOPassCnts  = pass;
        OSSemPost(AIOSem);                                             /* Release AI channel         */
        return (0);
    } else {
        return (1);
    }
}

/*$PAGE*/
```

10

Listing 10.1 (continued) `AIO.C`

```
/*
*********************************************************************************************************
*                         CONFIGURE THE SCALING PARAMETERS OF AN ANALOG INPUT CHANNEL
*
* Description : This function is used to configure the scaling parameters associated with an analog
*               input channel.
* Arguments   : n        is the analog input channel to configure (0..AIO_MAX_AI-1).
*               arg      is a pointer to arguments needed by the scaling function
*               fnct     is a pointer to a scaling function
* Returns     : 0        if successfull.
*               1        if you specified an invalid analog input channel number.
*********************************************************************************************************
*/

INT8U  AICfgScaling (INT8U n, void (*fnct)(AIO *paio), void *arg)
{
    AIO *paio;


    if (n < AIO_MAX_AI) {
        paio                  = &AITbl[n];               /* Faster to use a pointer to the structure     */
        OS_ENTER_CRITICAL();
        paio->AIOScaleFnct    = (void (*)())fnct;
        paio->AIOScaleFnctArg = arg;
        OS_EXIT_CRITICAL();
        return (0);
    } else {
        return (1);
    }
}


/*$PAGE*/
```

Listing 10.1 (continued) `AIO.C`

```
/*
*********************************************************************************************
*                            GET THE VALUE OF AN ANALOG INPUT CHANNEL
*
* Description : This function is used to get the currect value of an analog input channel (in engineering
*               units).
* Arguments   : n     is the analog input channel (0..AIO_MAX_AI-1).
*               pval  is a pointer to the destination engineering units of the analog input channel
* Returns     : 0        if successfull.
*               1        if you specified an invalid analog input channel number.
*                        In this case, the destination is not changed.
*********************************************************************************************
*/

INT8U  AIGet (INT8U n, FP32 *pval)
{
    AIO  *paio;


    if (n < AIO_MAX_AI) {
        paio = &AITbl[n];
        OS_ENTER_CRITICAL();                /* Obtain exclusive access to AI channel                     */
        *pval = paio->AIOEU;                /* Get the engineering units of the analog input channel     */
        OS_EXIT_CRITICAL();                 /* Release AI channel                                        */
        return (0);
    } else {
        return (1);
    }
}

/*$PAGE*/
```

Listing 10.1 (continued) `AIO.C`

```
/*
*********************************************************************************************************
*                                  ANALOG INPUTS INITIALIZATION
*
* Description : This function initializes the analog input channels.
* Arguments   : None
* Returns     : None.
*********************************************************************************************************
*/

static  void  AIInit (void)
{
    INT8U   i;
    AIO     *paio;


    paio = &AITbl[0];
    for (i = 0; i < AIO_MAX_AI; i++) {
        paio->AIOBypassEn     = FALSE;         /* Analog channel is not bypassed            */
        paio->AIORaw          = 0x0000;        /* Raw counts of ADC or DAC                  */
        paio->AIOEU           = (FP32)0.0;     /* Engineering units of AI channel           */
        paio->AIOGain         = (FP32)1.0;     /* Total gain                                */
        paio->AIOOffset       = (FP32)0.0;     /* Total offset                              */
        paio->AIOLim          =      0;
        paio->AIOPassCnts     =      1;        /* Pass counts                               */
        paio->AIOPassCtr      =      1;        /* Pass counter                              */
        paio->AIOCalGain      = (FP32)1.0;     /* Calibration gain                          */
        paio->AIOCalOffset    = (FP32)0.0;     /* Calibration offset                        */
        paio->AIOConvGain     = (FP32)1.0;     /* Conversion gain                           */
        paio->AIOConvOffset   = (FP32)0.0;     /* Conversion offset                         */
        paio->AIOScaleIn      = (FP32)0.0;     /* Input  to scaling function                */
        paio->AIOScaleOut     = (FP32)0.0;     /* Output of scaling function                */
        paio->AIOScaleFnct    = (void *)0;     /* No function to execute                    */
        paio->AIOScaleFnctArg = (void *)0;     /* No arguments to scale function            */
        paio++;
    }
}

/*$PAGE*/
```

Listing 10.1 (continued) `AIO.C`

```
/*
*********************************************************************************************************
*                                  ANALOG I/O MANAGER INITIALIZATION
*
* Description : This function initializes the analog I/O manager module.
* Arguments   : None
* Returns     : None.
*********************************************************************************************************
*/

void  AIOInit (void)
{
    INT8U   err;


    AIInit();
    AOInit();
    AIOInitIO();
    AIOSem = OSSemCreate(1);                        /* Create a mutual exclusion semaphore for AIOs       */
    OSTaskCreate(AIOTask, (void *)0, &AIOTaskStk[AIO_TASK_STK_SIZE], AIO_TASK_PRIO);
}


/*$PAGE*/

/*
*********************************************************************************************************
*                                     ANALOG I/O MANAGER TASK
*
* Description : This task is created by AIOInit() and is responsible for updating the analog inputs and
*               analog outputs.
*               AIOTask() executes every AIO_TASK_DLY milliseconds.
* Arguments   : None.
* Returns     : None.
*********************************************************************************************************
*/

void  AIOTask (void *data)
{
    INT8U err;


    data = data;                               /* Avoid compiler warning                            */
    for (;;) {
        OSTimeDlyHMSM(0, 0, 0, AIO_TASK_DLY);  /* Delay between execution of AIO manager             */

        OSSemPend(AIOSem, 0, &err);            /* Obtain exclusive access to AI channels             */
        AIUpdate();                            /* Update all AI channels                             */
        OSSemPost(AIOSem);                     /* Release AI channels (Allow high prio. task to run) */

        OSSemPend(AIOSem, 0, &err);            /* Obtain exclusive access to AO channels             */
        AOUpdate();                            /* Update all AO channels                             */
        OSSemPost(AIOSem);                     /* Release AO channels (Allow high prio. task to run) */
    }
}


/*$PAGE*/
```

10

Listing 10.1 (continued) AIO.C

```
/*
*********************************************************************************************
*                           SET THE STATE OF THE BYPASSED ANALOG INPUT CHANNEL
*
* Description : This function is used to set the engineering units of a bypassed analog input channel.
*               This function is used to simulate the presense of the sensor.  This function is only
*               valid if the bypass 'switch' is open.
* Arguments   : n     is the analog input channel (0..AIO_MAX_AI-1).
*               val   is the value of the bypassed analog input channel:
* Returns     : 0         if successfull.
*               1         if you specified an invalid analog input channel number.
*               2         if AIOBypassEn was not set to TRUE
*********************************************************************************************
*/

INT8U  AISetBypass (INT8U n, FP32 val)
{
    AIO    *paio;


    if (n < AIO_MAX_AI) {
        paio = &AITbl[n];                        /* Faster to use a pointer to the structure     */
        if (paio->AIOBypassEn == TRUE) {         /* See if the analog input channel is bypassed  */
            OS_ENTER_CRITICAL();
            paio->AIOEU = val;                    /* Yes, then set the new value of the channel   */
            OS_EXIT_CRITICAL();
            return (0);
        } else {
            return (2);
        }
    } else {
        return (1);
    }
}

/*$PAGE*/
```

Listing 10.1 *(continued)* AIO.C

```
/*
*********************************************************************************************
*                           SET THE STATE OF THE BYPASS SWITCH
*
* Description : This function is used to set the state of the bypass switch.  The analog input channel is
*               bypassed when the 'switch' is open (i.e. AIOBypassEn is set to TRUE).
* Arguments   : n      is the analog input channel (0..AIO_MAX_AI-1).
*               state  is the state of the bypass switch:
*                       FALSE disables the bypass (i.e. the bypass 'switch' is closed)
*                       TRUE  enables  the bypass (i.e. the bypass 'switch' is open)
* Returns     : 0      if successfull.
*               1      if you specified an invalid analog input channel number.
*********************************************************************************************
*/

INT8U  AISetBypassEn (INT8U n, BOOLEAN state)
{
    if (n < AIO_MAX_AI) {
        AITbl[n].AIOBypassEn = state;
        return (0);
    } else {
        return (1);
    }
}

/*$PAGE*/
```

Listing 10.1 (continued) `AIO.C`

```
/*
*********************************************************************************************
*                          UPDATE ALL ANALOG INPUT CHANNELS
*
* Description : This function processes all of the analog input channels.
* Arguments   : None.
* Returns     : None.
*********************************************************************************************
*/

static  void  AIUpdate (void)
{
    INT8U   i;
    AIO    *paio;


    paio = &AITbl[0];                        /* Point at first analog input channel         */
    for (i = 0; i < AIO_MAX_AI; i++) {       /* Process all analog input channels           */
        if (paio->AIOBypassEn == FALSE) {    /* See if analog input channel is bypassed     */
            paio->AIOPassCtr--;              /* Decrement pass counter                      */
            if (paio->AIOPassCtr == 0) {     /* When pass counter reaches 0, read and scale AI */
                paio->AIOPassCtr = paio->AIOPassCnts;    /* Reload pass counter             */
                paio->AIORaw     = AIRd(i);              /* Read ADC for this channel      */
                paio->AIOScaleIn = ((FP32)paio->AIORaw + paio->AIOOffset) * paio->AIOGain;
                if ((void *)paio->AIOScaleFnct != (void *)0) {    /* See if function defined */
                    (*paio->AIOScaleFnct)(paio);         /* Yes, execute function          */
                } else {
                    paio->AIOScaleOut = paio->AIOScaleIn;    /* No, just copy data          */
                }
                paio->AIOEU = paio->AIOScaleOut;         /* Output of scaling fnct to E.U. */
            }
        }
        paio++;                              /* Point at next AI channel                    */
    }
}

/*$PAGE*/
```

Listing 10.1 (continued) AIO.C

```
/*
*********************************************************************************************
*                    CONFIGURE THE CALIBRATION PARAMETERS OF AN ANALOG OUTPUT CHANNEL
*
* Description : This function is used to configure an analog output channel.
* Arguments   : n        is the analog output channel to configure (0..AIO_MAX_AO-1)
*               gain     is the calibration gain
*               offset   is the calibration offset
* Returns     : 0        if successfull.
*               1        if you specified an invalid analog output channel number.
*********************************************************************************************
*/

INT8U  AOCfgCal (INT8U n, FP32 gain, FP32 offset)
{
    INT8U  err;
    AIO    *paio;


    if (n < AIO_MAX_AO) {
        paio              = &AOTbl[n];                    /* Point to Analog Output structure        */
        OSSemPend(AIOSem, 0, &err);                       /* Obtain exclusive access to AO channel    */
        paio->AIOCalGain   = gain;                        /* Store new cal. gain and offset into struct  */
        paio->AIOCalOffset = offset;
        paio->AIOGain      = paio->AIOCalGain   * paio->AIOConvGain;     /* Compute overall gain    */
        paio->AIOOffset    = paio->AIOCalOffset + paio->AIOConvOffset;   /* Compute overall offset  */
        OSSemPost(AIOSem);                                               /* Release AO channel      */
        return (0);
    } else {
        return (1);
    }
}

/*$PAGE*/
```

10

Listing 10.1 (continued) AIO.C

```c
/*
*********************************************************************************************
*                   CONFIGURE THE CONVERSION PARAMETERS OF AN ANALOG OUTPUT CHANNEL
*
* Description : This function is used to configure an analog output channel.
* Arguments   : n        is the analog channel to configure (0..AIO_MAX_AO-1).
*               gain     is the conversion gain
*               offset   is the conversion offset
*               pass     is the value for the pass counts
* Returns     : 0        if successfull.
*               1        if you specified an invalid analog output channel number.
*********************************************************************************************
*/

INT8U  AOCfgConv (INT8U n, FP32 gain, FP32 offset, INT16S lim, INT8U pass)
{
    INT8U err;
    AIO  *paio;


    if (n < AIO_MAX_AO) {
        paio                = &AOTbl[n];            /* Point to Analog Output structure         */
        OSSemPend(AIOSem, 0, &err);                 /* Obtain exclusive access to AO channel     */
        paio->AIOConvGain   = gain;                 /* Store new conv. gain and offset into struct  */
        paio->AIOConvOffset = offset;
        paio->AIOGain       = paio->AIOCalGain   * paio->AIOConvGain;    /* Compute overall gain     */
        paio->AIOOffset     = paio->AIOCalOffset + paio->AIOConvOffset;  /* Compute overall offset   */
        paio->AIOLim        = lim;
        paio->AIOPassCnts   = pass;
        OSSemPost(AIOSem);                                               /* Release AO channel       */
        return (0);
    } else {
        return (1);
    }
}
/*$PAGE*/
```

Listing 10.1 (continued) AIO.C

```
/*
*********************************************************************************************************
*                        CONFIGURE THE SCALING PARAMETERS OF AN ANALOG OUTPUT CHANNEL
*
* Description : This function is used to configure the scaling parameters associated with an analog
*               output channel.
* Arguments   : n        is the analog output channel to configure (0..AIO_MAX_AO-1).
*               arg      is a pointer to arguments needed by the scaling function
*               fnct     is a pointer to a scaling function
* Returns     : 0        if successfull.
*               1        if you specified an invalid analog output channel number.
*********************************************************************************************************
*/

INT8U  AOCfgScaling (INT8U n, void (*fnct)(AIO *paio), void *arg)
{
    AIO *paio;


    if (n < AIO_MAX_AO) {
        paio                  = &AOTbl[n];                 /* Faster to use a pointer to the structure     */
        OS_ENTER_CRITICAL();
        paio->AIOScaleFnct    = (void (*)())fnct;
        paio->AIOScaleFnctArg = arg;
        OS_EXIT_CRITICAL();
        return (0);
    } else {
        return (1);
    }
}
/*$PAGE*/
```

10

Listing 10.1 (continued) AIO.C

```
/*
*********************************************************************************************
*                               ANALOG OUTPUTS INITIALIZATION
*
* Description : This function initializes the analog output channels.
* Arguments   : None
* Returns     : None.
*********************************************************************************************
*/

static void AOInit (void)
{
    INT8U  i;
    AIO    *paio;


    paio = &AOTbl[0];
    for (i = 0; i < AIO_MAX_AO; i++) {
        paio->AIOBypassEn      = FALSE;         /* Analog channel is not bypassed              */
        paio->AIORaw           = 0x0000;        /* Raw counts of ADC or DAC                    */
        paio->AIOEU            = (FP32)0.0;     /* Engineering units of AI channel             */
        paio->AIOGain          = (FP32)1.0;     /* Total gain                                  */
        paio->AIOOffset        = (FP32)0.0;     /* Total offset                                */
        paio->AIOLim           =     0;         /* Maximum count of an analog output channel   */
        paio->AIOPassCnts      =     1;         /* Pass counts                                 */
        paio->AIOPassCtr       =     1;         /* Pass counter                                */
        paio->AIOCalGain       = (FP32)1.0;     /* Calibration gain                            */
        paio->AIOCalOffset     = (FP32)0.0;     /* Calibration offset                          */
        paio->AIOConvGain      = (FP32)1.0;     /* Conversion gain                             */
        paio->AIOConvOffset    = (FP32)0.0;     /* Conversion offset                           */
        paio->AIOScaleIn       = (FP32)0.0;     /* Input  to scaling function                  */
        paio->AIOScaleOut      = (FP32)0.0;     /* Output of scaling function                  */
        paio->AIOScaleFnct     = (void *)0;     /* No function to execute                      */
        paio->AIOScaleFnctArg  = (void *)0;     /* No arguments to scale function              */
        paio++;
    }
}
/*$PAGE*/
```

Listing 10.1 (continued) `AIO.C`

```
/*
*********************************************************************************************************
*                               SET THE VALUE OF AN ANALOG OUTPUT CHANNEL
*
* Description : This function is used to set the currect value of an analog output channel
*               (in engineering units).
* Arguments   : n     is the analog output channel (0..AIO_MAX_AO-1).
*               val   is the desired analog output value in Engineering Units
* Returns     : 0     if successfull.
*               1     if you specified an invalid analog output channel number.
*********************************************************************************************************
*/

INT8U  AOSet (INT8U n, FP32 val)
{
    if (n < AIO_MAX_AO) {
        OS_ENTER_CRITICAL();
        AOTbl[n].AIOEU = val;              /* Set the engineering units of the analog output channel     */
        OS_EXIT_CRITICAL();
        return (0);
    } else {
        return (1);
    }
}

/*$PAGE*/
```

10

Listing 10.1 (continued) AIO.C

```
/*
*********************************************************************************************************
*                           SET THE STATE OF THE BYPASSED ANALOG OUTPUT CHANNEL
*
* Description : This function is used to set the engineering units of a bypassed analog output channel.
* Arguments   : n     is the analog output channel (0..AIO_MAX_AO-1).
*               val   is the value of the bypassed analog output channel:
* Returns     : 0     if successfull.
*               1     if you specified an invalid analog output channel number.
*               2     if AIOBypassEn is not set to TRUE
*********************************************************************************************************
*/

INT8U  AOSetBypass (INT8U n, FP32 val)
{
    AIO *paio;


    if (n < AIO_MAX_AO) {
        paio = &AOTbl[n];                        /* Faster to use a pointer to the structure        */
        if (paio->AIOBypassEn == TRUE) {         /* See if the analog output channel is bypassed    */
            OS_ENTER_CRITICAL();
            paio->AIOScaleIn = val;              /* Yes, then set the new value of the channel      */
            OS_EXIT_CRITICAL();
            return (0);
        } else {
            return (2);
        }
    } else {
        return (1);
    }
}

/*$PAGE*/
```

Listing 10.1 (continued) `AIO.C`

```
/*
*********************************************************************************************************
*                                 SET THE STATE OF THE BYPASS SWITCH
*
* Description : This function is used to set the state of the bypass switch.  The analog output channel
*               is bypassed when the 'switch' is open (i.e. AIOBypassEn is set to TRUE).
* Arguments   : n      is the analog output channel (0..AIO_MAX_AO-1).
*               state  is the state of the bypass switch:
*                      FALSE disables the bypass (i.e. the bypass 'switch' is closed)
*                      TRUE  enables  the bypass (i.e. the bypass 'switch' is open)
* Returns     : 0      if successfull.
*               1      if you specified an invalid analog output channel number.
*********************************************************************************************************
*/

INT8U  AOSetBypassEn (INT8U n, BOOLEAN state)
{
    INT8U err;


    if (n < AIO_MAX_AO) {
        AOTbl[n].AIOBypassEn = state;
        return (0);
    } else {
        return (1);
    }
}

/*$PAGE*/
```

10

Listing 10.1 (continued) AIO.C

```
/*
*********************************************************************************************
*                             UPDATE ALL ANALOG OUTPUT CHANNELS
*
* Description : This function processes all of the analog output channels.
* Arguments   : None.
* Returns     : None.
*********************************************************************************************
*/

static  void  AOUpdate (void)
{
    INT8U    i;
    AIO      *paio;
    INT16S   raw;


    paio = &AOTbl[0];                       /* Point at first analog output channel         */
    for (i = 0; i < AIO_MAX_AO; i++) {      /* Process all analog output channels           */
        if (paio->AIOBypassEn == FALSE) {   /* See if analog output channel is bypassed     */
            paio->AIOScaleIn = paio->AIOEU; /* No                                           */
        }
        paio->AIOPassCtr--;                 /* Decrement pass counter                       */
        if (paio->AIOPassCtr == 0) {        /* When pass counter reaches 0, read and scale AI */
            paio->AIOPassCtr = paio->AIOPassCnts;        /* Reload pass counter             */
            if ((void *)paio->AIOScaleFnct != (void *)0) { /* See if function defined       */
                (*paio->AIOScaleFnct)(paio);             /* Yes, execute function           */
            } else {
                paio->AIOScaleOut = paio->AIOScaleIn;    /* No,  bypass scaling function    */
            }
            raw = (INT16S)(paio->AIOScaleOut * paio->AIOGain + paio->AIOOffset);
            if (raw > paio->AIOLim) {                    /* Never output > maximum DAC counts */
                raw = paio->AIOLim;
            } else if (raw < 0) {                        /* DAC counts must always be >= 0  */
                raw = 0;
            }
            paio->AIORaw = raw;
            AOWr(i, paio->AIORaw);          /* Write counts to DAC                          */
        }
        paio++;                             /* Point at next AO channel                     */
    }
}

/*$PAGE*/
```

Listing 10.1 (continued) `AIO.C`

```c
#ifndef CFG_C
/*
*********************************************************************************************************
*                                    INITIALIZE PHYSICAL I/Os
*
* Description : This function is called by AIOInit() to initialize the physical I/O used by the AIO
*               driver.
* Arguments   : None.
* Returns     : None.
*********************************************************************************************************
*/

void  AIOInitIO (void)
{
    /* This is where you will need to put you initialization code for the ADCs and DACs          */
    /* You should also consider initializing the contents of your DAC(s) to a known value.       */
}


/*
*********************************************************************************************************
*                                    READ PHYSICAL INPUTS
*
* Description : This function is called to read a physical ADC channel.  The function is assumed to
*               also control a multiplexer if more than one analog input is connected to the ADC.
* Arguments   : ch     is the ADC logical channel number (0..AIO_MAX_AI-1).
* Returns     : The raw ADC counts from the physical device.
*********************************************************************************************************
*/

INT16S  AIRd (INT8U ch)
{
    /* This is where you will need to provide the code to read your ADC(s).                        */
    /* AIRd() is passed a 'LOGICAL' channel number.  You will have to convert this logical channel */
    /* number into actual physical port locations (or addresses) where your MUX. and ADCs are located. */
    /* AIRd() is responsible for:                                                                  */
    /*      1) Selecting the proper MUX. channel,                                                  */
    /*      2) Waiting for the MUX. to stabilize,                                                  */
    /*      3) Starting the ADC,                                                                   */
    /*      4) Waiting for the ADC to complete its conversion,                                     */
    /*      5) Reading the counts from the ADC and,                                                */
    /*      6) Returning the counts to the calling function.                                       */

    return (ch);
}

/*$PAGE*/
```

10

Listing 10.1 (continued) `AIO.C`

```
/*
*********************************************************************************************
*                                 UPDATE PHYSICAL OUTPUTS
*
* Description : This function is called to write the 'raw' counts to the proper analog output device
*               (i.e. DAC).  It is up to this function to direct the DAC counts to the proper DAC if more
*               than one DAC is used.
* Arguments   : ch     is  the DAC logical channel number (0..AIO_MAX_AO-1).
*               cnts   are the DAC counts to write to the DAC
* Returns     : None.
*********************************************************************************************
*/

void  AOWr (INT8U ch, INT16S cnts)
{
    ch   = ch;
    cnts = cnts;

    /* This is where you will need to provide the code to update your DAC(s).                 */
    /* AOWr() is passed a 'LOGICAL' channel number.  You will have to convert this logical channel  */
    /* number into actual physical port locations (or addresses) where your DACs are located.  */
    /* AOWr() is responsible for writing the counts to the selected DAC based on a logical number.  */
}
#endif
```

Listing 10.2 `AIO.H`

```
/*
*********************************************************************************************
*                                    Analog I/O Module
*
*                          (c) Copyright 1999, Jean J. Labrosse, Weston, FL
*                                    All Rights Reserved
*
* Filename   : AIO.H
* Programmer : Jean J. Labrosse
*********************************************************************************************
*/

#ifdef   AIO_GLOBALS
#define  AIO_EXT
#else
#define  AIO_EXT extern
#endif


/*
*********************************************************************************************
*                                 CONFIGURATION CONSTANTS
*********************************************************************************************
*/

#ifndef  CFG_H

#define  AIO_TASK_PRIO          40
#define  AIO_TASK_DLY           100
#define  AIO_TASK_STK_SIZE      512

#define  AIO_MAX_AI             8        /* Maximum number of Analog Input  Channels (1..250)    */
#define  AIO_MAX_AO             8        /* Maximum number of Analog Output Channels (1..250)    */

#endif
/*$PAGE*/
```

10

Listing 10.2 (continued) `AIO.H`

```
/*
************************************************************************************************************
*                                            DATA TYPES
************************************************************************************************************
*/

typedef struct aio {                        /* ANALOG I/O CHANNEL DATA STRUCTURE                          */
    BOOLEAN     AIOBypassEn;                 /* Bypass enable switch (Bypass when TRUE)                   */
    INT16S      AIORaw;                      /* Raw counts of ADC or DAC                                  */
    FP32        AIOEU;                       /* Engineering units of AI channel                           */
    FP32        AIOGain;                     /* Total gain    (AIOCalGain   * AIOConvGain)                */
    FP32        AIOOffset;                   /* Total offset (AIOCalOffset + AIOConvOffset)               */
    INT16S      AIOLim;                      /* Maximum count of an analog output channel                 */
    INT8U       AIOPassCnts;                 /* Pass counts                                               */
    INT8U       AIOPassCtr;                  /* Pass counter (loaded from PassCnts)                       */
    FP32        AIOCalGain;                  /* Calibration gain                                          */
    FP32        AIOCalOffset;                /* Calibration offset                                        */
    FP32        AIOConvGain;                 /* Conversion gain                                           */
    FP32        AIOConvOffset;               /* Conversion offset                                         */
    FP32        AIOScaleIn;                  /* Input  to   scaling function                              */
    FP32        AIOScaleOut;                 /* Output from scaling function                              */
    void    (*AIOScaleFnct)(struct aio *paio);   /* Function to execute for further processing            */
    void     *AIOScaleFnctArg;                   /* Pointer to argument to pass to 'AIOScaleFnct'         */
} AIO;


/*
************************************************************************************************************
*                                          GLOBAL VARIABLES
************************************************************************************************************
*/

AIO_EXT    AIO      AITbl[AIO_MAX_AI];
AIO_EXT    AIO      AOTbl[AIO_MAX_AO];

/*$PAGE*/
```

Listing 10.2 (continued) `AIO.H`

```
/*
*********************************************************************************************************
*                                         FUNCTION PROTOTYPES
*********************************************************************************************************
*/

void      AIOInit(void);

INT8U     AICfgCal(INT8U n, FP32 gain, FP32 offset);
INT8U     AICfgConv(INT8U n, FP32 gain, FP32 offset, INT8U pass);
INT8U     AICfgScaling(INT8U n, void (*fnct)(AIO *paio), void *arg);
INT8U     AISetBypass(INT8U n, FP32 val);
INT8U     AISetBypassEn(INT8U n, BOOLEAN state);
INT8U     AIGet(INT8U n, FP32 *pval);

INT8U     AOCfgCal(INT8U n, FP32 gain, FP32 offset);
INT8U     AOCfgConv(INT8U n, FP32 gain, FP32 offset, INT16S lim, INT8U pass);
INT8U     AOCfgScaling(INT8U n, void (*fnct)(AIO *paio), void *arg);
INT8U     AOSet(INT8U n, FP32 val);
INT8U     AOSetBypass(INT8U n, FP32 val);
INT8U     AOSetBypassEn(INT8U n, BOOLEAN state);

void      AIOInitIO(void);                    /* Hardware dependant functions              */
INT16S    AIRd(INT8U ch);
void      AOWr(INT8U ch, INT16S cnts);
```

10

## Chapter 11

Asynchronous Serial Communications

The world of data communications is very complex. A single book (let alone a chapter) cannot cover everything. Data communication is concerned specifically with the issues that must be considered when communicating data between two devices (generally computers). When computing elements are distant from one another, in most cases data is transmitted serially. Because data in a computer is handled in parallel (8 bits or more), it is necessary to convert this information from parallel to serial (when sending) and from serial to parallel (when receiving). There are basically three modes of communication, as shown in Figure 11.1:

1. *Simplex*: Data travels in one direction (from A to B). An example of a simplex link would be scoreboards such as those used in hockey, basketball, or other sports. The information is entered at a console by the score/timekeeper and sent serially to large displays that everybody can see.

2. *Half-duplex*: Data travels in one direction (from A to B) and then the other direction (from B to A) but not at the same time. The RS-485 interface (discussion starts on page 408) is half-duplex.

3. *Full-duplex*: Data can travel in both directions at the same time.

Figure 11.1   Communication modes.

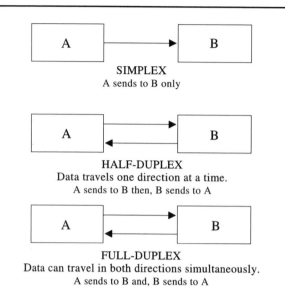

SIMPLEX
A sends to B only

HALF-DUPLEX
Data travels one direction at a time.
A sends to B then, B sends to A

FULL-DUPLEX
Data can travel in both directions simultaneously.
A sends to B and, B sends to A

In this chapter, I will briefly discuss *asynchronous communications*, the RS-232C standard, the RS-485 standard, the serial ports on a PC, and how data is sent and received on an asynchronous communication port. This chapter is not concerned with what is actually sent and received. In other words, in this chapter, I will not cover *data communication protocols*. This chapter provides three software modules:

1. A low-level driver that allows characters to be sent and received on either of the two serial I/O ports on a PC. The driver is called COMM_PC and is interrupt-driven.

2. An interface to the low-level driver (described previously) which allows bytes sent and received to be buffered. This interface allows you to use buffered serial I/O without requiring a real-time operating system. This software module is called COMMBGND and is applicable to just about any Foreground/Background system.

3. An interface to the low-level driver which assumes the presence of a real-time operating system. This software module (called COMMRTOS) allows you to use buffered serial I/O in a multitasking environment.

The code provided in this chapter doesn't make any assumption about the communication mode, i.e., simplex, half-duplex, or full-duplex.

11.00 Asynchronous Communications

You can find just about everything there is to know about asynchronous serial communications in the excellent book from Joe Campbell, *C Programmer's Guide to Serial Communications*, which is now in its second edition (see "Bibliography" on page 455). If you are further interested in the world of data communications, you should also add the books from Andrew S. Tanenbaum and Fred Halsall to your collection.

In asynchronous communication systems, the receiver clock is not synchronized to the transmitter clock when data is being transmitted between two devices. Generally speaking, asynchronous transmission is used to indicate that data is being transmitted as individual bytes. Each byte is preceded by a *start signal* and terminated by one or more *stop signals*. The start and stop signals are used by the receiver for synchronization purposes. As shown in Figure 11.2, the transmission line is in a *mark* (binary 1) condition in its idle state. As each byte is transmitted, it is preceded by a start bit which is a transition from a mark to a *space* (binary 0). This transition indicates to the receiving device that a byte is being transmitted. The receiving device detects the start bit and the data bits that make up the byte. At the end of the byte transmission, the line is returned to a mark condition by one or more stop bit(s). At this point, the transmitter is ready to send the next byte. The start and stop bits permit the receiving device to synchronize itself to the transmitter on a byte-by-byte basis. From Figure 11.2, you should note that bytes are transmitted least-significant bit first. Also, each byte of data being transmitted requires at least two bits which are used for synchronization purpose. The synchronization bits thus impose an overhead of 20 percent.

Figure 11.2 *Asynchronous communications timing diagram.*

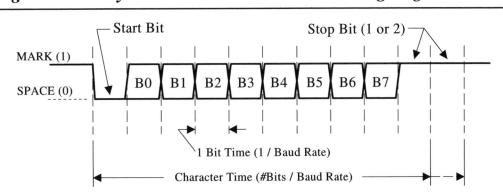

It is assumed that the receiver knows how fast each bit is being transmitted. This transmission rate is known as the *baud rate*. As long as the sender and the receiver agree to use the same baud rate, the actual rate used is not important. The industry has, however, standardized baud rates, as shown in Table 11.1.

Table 11.1 *Standard baud rates*

Baud rate	Bit time (μS)	#Bytes/sec. (note 1)	Time between bytes (μS) (note 1)
300	3,333.3	30	33,333
600	1,666.6	60	16,667
1200	833.3	120	8,333
2400	4166.7	240	4,167
4800	208.3	480	2,083
9600	104.2	960	1,042
19200	52.1	1920	521
38400	26.0	3840	260
56000	17.9	5600	179

Note 1: Assuming 1 start, 8 data bits, and 1 stop.

11

Asynchronous communications is performed almost transparently by a device called a *UART* (Universal Asynchronous Receiver Transmitter). To send and receive data, your program simply writes and reads bytes to and from the UART. UARTs are generally capable of sending and receiving data at the same time (i.e., they support full-duplex communication). A UART appears to the microprocessor as one or more memory locations or I/O ports. UARTs generally contain one or more *status register(s)*, which are used to verify the progress and state of data transmission and reception. The microprocessor can thus know when a byte has been received, whether a communication error occurred, or when a byte has been sent. UARTs can also be configured through one or more *control registers*. Configuration of a UART consists of setting the baud rate, setting the number of stop bits (1, 1-1/2 or 2), enabling interrupts when bytes are sent or received, etc.

Probably the most popular UART is the National Semiconductor NS16550 (see `16450.pdf` on the companion CD-ROM). There are many other UARTs available on the market and some of the more popular ones are: the AMD Z8530, the Motorola 6850 ACIA, the Zilog Z-80 SIO, etc. The NS16550 contains all the required functionality to send and receive characters, but the NS16550 also is equipped with an internal *Baud Rate Generator*, which makes it especially easy to interface to most microprocessors. What is nice about UARTs is that they also are available on a large number of single chip CPUs. Embedded systems can thus benefit from the capability of communicating with terminals, computers or even other embedded microprocessors.

Data sent and received by UARTs can consist of anything that can be represented by eight bits (or less) or any multiple of eight bits. You can thus send binary data, *ASCII* (American Standard Code for Information Interchange) characters, *EBCDIC* (Extended Binary Coded Decimal Interchange Code), BCD (Binary Coded Decimal) digits, etc. By far the most important character set used by the English-speaking world is ASCII. ASCII is a 7-bit code. The mapping of a 7-bit binary value to an ASCII code is shown in Figure 11.3. ASCII characters are used to represent strings in C. For example, the string "HELLO" is represented by the following ASCII codes:

ASCII:	H	E	L	L	O	\0
Binary	0x48	0x45	0x4C	0x4C	0x4F	0x00

The ASCII chart contains two columns of "special" characters. Some of these ASCII characters are well known to C programmers: NUL (Nul character, 0x00), BEL (Bell, 0x07), BS (Back Space, 0x08), LF (Line Feed, 0x0A), CR (Carriage Return, 0x0C), FF (Form Feed, 0x0F), ESC (Escape, 0x1B), and SP (Space, 0x20). The first two columns also contain character codes that can be used in data communication protocols (beyond the scope of this book).

Figure 11.3   ASCII character set (7-bit code).

LSD		MSD							
		0 000	1 001	2 010	3 011	4 100	5 101	6 110	7 111
0	0000	NUL	DLE	SP	0	@	P	`	p
1	0001	SOH	DC1	!	1	A	Q	a	q
2	0010	STX	DC2	"	2	B	R	b	r
3	0011	ETX	DC3	#	3	C	S	c	s
4	0100	EOT	DC4	$	4	D	T	d	t
5	0101	ENQ	NAK	%	5	E	U	e	u
6	0110	ACK	SYN	&	6	F	V	f	v
7	0111	BEL	ETB	'	7	G	W	g	w
8	1000	BS	CAN	(	8	H	X	h	x
9	1001	HT	EM	)	9	I	Y	i	y
A	1010	LF	SUB	*	:	J	Z	j	z
B	1011	VT	ESC	+	;	K	[	k	{
C	1100	FF	FS	,	<	L	\	l	|
D	1101	CR	GS	-	=	M	]	m	}
E	1110	SO	RS	.	>	N	^	n	~
F	1111	SI	US	/	?	O	_	o	DEL

11

11.01 RS-232C

Dating all the way back to 1969, the RS-232C standard is probably the most widely used communication interface in the world. RS-232C was defined by the *Electronic Industries Association* (EIA) and is formally known as: "Interface between data terminal equipment and data communication equipment employing serial binary data interchange." As shown in Figure 11.4, the RS-232C standard is a hardware protocol used to interface between two devices: one is called the *Data Terminal Equipment* (DTE) and the other, the *Data Communication Equipment* (DCE). The RS-232C standard defines:

1. The mechanical aspects of the interface.
2. The characteristics of the electrical signals.
3. The functional aspects of the interchange.

Figure 11.4   RS-232C interface.

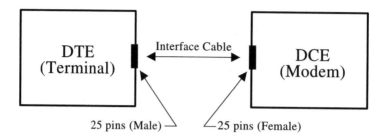

The RS-232C standard says that there should be two 25-pin connectors: the male connector is used on the DTE while the female connector is used on the DCE. The actual type of connector is not defined by the standard. The industry has, however, standardized on 25 pins D-shell type connectors.

Electrically speaking, the RS-232C standard specifies that:

- the load capacitance on a driver is not to exceed 2500 picofarads (pF),

- the load resistance on a driver must be between 3000 and 7000 ohms,

- the data signaling rate (or baud rate) must be below 20,000 bits per second (bps) under the specified load,

- the maximum levels on the RS-232C lines are not to exceed 15 volts (with respect to signal ground),

- drivers must be able to produce between +5 and +15 volts (logic 1) and –5 to –15 volts (logic 0),

- inputs must be able to accept signals from +3 to +15 volts (logic 1) and –3 to –15 volts (logic 0).

Under the maximum load suggested by the RS-232C standard, the distance between the DTE and the DCE should not exceed 50 feet. Simple math would have you conclude that at a distance of 25 feet (half the capacitance) you should be able to increase the signaling rate to 40,000 bps, 80,000 bps at 12.5 feet, and about 160,000 bps at 6 feet. In fact, many communication packages allow you to interface two computers at a data signaling rate of up to 115,200 bps. You should note that the RS-232C standard does not define "standard" baud rates. The RS-232C standard allows data to be sent and received at the same time (i.e., full-duplex).

From the 25 pins defined by the RS-232C standard only nine (9) lines are actually used in "real-world" applications. Probably for that reason and to reduce cost, IBM started to use 9-pin connectors for RS-232C communication when they introduced the IBM PC/AT back in the mid-1980s. The nine pins that are retained for RS-232C communications are shown in Table 11.2. You should note that communication ports on PCs are generally connected as DTEs (i.e., male connectors).

Table 11.2   RS-232C connections.

Description	Acronym	DTE DB-25M Pin#	DTE DB-9M Pin#	Direction	DCE DB-9F Pin#	DCE DB-25F Pin#
Transmit	TxD	2	3	->	2	3
Receive Data	RxD	3	2	<--	3	2
Request To Send	RTS	4	7	->	8	5
Clear To Send	CTS	5	8	<--	7	4
Data Set Ready	DSR	6	6	<--	4	20
Data Carrier Detect	DCD	8	1	<--	1	8
Data Terminal Ready	DTR	20	4	->	6	6
Ring Indicator	RI	22	9	<--	9	22
Signal Ground	SG	7	5	-	5	7

A full description of the use of each of the pins is beyond the scope of this chapter because the code presented in this chapter only assumes the presence of the TxD, RxD, and SG lines. You will find, however, detailed information about these lines in Joe Campbell's book.

An RS-232C communications port generally consists of a UART and what are called *EIA drivers/receivers*. The EIA drivers and receivers are used to convert microprocessor levels (typically 0 to 5 volts) to RS-232C compatible levels: –3 to –15 volts (logic 0) to +3 to +15 volts (logic 1). An RS-232C DTE using an NS16550 and EIA drivers/receivers is shown in Figure 11.5. Inverters are used for electrical reasons. For your convenience, Figure 11.5 shows the pinout for both the DB25 and DB9 connectors. (Note that the "M" in DB-25M and DB-9M stands for "Male.") Only one of the two connectors, however, would actually be used.

Figure 11.5   RS-232C connections (DTE).

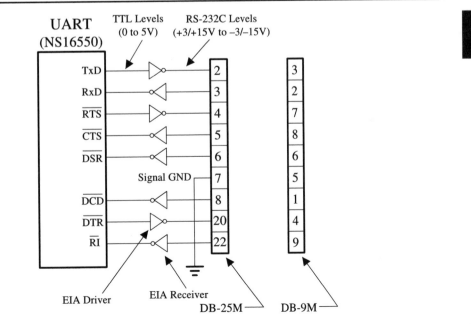

Connection between a DTE and a DCE is quite straightforward and is shown in Figure 11.6. A readily available DB25F to DB25M (or DB9F to DB9M) cable is typically all that is required.

Figure 11.6   RS-232C connections (DTE to DCE).

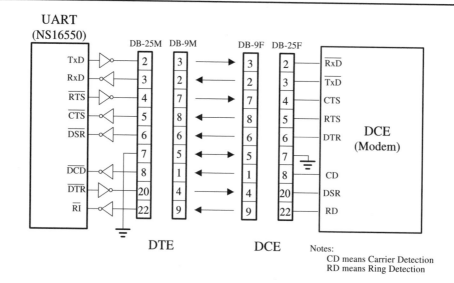

There might be situations where you would need to connect two DTEs together. For example, you may want to connect a terminal to a PC or even interface two PCs. Connecting two DTEs together is a little tricky because:

• Both DTEs have male connectors and,

• outputs would be connected to outputs and, inputs would be connected to inputs on each DTE.

This situation can be resolved by using what is called a *Null Modem adapter* (also known as a *Gender Changer*) or by using two female connectors and making the connections shown in Figure 11.7.

Figure 11.7   RS-232C NULL Model (DTE to DTE).

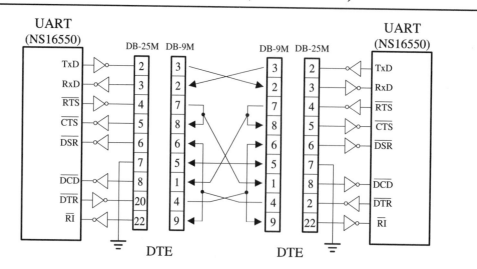

Communication between DTEs is also possible by using only three wires as shown in Figure 11.8. The unused inputs must be asserted to satisfy the UART (specifically, the TxD output line typically is disabled when CTS is negated). This can be accomplished by asserting the DTR output on each DTE. The software modules presented in this chapter assume that you are using a three-wire interface.

Figure 11.8   RS-232C 3-wire DTE to DTE.

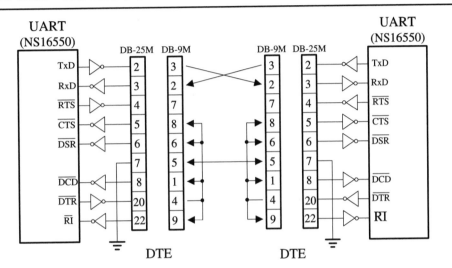

11.02 RS-485

The RS-232C standard requires that a direct connection be made between two devices. This is known as a *point-to-point interface*. If, for example, you need to communicate with many embedded microprocessors, you would need to dedicate an RS-232C port for each embedded processor, as shown in Figure 11.9. This situation can become expensive if the embedded processors are located far from the PC. Also, RS-232C is fairly susceptible to noise because of its common ground arrangement.

11

Figure 11.9   PC interfacing to multiple embedded processors.

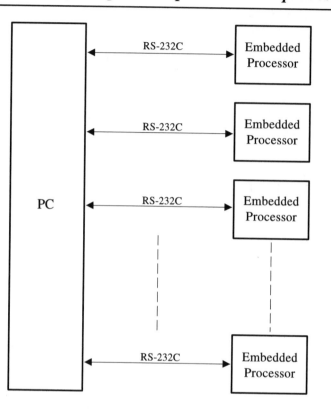

The RS-485 interface has been created to allow multiple (up to 32) processors to communicate with each other on a common line. RS-485 is sometimes called a *party-line* or a *multi-drop* interface and is shown in Figure 11.10. The RS-485 interface uses differential line driver/receiver chips (such as the Texas Instruments SN75176A Differential Bus Transceiver) and only requires a single twisted pair of wires. Communication on an RS-485 interface is, however, half-duplex. Each communicating element on an RS-485 interface is called a *node* and communication generally follows a *MASTER/SLAVE* protocol (but doesn't have to). One of the nodes is called the *MASTER* while all other nodes are called *SLAVEs*. In a MASTER/SLAVE arrangement, all communication occurs between the MASTER and a SLAVE (not between SLAVEs). Each node on an RS-485 is assigned a unique *node I.D.* number. Node #0 is generally assigned to the MASTER. The MASTER selectively communicates with one of the SLAVEs at any given time. An RS-485 interface has the following features:

- very noise immune,
- maximum cable length of 4000 feet,
- data signaling rate up to 10 Mbps (mega-bits per second),
- capable of supporting up to 32 nodes, and
- capable of supporting a multi-MASTER configuration.

Figure 11.10 RS-485 interface.

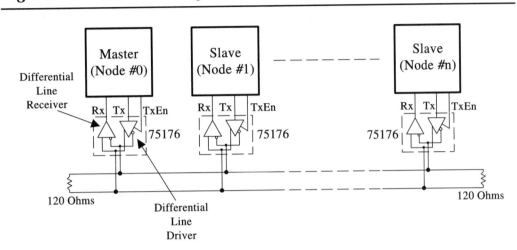

Communication on an RS-485 interface proceeds as shown in Figure 11.11. The MASTER enables its transmit line driver and sends a command or data to a SLAVE (①). The desired SLAVE I.D. number is typically sent as one of the first bytes in the message from the MASTER. When all bytes of the command or data are sent, the MASTER disables its transmit line driver (②) and waits for a reply from the SLAVE. The SLAVE processes the command or data received and formulates a response for the MASTER (③). The SLAVE enables its transmit line driver (④) and sends the response back to the MASTER. When all bytes which make up the response from the SLAVE are sent, the SLAVE disables its transmit line driver (⑤). The MASTER analyzes the response from the SLAVE (⑥) and performs whatever action is needed. The MASTER is then ready to initiate the next command or data transfer. You should note that when either the MASTER or the SLAVE is sending data the respected receivers are monitoring what is being sent. The data sent can be verified by the sender to ensure the integrity of the line, or the sender can simply discard the same number of bytes received as sent. The sender can also ignore any received data until it is done with the transmission.

11

Figure 11.11 RS-485 timing diagram.

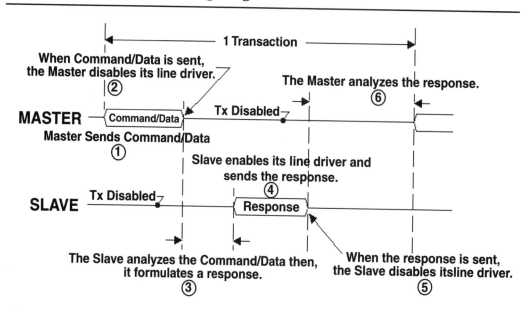

The NS16550 is not a good UART to use for RS-485 communication because it doesn't provide an interrupt when the last byte has been transmitted. Instead, the NS16550 only tells you when it is ready to send another byte. Figure 11.12(a) will help illustrate what happens. The NS16550 contains two registers for data transmission: a Transmitter Holding Register (THR) and a Transmitter Shift Register (TSR). When you write a data byte to the NS16550, the byte is actually deposited into the THR (①) and is then automatically transferred to the TSR (②). At this point, the bits in the TSR are shifted out at the baud rate that you selected (③) and an interrupt is generated by the NS16550 to indicate that the THR can accept another byte (④); the THR *holds* the byte while the previous byte is being transmitted. If you disable the RS-485 line driver in the THR Interrupt Service Routine (ISR), you will actually prevent the last byte from being sent because it is still in the process of being shifted out.

Figure 11.12 Disabling the RS-485 line driver.

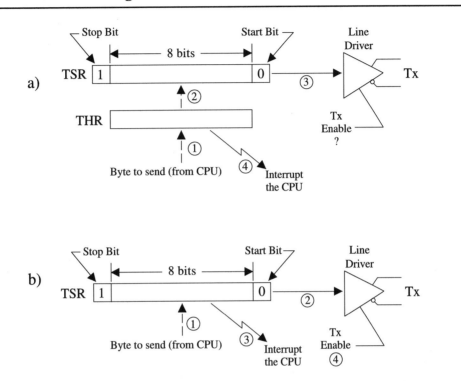

What you actually need is a UART that interrupts the processor when the STOP bit of the last byte has been shifted out, as illustrated in Figure 11.12(b). In this case, there is no need for a THR. The CPU writes a byte to the TSR (①), which then gets shifted out by the UART (②). When the start bit, the byte, and the stop bit are sent, the UART interrupts the CPU (③). If there are no more bytes to send, the ISR disables the line driver (④).

The low-level code provided in this chapter is designed to work with the NS16550 and so it does not support RS-485. It should, however, be fairly easy to port the code to another UART which supports the scheme described in Figure 11.12(b).

11.03 Sending and Receiving Data

As previously mentioned, data is sent and received by a UART by writing and reading from memory or I/O port locations. A bit in the UART's status register can be monitored to determine when a byte has been received. Similarly, another bit can be examined to see when a byte has been transmitted through the interface. This method of monitoring the UART status is called *polling the I/O device* and generally is used when the microprocessor can monitor the status register faster than bytes are sent and received. Polling has serious shortcomings, especially for input, because bytes can be missed while the processor is occupied with other duties. Because microprocessors have other things to do besides wait for serial I/O ports, it is common to resort to an *interrupt-driven* scheme to handle data reception and transmission.

11

11.03.01 Receiving Data

When using an interrupt-driven scheme, an interrupt is generated when a byte arrives through the serial port. The interrupt handler reads the byte from the port, which generally clears the interrupt source. At this point you have a choice of either processing the byte received in the ISR or putting the byte into some sort of buffer to let a background process handle the data. When you use a buffer, the size of the buffer depends on how quickly your background process can get control of the CPU to process the information. For example, if the worst case latency of your background process is 200 mS, you should plan for a buffer of at least 192 bytes if your serial port receives bytes at 9600 baud (960 bytes/sec. × 200 mS). A special type of buffer called a *Ring Buffer* (also called a *Circular Buffer*) is often used to capture data from a serial port.

To avoid allocating very large buffers, you can resort to what is called *flow control*. Basically, the interrupt receiving data can notify the sender that the receiver's buffer is getting full. The sender would then hold off with its transmission until the receiver empties out the buffer and notifies the sender that it can proceed. The most common flow control scheme is called *XON-XOFF* and it uses the ASCII characters DC1 (0x11) for XON (i.e., "send me more") and DC3 (0x13) for XOFF (i.e., "don't send me any more"). Using the XON-XOFF scheme precludes you from sending binary data because the data you are sending could happen to be one of these two characters.

Flow control can also be performed by using some of the RS-232C lines. This would allow you to send and receive binary data. Unfortunately, the RS-232C standard doesn't specify which lines to use when you are not interfacing to a modem. Nothing prevents you from using the modem control lines RTS, CTS, DSR, and DTR, but you will have to establish how flow control will work between your devices.

Input buffering using a ring buffer is shown in Figure 11.3. When bytes are received, the ISR reads the byte from the serial port (①) and places the byte into the ring buffer (②). Your application code (background) then monitors the ring buffer to see if bytes have been received (③). If the ring buffer is not empty, the "oldest" byte (least recent byte) is extracted from the ring buffer.

Figure 11.13 Buffered serial I/O, receiving bytes.

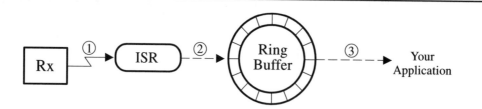

The following pseudocode for both the ISR and the interface function to your application follow. Actual code for the ISR and the interface function will be described later.

```
ISR CommRxISR (void)
{
    INT8U c;


    Save processor context;
    c = Get byte from RX port;
    if (Rx Ring Buffer not full) {
        Put byte received into ring buffer;
    }
    Restore processor context;
    Return from Interrupt;
}


INT8U CommGetChar (void)
{
    INT8U c;


    c = NUL;
    Disable interrupts;                 /* Prevent INTs during access    */
    if (Rx Ring Buffer not empty) {
        c = Get byte from ring buffer;
    }
    Enable interrupts;
    return (c);
}
```

You should note that interrupts are disabled when your application accesses the ring buffer to ensure exclusive access to the ring buffer from either the ISR or the interface function.If your application doesn't extract bytes from the ring buffer in time, the ring buffer will become full and received bytes will be lost.

The response to incoming data depends on how soon your background process gets to execute. If you are using a real-time kernel, you can process incoming data almost as quickly as you receive it without doing so in an ISR. To accomplish this, a semaphore is added to the management of the ring buffer as shown in Figure 11.4. In this case, your application waits on the semaphore (①). When a byte is received, the ISR reads the byte from the serial port (②) and deposits it in the ring buffer (③). The ISR then signals the semaphore to indicate to the waiting task that a byte was received (④). Signaling the semaphore makes the waiting task ready to run. When the ISR completes, the kernel determines if your waiting task is now the highest-priority task ready to get the CPU. If it is, the ISR resumes the task waiting for the byte (assuming a preemptive kernel). Your application code then extracts the byte from the ring buffer and performs whatever processing is required.

Figure 11.14 Buffered serial I/O with semaphore, receiving bytes.

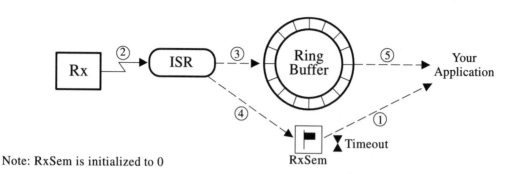

Note: RxSem is initialized to 0

The following pseudocode for both the ISR and the interface function to your application follow. Actual code for the ISR and the interface function will be described later. As with the previous scheme, if your application doesn't extract bytes from the ring buffer in time, the ring buffer will become full and bytes received will be lost. The use of a real-time kernel, however, reduces the chance of this situation from happening.

Most real-time kernels allow you to specify the maximum amount of time your task is willing to wait for a byte to be received. This gives your task a chance to take corrective action in case something happened to the communication link. For example, a task can send a message and then wait for a response. If the response doesn't arrive within a certain amount of time, the sender can conclude either that there is nobody listening or that something happened to the transmission medium.

```
ISR CommRxISR (void)
{
    INT8U c;


    Save processor context;
    Tell OS that we are processing an ISR;
    c = Get byte from RX port;
    if (Rx Ring Buffer is not Full) {
        Put received byte into Ring Buffer;
        Signal Rx Semaphore;
    }
    Tell OS that we are exiting an ISR;
    Restore processor context;
    Return from Interrupt;
}


INT8U CommGetChar (INT8U *err)
{
    INT8U c;


    Wait for byte to be received (using semaphore with T.O.);
    if (timed out) {
        *err = Time out error;
        return (0);
    }
    Disable interrupts;
    c    = Get byte from Ring Buffer;
    Enable interrupts;
    *err = No error;
    return (c);
}
```

Signalling the semaphore everytime a character is received can consume valuable CPU time. An alternate method is to only signal the semaphore when a special character is received. For example, you can signal the semaphore when a carriage return character (i.e., CR or 0x0D) is received. You application can thus be notified once a full command is received which reduces the overhead. Of course, your buffer needs to have sufficient storage to hold one or more commands. This alternate method is shown in the following pseudocode.

```
ISR CommRxISR (void)
{
    INT8U c;


    Save processor context;
    Tell OS that we are processing an ISR;
    c = Get byte from RX port;
    if (Rx Ring Buffer is not Full) {
        Put received byte into Ring Buffer;
        if (received byte is the end-of-command byte) {
            Signal Rx Semaphore;
        }
    }
    Tell OS that we are exiting an ISR;
    Restore processor context;
    Return from Interrupt;
}
INT8U CommGetCommand (INT8U *command, INT8U *nbytes)
{
    INT8U c;
    INT8U nrx;


    Wait for command to be received (using semaphore with T.O.);
    if (timed out) {
        *nbytes = 0;
        return (Timeout error);
    }
    nrx = 0;                  /* Clear number of bytes received counter */
    Disable interrupts;
    c   = Get byte from Ring Buffer;
    while (c != end-of-command byte) {
        *command++ = c;   /* Save command byte                        */
        nrx++;                /* Clear number of bytes received counter */
        c         = Get byte from Ring Buffer;
    }
    Enable interrupts;
    *nbytes = nrx error;   /* Set number of bytes received            */
    return (No error);
}
```

11.03.02 Transmitting Data

Transmission of bytes works somewhat like byte reception. Your background process deposits bytes in an output buffer. When the transmitter on the UART is ready to send a byte, an interrupt is generated, the byte is extracted from the buffer, and the ISR outputs the byte. There is, however, one small complication: The serial port generates an interrupt only AFTER the port has finished sending the byte. The most elegant way I found to resolve this dilemma is to disable interrupts from the transmitter until you need to send bytes. Interrupts are enabled AFTER the output buffer is loaded with at least one byte. As soon as you allow the transmitter to interrupt, the first byte to send will be removed by the transmit ISR and output to the UART. The ISR then examines the buffer and, if there are no more bytes to send, the ISR disables the transmit interrupt.

Buffering of data makes a lot of sense when you have to transmit a relatively large amount of data on the serial port, such as the contents of a disk file. Output buffering using a ring buffer is shown in Figure 11.15. When one or more bytes need to be sent, they are placed in the ring buffer (①). Transmit interrupts are enabled after putting a byte into the buffer (②). If the UART is ready to send a byte, an interrupt occurs and the ISR extracts the "oldest" (least recent) byte from the ring buffer (③). The byte is then output to the serial port (④). Transmit interrupts will be inhibited if the byte extracted from the buffer makes the ring buffer empty.

Figure 11.15 Buffered serial I/O, transmitting bytes.

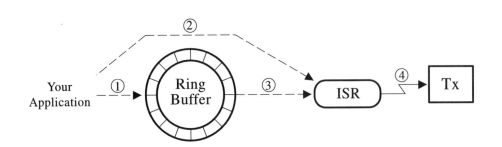

The following pseudocode for both the ISR and the interface function to your application follows. Actual code for the ISR and the interface function will be described later.

11

```
void CommPutChar (INT8U c)
{
    Disable interrupts;              /* Prevent INTs during access    */
    if (Tx Ring Buffer is not Full) {
        Put byte to send into ring buffer;
        if (This is the first byte in the Ring Buffer) {
            Enable Tx Interrupts;
        }
    }
    Enable interrupts;               /* Allow CPU interruptions       */
}



ISR CommTxCharISR (void)
{
    INT8U c;


    Save processor context;
    if (Tx Ring Buffer not empty) {
        c = Get next byte to send from ring buffer;
        Output byte 'c' to TX port;
    } else {
        Disable Tx Interrupts;
    }
    Restore processor context;
    Return from Interrupt;
}
```

Figure 11.16 shows how you can make use of a real-time kernel's facilities. The semaphore is used as a *traffic light* pausing the sending task when the ring buffer is full. To send data, the task waits for the semaphore (①). If the ring buffer is not full, the task proceeds to deposit the byte into the ring buffer (②). Transmitter interrupts are enabled if the byte deposited is the first byte in the ring buffer (③). The transmit interrupt ISR extracts the "oldest" byte from the ring buffer (④) and signals the semaphore (⑤) to indicate that the ring buffer has room to accept another character. The ISR then outputs the byte to the UART.

Figure 11.16 Buffered serial I/O with semaphore, transmitting bytes.

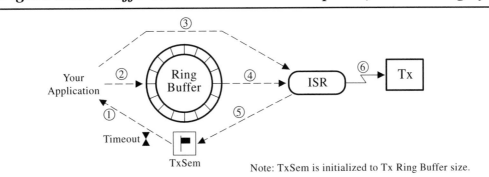

Note: TxSem is initialized to Tx Ring Buffer size.

It is important to note that TxSem needs to be a counting semaphore, and the semaphore must be initialized to the size of the ring buffer. The pseudocode for both the interface function to your application and the ISR follows. Actual code for the ISR and the interface function will be described later.

```
void CommPutChar (INT8U c, INT8U *err)
{
    Wait for space in the Tx Ring Buffer (using semaphore T.O.);
    if (timed out) {
        *err = Time out error;
        return;
    }
    Disable interrupts;
    Put byte to send (c) into the Tx Ring Buffer;
    if (This is the first byte in the Tx Ring Buffer) {
        Enable TX Interrupts;
    }
    Enable interrupts;
    *err = No error;
}
```

11

```
ISR CommTxCharISR (void)
{
    INT8U c;


    Save processor context;
    if (Tx Ring Buffer is not empty) {
        c = Get next character to send from Tx Ring Buffer;
        Output character 'c' to TX port;
        Signal Tx semaphore;
    } else {
        Disable TX Interrupts;
    }
    Restore processor context;
    Return from Interrupt;
}
```

11.04 Serial Ports on a PC

The software modules provided in this chapter allow you to use both serial ports on an IBM-PC/AT compatible computer although it can be easily altered to support different hardware. A review of the PC's architecture relating to the serial ports available on PCs is thus necessary in order to better understand the code.

PCs are typically equipped with two RS-232C communication ports that are referred to as COM1 and COM2. Both ports generally consist of a National Semiconductor NS16550 or equivalent UART and are capable of communicating at baud rates up to 115200 bps. The PC provides services through its BIOS (Basic Input/Output System) but unfortunately, communications using the BIOS must be done by *polling* (monitoring the port to see if bytes have been received or sent). This limitation means that communication effectively cannot exceed about 1200 baud. This shortcoming can be corrected by replacing the BIOS services with interrupt-driven functions.

An IBM-PC/AT computer contains two interrupt controllers (Intel 82C59A PIC) providing 15 sources of interrupts to the PC's microprocessor. Interrupts are labeled IRQ0 through IRQ15, as shown in Figure 11.17. IRQ2 of the first i82C59A is actually the output of the second i82C59A interrupt controller.

Figure 11.17 PC/AT interrupt controllers.

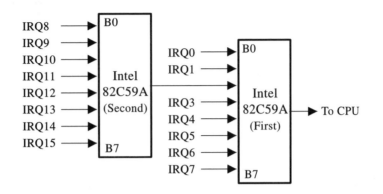

Table 11.3 shows what devices are typically connected to the interrupt controllers. The table lists the interrupt sources in priority order (IRQ0 has the highest priority). Table 11.3 also shows that each serial I/O port is connected to its own IRQ line: COM1 is connected to IRQ4 while COM2 is connected to IRQ3.

11

Table 11.3   PC/AT interrupts summary.

IRQ#	Description	Interrupt vector #	Interrupt vector address	Mask register address	Mask bit#	Clear IRQ
IRQ0	Timer (i.e., ticker, 18.2 Hz)	0x08	0x0000:0x0020	0x0021	0	0x0020
IRQ1	Keyboard	0x09	0x0000:0x0024	0x0021	1	0x0020
IRQ2	(Interrupts 8-15 shown below)	0x0A	0x0000:0x0028	0x0021	2	0x0020
IRQ8	Real-Time Clock	0x70	0x0000:0x01C0	0x00A1	0	0x00A0 then 0x0020
IRQ9	Redirected to IRQ2	0x71	0x0000:0x01C0	0x00A1	1	0x00A0 then 0x0020
IRQ10	Unassigned	0x72	0x0000:0x01C8	0x00A1	2	0x00A0 then 0x0020
IRQ11	Unassigned	0x73	0x0000:0x01CC	0x00A1	3	0x00A0 then 0x0020
IRQ12	Unassigned	0x74	0x0000:0x01D0	0x00A1	4	0x00A0 then 0x0020
IRQ13	80x87 co-processor	0x75	0x0000:0x01D4	0x00A1	5	0x00A0 then 0x0020
IRQ14	Hard Disk	0x76	0x0000:0x01D8	0x00A1	6	0x00A0 then 0x0020
IRQ15	Unassigned	0x77	0x0000:0x01DC	0x00A1	7	0x00A0 then 0x0020
IRQ3	COM2	0x0B	0x0000:0x002C	0x0021	3	0x0020
IRQ4	COM1	0x0C	0x0000:0x0030	0x0021	4	0x0020
IRQ5	LPT2	0x0D	0x0000:0x0034	0x0021	5	0x0020
IRQ6	Floppy Disk	0x0E	0x0000:0x0038	0x0021	6	0x0020
IRQ7	LPT1	0x0F	0x0000:0x003C	0x0021	7	0x0020

IRQ4 is asserted whenever a byte is either received on COM1 or whenever COM1 has completed the transmission of a byte. When an interrupt occurs, the CPU automatically vectors to the *Interrupt Vector Address* shown in Table 11.3. The Interrupt Vector Address points to the Interrupt Service Routine (ISR) responsible for handling the source of the interrupt: either a byte was received, a byte was sent, or both. IRQ3 works just like IRQ4 except that it uses a different vector.

As shown in Figure 11.18, COM port interrupts have to travel through many "doors" (gates) in order to actually interrupt the CPU. First, interrupts must be allowed by the CPU by setting the IF bit in the PSW (*Processor Status Word*). Second, the interrupt controller can inhibit interrupts from any device connected to it through the i82C59A Interrupt Mask Register. Finally, the NS16550 UART is capable of inhibiting either the Rx (byte received) or the Tx (byte sent) interrupts through its Interrupt Enable Register.

Figure 11.18 COM ports interrupt path.

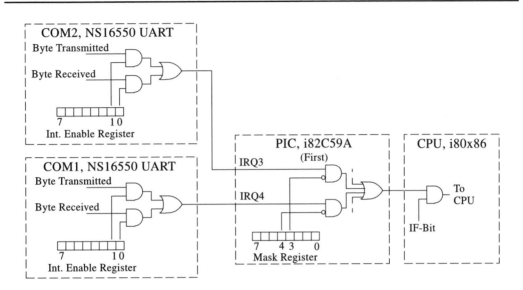

11.05 Low-Level PC Serial I/O Module (COMM_PC)

This section describes a driver that I wrote which makes much better use of the serial I/O ports provided on a PC. The code and the functionality of the driver easily can be ported to other environments. Your application actually interfaces with two modules, as shown in Figure 11.19. Note that the term PC is used generically to mean any PC having either an Intel 80286, 80386, 80486, or Pentium microprocessor.

The low-level driver is responsible for interfacing with the National Semiconductor NS16550 UART. Functions are provided to your application to configure the two ports (COM1 or COM2), enable/disable communication interrupts, and acquire/release the COM port interrupt vectors. The interface functions will be described later.

Your application also interfaces to either one of two buffered serial I/O modules: COMMBGND or COMMRTOS. You would use COMMBGND in a foreground/background application and COMMRTOS if you are running a real-time kernel such a μC/OS-II.

This section specifically describes the low-level driver interface functions. The source code for the low-level code is found in the \SOFTWARE\BLOCKS\COMM\SOURCE directory, and specifically, in the following files:

- COMM_PCA.ASM (Listing 11.1)
- COMM_PC.C (Listing 11.2)
- COMM_PC.H (Listing 11.3)

11

Figure 11.19 PC/AT buffered serial I/O block diagram.

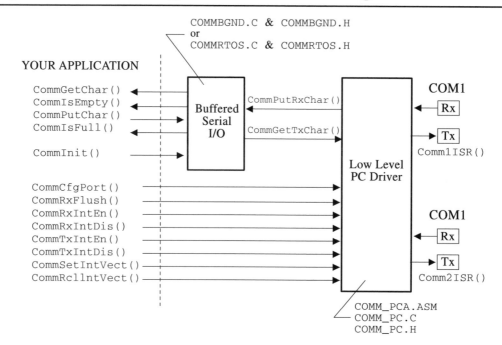

As a convention, all functions and variables related to the low-level serial I/O module start with Comm while all #define constants start with COMM_.

Comm1ISR() and Comm2ISR() (COMM_PCA.ASM) are the functions that are executed when an interrupt occurs on the PC's COM1 or COM2, respectively. These functions start by saving the CPU registers onto the current task stack or the background stack in a foreground/background system. If you are using COMMRTOS, Comm1ISR() needs to increment the µC/OS-II global variable OSIntNesting after saving the CPU registers and call OSIntExit() prior to restoring the registers. After incrementing OSIntNesting, the ISRs call CommISRHandler().

CommISRHandler() is responsible for doing most of the ISR processing and knows about the NS16550 UART internals. You can easily expand this function to support more than just two serial ports. CommISRHandler() determines whether the interrupt was caused by the reception of a byte, the completion of a byte transmission, or both.

If a byte is received, CommISRHandler() reads the UART's receive data register and calls CommPutRxChar(). CommPutRxChar() (described later) is a function that knows what to do with the byte just received. In our case, the byte received is placed in a ring buffer.

If the interrupt is caused by the completion of byte transmission, CommISRHandler() calls CommGetTxChar() (described later) to see if anything else needs to be sent. When all bytes have been sent, CommISRHandler() disables further transmit interrupts from the UART. The interrupt source is not cleared because CommISRHandler() does not actually write to the UART's transmit data register (there is nothing to send). The next time your application code puts something in the ring buffer the transmit interrupt will be re-enabled and an interrupt will occur immediately. The ISR will then extract the byte to send from the ring buffer and satisfy the UART.

Before returning to Comm1ISR() or Comm2ISR(), CommISRHandler() clears the interrupt from the PC's i82C59A interrupt controller.

CommCfgPort()

```
INT8U CommCfgPort(INT8U ch, INT16U baud, INT8U bits, INT8U parity, INT8U stops);
                              (COMM_PC.C)
```

CommCfgPort() is used to establish the characteristics of a serial port. You will need to call this function before calling any of the other services provided by this module for the specific port.

Arguments

ch specifies the channel and can be either COMM1 (for the PC's COM1) or COMM2 (for the PC's COM2).

baud specifies the desired baud rate. The NS16550 sets the baud rate (i.e., baud) according to the following equation:

[11.1] $baud\_rate\_divisor = 115200 / baud$;

You can specify just about any baud rate except that the baud rate divisor will be truncated to a 16-bit integer. For example, you can specify 7500 baud, but you will actually get 7680, as shown:

```
115200 / 7500 = 15.36
```

Truncation produces a baud rate divisor of 15 and the NS16550 UART will actually be set to a baud rate of 115200 / 15 = 7680.

bits specifies the number of bits used. The NS16550 supports either 5, 6, 7, or 8. Generally, you would specify 7 bits with either ODD or EVEN parity or 8 bits with NO parity.

parity specifies the type of parity checking used by the serial port. You can specify either:
COMM_PARITY_NONE for no parity
COMM_PARITY_ODD for odd parity
COMM_PARITY_EVEN for even parity

stops specifies the number of stop bits used. The NS16550 supports either 1 or 2. You would typically specify 1 stop bit, though.

Return Value

CommCfgPort() returns either COMM_NO_ERR (if the channel you specified was either COMM1 or COMM2) or COMM_BAD_CH.

Notes/Warnings

In the previous edition of this book, CommCfgPort() only allowed you to configure the baud rate. The number of bits was always assumed to be 8, the parity was always set to NONE, and the number of stop bits 1.

11

Example

```
void main (void)
{
    INT8U err;
    .
    .
    CommCfgPort(COMM1, 9600, 8, COMM_PARITY_NONE, 1);
    .
    .
}
```

CommRxFlush()

```
void CommRxFlush(INT8U ch);
              (COMM_PC.C)
```

CommRxFlush() allows your application to clear the contents of the UART's receive register. The receive register on the NS16550 UART can receive a byte while another byte waits for the CPU to be processed. CommRxFlush() simply discards the last received byte. If you use the more powerful NS 16550 UART then you would set COMM_MAX_RX (in COMM_PC.H or CFG.G) to 16 because this chip can buffer up to 16 characters.

Arguments

ch specifies the channel and can be either COMM1 (for the PC's COM1) or COMM2 (for the PC's COM2).

Return Value

None

Notes/Warnings

None

Example

The following code example assumes the presence of an RTOS but the function can just as easily be used in a foreground/background environment.

```
void Task (void *pdata)
{
    .
    .
    .
    for (;;) {
        .
        CommRxFlush(COMM2);
        .
        .
    }
}
```

11

CommRxIntDis()

void CommRxIntDis(INT8U ch);
(COMM_PC.C)

CommRxIntDis() is used to prevent interrupts from the desired serial port when bytes are received. CommRxIntDis() hides the details of disabling interrupts for the selected serial port from your application. Note that CommRxIntDis() will ensure that the interrupt controller bit will not be cleared (disabling the port's interrupts) if the UART's transmit interrupt is enabled.

Arguments

ch specifies the channel and can be either COMM1 (for the PC's COM1) or COMM2 (for the PC's COM2).

Return Value

None

Notes/Warnings

None

Example

The following code example assumes the presence of an RTOS but the function can just as easily be used in a foreground/background environment.

```
void Task (void *pdata)
{
    .
    .
    .
    for (;;) {
        .
        CommRxIntDis(COMM2);
        .
        .
        .
    }
}
```

CommRxIntEn()

```
void CommRxIntEn(INT8U ch);
                (COMM_PC.C)
```

CommRxIntEn() is used to enable interrupts from the desired serial port when bytes are received. CommRxIntEn() hides the details of enabling interrupts for the selected serial port from your application. Enabling interrupts consist of setting bit 0 of the UART's Interrupt Enable Register (IER) and clearing the appropriate bit on the PC's i82C59A interrupt controller.

Arguments

ch specifies the channel and can be either COMM1 (for the PC's COM1) or COMM2 (for the PC's COM2).

Return Value

None

Notes/Warnings

None

Example

The following code example assumes the presence of an RTOS but the function can just as easily be used in a foreground/background environment.

```
void Task (void *pdata)
{
    .
    .
    for (;;) {
        .
        CommRxIntEn(COMM2);
        .
        .
    }
}
```

11

CommTxIntDis()

void CommTxIntDis(INT8U ch);
(COMM_PC.C)

CommTxIntDis() is used to prevent interrupts from the desired serial port when bytes are sent. CommTxIntDis() hides the details of disabling interrupts for the selected serial port from your application. Note that CommTxIntDis() will ensure that the interrupt controller bit will not be cleared (disabling the port's interrupts) if the UART's receive interrupt is enabled.

Arguments

ch specifies the channel and can be either COMM1 (for the PC's COM1) or COMM2 (for the PC's COM2).

Return Value

None

Notes/Warnings

None

Example

The following code example assumes the presence of an RTOS but the function can just as easily be used in a foreground/background environment.

```
void Task (void *pdata)
{
    .
    .
    for (;;) {
        .
        CommTxIntDis(COMM2);
        .
        .
    }
}
```

CommTxIntEn()

void CommTxIntEn(INT8U ch);
(COMM_PC.C)

CommTxIntEn() is used to enable interrupts when a byte is sent by the UART. CommTxIntEn() hides the details of enabling interrupts for the selected serial port from your application. Enabling transmission interrupts consist of setting bit 1 of the UART's Interrupt Enable Register (IER) and clearing the appropriate bit on the PC's i82C59A interrupt controller.

Arguments

ch specifies the channel and can be either COMM1 (for the PC's COM1) or COMM2 (for the PC's COM2).

Return Value

None

Notes/Warnings

None

Example
The following code example assumes the presence of an RTOS but the function can just as easily be used in a foreground/background environment.

```
void Task (void *pdata)
{
    .
    .
    .
    for (;;) {
        .
        CommTxIntEn(COMM2);
        .
        .
    }
}
```

11

CommSetIntVect()

void CommSetIntVect(INT8U ch);
(COMM_PC.C)

CommSetIntVect() is used to set the contents of the Interrupt Vector Table (IVT) for the desired serial port (see Table 11.3). CommSetIntVect() saves the old contents of the IVT (i.e., a pointer to the BIOS communication handler) so that it can be recovered when your application code returns to DOS.

Arguments

ch is the serial channel to process and can either be COMM1 or COMM2. When you specify COMM1, CommSetIntVect() places a pointer to Comm1ISR() at address 0x0000:0x0030 (see Table 11.3). Similarly, when you specify COMM2, CommSetIntVect() places a pointer to Comm2ISR() at address 0x0000:0x002C (see Table 11.3).

Return Value

None

Notes/Warnings

None

Example

```
void main (void)
{
    INT8U err;
    .
    .
    CommCfgPort(COMM1, 9600, COMM_PARITY_NONE, 8, 1);
    CommSetIntVect(COMM1);
    CommRxIntEn(COMM1);
    .
    .
}
```

CommRclIntVect()

```
void CommRclIntVect(INT8U ch);
(COMM_PC.C)
```

CommRclIntVect() is used to restore the original interrupt vectors of the desired serial port in the IVT (Interrupt Vector Table).

Arguments

ch is the serial channel to process and can either be COMM1 or COMM2. When you specify COMM1, CommRclIntVect() places the previous vector for the PC's COM1 at address 0x0000:0x0030 (see Table 11.3). Similarly, when you specify COMM2, CommRclIntVect() places the previous vector for the PC's COM2 at address 0x0000:0x002C (see Table 11.3).

Return Value

None

Notes/Warnings

None

Example

The following code example assumes the presence of an RTOS but the function can just as easily be used in a foreground/background environment.

```
void Task (void *pdata)
{
    .
    .
    .
    for (;;) {
        .
        if (done with serial port #1 and returning to DOS) {
            CommRclIntVect(COMM1);
        }
        .
        .
    }
}
```

11

11.06 Buffered Serial I/O Module (COMMBGND)

The COMMBGND module allows data received from and sent to a UART to be buffered. Specifically, you would use the COMMBGND module if you write an application destined for a foreground/background environment. The COMMBGND module is designed to work in conjunction with the COMM_PC module described in the previous section. COMMBGND allows you to do full-duplex communication on either serial port (concurrently). The source code for the COMMBGND module is found in the \SOFTWARE\BLOCKS\COMM\SOURCE directory and specifically, in COMMBGND.C (Listing 11.4) and COMMBGND.H (Listing 11.5).

WARNING

In the previous edition of this book, COMMBGND was called COMMBUF1. The file COMMBUF1.C is now COMMBGND.C and, COMMBUF1.H is now COMMBGND.H.

As a convention, all functions and variables related to the COMMBGND module start with Comm while all #define constants start with COMM_.

Each serial port is assigned two ring buffers: one for byte reception and another for byte transmission. Both ring buffers are stored in a structure called COMM_RING_BUF (see COMMBGND.C on page 473). Each ring buffer consists of four elements:

1. storage for data (an array of INT8Us)

2. a counter containing the number of bytes in the ring buffer

3. a pointer where the next byte will be placed in the ring buffer

4. a pointer where the next byte will be extracted from the ring buffer

Figure 11.20 shows a flow diagram for data reception using the COMMBGND module and how it interfaces with the COMM_PC module. .RingBuf??? are elements of the COMM_RING_BUF data structure. An interrupt occurs when a byte is received by the UART (①). If interrupts are enabled, the CPU vectors to the appropriate ISR, i.e., Comm?ISR(). Comm?ISR() saves the CPU's context (its registers), and calls CommISRHandler() (②). CommISRHandler() gets the byte from the UART and calls CommPutRxChar() in order to save the byte into the ring buffer (③). Reading the byte from the UART clears the interrupt from the UART. If the buffer is not already full, a counter, which keeps track of how many bytes are in the buffer is incremented (.RingBufRxCtr). Next, the byte retrieved from the UART is stored at the location pointed to by .RingBufRxInPtr (④). The pointer is then incremented and checked to make sure that it still points somewhere in .RingBufRx[]. If .RingBufRxInPtr points past the array, it is re-initialized to point at .RingBufRx[0].

Figure 11.20 Buffered serial I/O, receiving bytes.

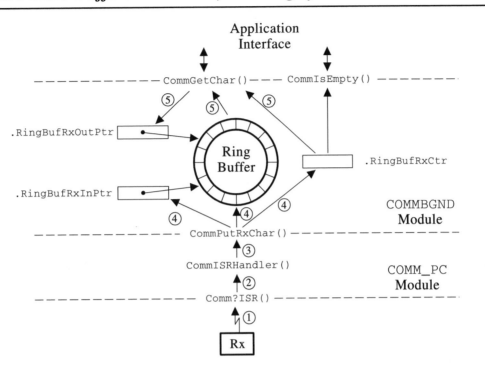

CommPutRxChar() is an interface function between the COMMBGND module and the COMM_PC module. The COMM_PC module calls this function when a byte is received. CommPutRxChar() deposits the byte into the receive ring buffer — but only if the buffer is not already full. The byte is discarded if the buffer is full.

Your application code can find out whether there are bytes in the ring buffer by calling CommIsEmpty(). CommIsEmpty() only needs to check the byte count to determine the state of the ring buffer. When data is available, it is extracted from the ring buffer by calling CommGetChar() (⑤).

Figure 11.21 shows a flow diagram for data transmission using the COMMBGND module and how COMMBGND interfaces with the COMM_PC module. Your application code inserts data to be sent to the serial port into the ring buffer by calling CommPutChar(). If the buffer is not already full, a counter keeping track of how many bytes are in the buffer is incremented (.RingBufTxCtr). Next, the byte you are sending is stored at the location pointed to by .RingBufTxInPtr (①). The pointer is incremented and checked to make sure that it still points somewhere in .RingBufTx[]. If .RingBufTxInPtr points past the array, it is re-initialized to point at the beginning of the array, i.e., .RingBufTx[0]. If CommPutChar() inserted the first character in the buffer, the UART's transmit interrupt is enabled (②). Because you called CommPutChar() from the background, an interrupt will immediately occur (③). The CPU then vectors to the appropriate ISR (Comm?ISR()), saves the CPU's context, and calls CommISRHandler() (④). CommISRHandler() gets the byte from the ring buffer by calling CommGetTxChar() (⑤). Note that CommGetTxChar() obtains the byte from a different pointer than CommPutChar() (⑥). This allows the bytes to be sent in the same order as they were placed in the buffer (First In First Out, FIFO). Obviously, when a byte is removed from the buffer, the byte count is decremented. Writing a byte to the UART clears the interrupt, however, when there are no more bytes to send, CommISRHandler() will not write any-thing to the UART. Instead, further interrupts from the UART will be disabled. In this case, the interrupt status of the UART will remain active but will not make it to the processor until UART interrupts are again enabled.

Figure 11.21 Buffered serial I/O, transmitting bytes.

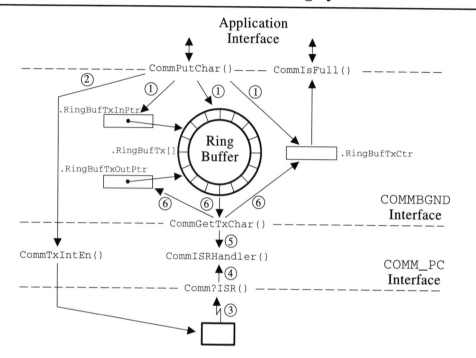

CommGetTxChar() is an interface function between the COMMBGND module and the COMM_PC mod-ule. The COMM_PC module calls this function when a byte has been sent by the UART. Basically, this func-tion says, "Give me the next byte to send." CommGetTxChar() returns the next byte to send from the transmit ring buffer if there is at least one byte in the ring buffer. If the buffer is empty, CommGetTxChar() returns the NUL character and tells the caller that there is no more data in the buffer. This allows the caller to disable further transmit interrupts until there is more data to send.

CommGetChar()

INT8U CommGetChar(INT8U ch, INT8U *err);
(COMMBGND.C)

CommGetChar() allows your application to extract data from the received data ring buffer.

Arguments

ch is the serial channel and can be either COMM1 or COMM2.

err is a pointer to a variable that will hold status about the outcome of the function. CommGetChar() sets *err to one of the following:

COMM_NO_ERR. if a byte is available from the ring buffer.
COMM_RX_EMPTY if the ring buffer is empty.
COMM_BAD_CH if you do not specify either COMM1 or COMM2.

Return Value

The function returns the oldest byte stored in the ring buffer if the buffer is not empty. If the buffer is empty, CommGetChar() returns the NUL character (i.e., 0x00).

Notes/Warnings

None

Example

```
void BgndFnct (void)
{
    INT8U err;


        .

        .

    c = CommGetChar(COMM1, &err);
    if (err == COMM_NO_ERR) {
        Process character;
    }

        .

        .

}
```

11

CommInit()

void CommInit(void);
(COMMBGND.C)

CommInit() is used to initialize the COMMBGND module. This function must be called before any other services provided by this module. CommInit() clears the number of bytes in the ring buffer counter and also initializes both the IN and OUT pointers of each ring buffer to point at the beginning of the data storage area.

Arguments

None

Return Value

None

Notes/Warnings

None

Example

```
void main (void)
{
       .
       .
   CommInit();
       .
       .
}
```

CommIsEmpty()

BOOLEAN CommIsEmpty(INT8U ch);
(COMMBGND.C)

CommIsEmpty() allows your application to determine if a byte was received on the serial port.

Arguments

ch is the serial channel and can be either COMM1 or COMM2.

Return Value

The function returns TRUE if no data was received and FALSE if data is available in the ring buffer.

Notes/Warnings

If you specify an incorrect channel number the function returns TRUE to prevent you from extracting data from an invalid serial port.

Example

```
void BgndFnct (void)
{
    INT8U err;

    .
    .

    if (!CommIsEmpty(COMM1)) {
        /* Characters have been received */
    }
    .
    .

}
```

11

CommIsFull()

BOOLEAN CommIsFull(INT8U ch);
(COMMBGND.C)

CommIsFull() allows your application code to check the status of the transmit ring buffer.

Arguments

ch is the serial channel and can be either COMM1 or COMM2.

Return Value

The function returns TRUE when the buffer is full and FALSE otherwise.

Notes/Warnings

If you specify an incorrect channel number, the function returns TRUE to prevent you from sending data to an invalid serial port.

Example

```
void BgndFnct (void)
{
    INT8U err;


        .

        .

    if (!CommIsFull(COMM1)) {
        /* Characters can be sent to serial port */
    }
        .

        .

}
```

CommPutChar()

UBYTE CommPutChar(INT8U ch, UBYTE ch);
(COMMBGND.C)

CommPutChar() allows your application to send data to a serial port (one byte at a time).

Arguments

ch is the serial channel and can be either COMM1 or COMM2.

c is the byte that your application sends to the serial port. The byte can have any value between 0x00 and 0xFF (i.e., you can send binary data).

Return Value

CommPutChar() returns a value representing the outcome of the function call as follows:

COMM_NO_ERR   the byte was placed in the ring buffer and will eventually be sent by the UART if a byte is available from the ring buffer.

COMM_BAD_CH   if you do not specify either COMM1 or COMM2.

COMM_TX_FULL   indicates that you tried to send a byte to an already-full buffer.

Notes/Warnings

If you configured the serial port to 7 data bits then you will not be able to send binary data.

Example

```
char Message[] = "Hello World!");



void BgndFnct (void)
{
    INT8U err;


    .

    .

    err  = COMM_NO_ERR;
    s    = &Message[0];
    while (*s && err == COMM_NO_ERR) {
        err = CommPutChar(COMM1, *s++);
    }
    .

    .

}
```

11

11.07 Buffered Serial I/O Module (COMMRTOS)

The COMMRTOS module works just like the COMMBGND module except that the COMMRTOS module uses semaphores to indicate when bytes are inserted into the buffer. Semaphores allow your task-level code to process incoming and outgoing data as quickly as possible. Furthermore, your application code no longer needs to poll the receive buffer to see if bytes are available. Similarly, your application code also will be suspended if the transmit buffer is full. This also prevents your code from having to check that the transmit buffer is not full when you are sending data on a serial port.

The source code for the COMMRTOS module is found in the \SOFTWARE\BLOCKS\COMM\SOURCE directory and, specifically, in COMMRTOS.C (Listing 11.6) and COMMRTOS.H (Listing 11.7). As a convention, all functions and variables related to the COMMRTOS module start with Comm while all #define constants start with COMM_.

WARNING

In the previous edition of this book, COMMBGND was called COMMBUF2. The file COMMBUF2.C is now COMMRTOS.C and, COMMBUF2.H is now COMMRTOS.H.

Along with the two ring buffers, each serial port now has two semaphores: one to signal that a byte was received and the other to signal that a byte was sent. The COMM_RING_BUF structure (see COMMRTOS.C on page 484) is identical to the COMMBGND structure except for the addition of the semaphores.

Figure 11.22 Buffered serial I/O, receiving bytes.

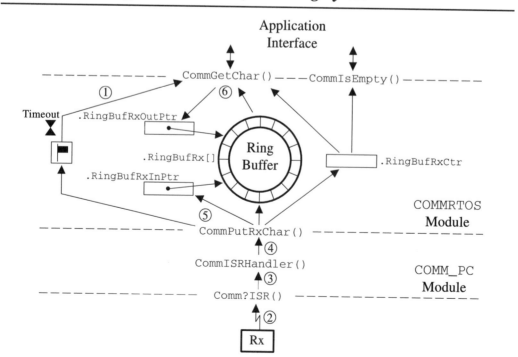

Figure 11.22 shows a flow diagram for data reception using the COMMRTOS module and how COMMRTOS interfaces with the COMM_PC module. Your application still calls CommGetChar() except that your task will be suspended if the buffer is empty. You can specify to CommGetChar() a time-out value to prevent suspending your application task forever. When a byte is received, your task will "wake-up" and will receive the byte from the serial port.

CommPutRxChar() is an interface function between the COMMRTOS module and the COMM_PC module. The COMM_PC module calls this function when a byte is received. CommPutRxChar() deposits the byte into the receive ring buffer but only if the buffer is not already full. The byte is discarded if the buffer is full. When the byte is inserted in the buffer, CommPutRxChar() signals the data reception semaphore to indicate to any pending task that data was received.

To prevent suspending your application code, you can find out whether there are bytes in the ring buffer by calling CommIsEmpty().

Figure 11.23 shows a flow diagram for data transmission using the COMMRTOS module and how it interfaces with the COMM_PC module. Again, everything is identical to the COMMBGND module except for the semaphore. When you want to send data to a serial port, CommPutChar() waits for the semaphore. Because the transmit semaphore is initialized to the size of the buffer when the COMMRTOS module is initialized, CommPutChar() will suspend your application code when there is no more room in the buffer. The suspended task will resume as soon as the UART catches up sending the bytes.

Figure 11.23 Buffered serial I/O, transmitting bytes.

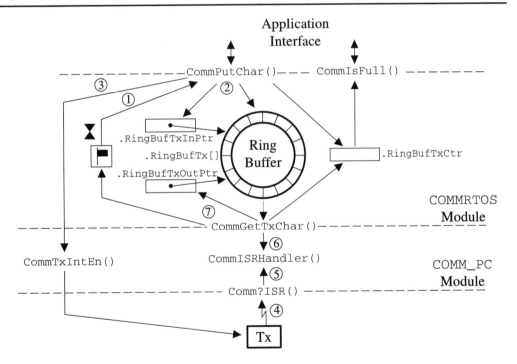

CommGetTxChar() is an interface function between the COMMRTOS module and the COMM_PC module. The COMM_PC module calls this function when a byte has been sent by the UART. Basically, this function says, "Give me the next byte to send." CommGetTxChar() returns the next byte to send from the transmit ring buffer if there is at least one byte in the ring buffer. If the buffer is empty, CommGetTxChar()

returns the NUL character and tells the caller that there is no more data in the buffer. This allows the caller to disable further transmit interrupts until there is more data to send. The data transmit semaphore is signaled when a byte is extracted from the buffer. This indicates to the sending task that there is more room in the transmit buffer.

CommGetChar()

INT8U CommGetChar(INT8U ch, INT16U to, INT8U *err);
(COMMRTOS.C)

CommGetChar() allows your application to extract data from the received data ring buffer.

Arguments

ch is the serial channel and can be either COMM1 or COMM2.

to specifies a timeout (in "clock ticks"). If a byte is not received on the serial port within this time, CommGetChar() will return to your application. Your task will wait for a byte forever when you specify a timeout of 0.

err is a pointer to a variable that will hold status about the outcome of the function. CommGetChar() sets *err to one of the following:

COMM_NO_ERR   if a byte is available from the ring buffer within the timeout period.

COMM_RX_TIMEOUT   if no data is received within the specified timeout.

COMM_BAD_CH   if you do not specify either COMM1 or COMM2.

Return Value

The function returns the oldest byte stored in the ring buffer if the buffer is not empty. If the function times out, CommGetChar() returns the NUL character (i.e., 0x00).

Notes/Warnings

None

11

Example

```
void Task (void *pdata)
{
    INT8U err;

        .
    for (;;) {
          .
        c = CommGetChar(COMM1, 0, &err);
        if (err == COMM_NO_ERR) {
            Process character;
        }
          .
    }
      .
      .
}
```

CommInit()

```
void CommInit(void);
(COMMRTOS.C)
```

CommInit() is used to initialize the COMMRTOS module. This function must be called before any other services provided by this module. CommInit() clears the number of bytes in the ring buffer counter and also initializes both the IN and OUT pointers of each ring buffer to point at the beginning of the data storage area. The data reception semaphore is initialized to 0, indicating that there is no data in the ring buffer. The data transmission semaphore is initialized with the size of the transmit buffer, indicating that the buffer is empty.

Arguments

None

Return Value

None

Notes/Warnings

None

Example

```
void main (void)
{
    .
    .
    CommInit();
    .
    .
}
```

11

CommIsEmpty()

```
BOOLEAN CommIsEmpty(INT8U ch);
(COMMRTOS.C)
```

CommIsEmpty() allows your application to determine if a byte was received on the serial port. This function allows you to avoid task suspension if no data is available.

Arguments

ch is the serial channel and can be either COMM1 or COMM2.

Return Value

The function returns TRUE if no data was received and FALSE if data is available in the ring buffer.

Notes/Warnings

If you specify an incorrect channel number, the function returns TRUE to prevent you from calling CommGetChar() thinking that data is available from an invalid port.

Example

```c
void Task (void *pdata)
{
    INT8U err;


    .

    for (;;) {

        .

        if (CommIsEmpty(COMM1) == FALSE) {
            c = CommGetChar(COMM1, 0, &err); /* Character available */
            Process character;
        }

        .

    }
    .

    .

}
```

CommIsFull()

```
BOOLEAN CommIsFull(INT8U ch);
(COMMRTOS.C)
```

CommIsFull() allows your application code to check the status of the transmit ring buffer. This function allows you to avoid task suspension if the buffer is already full.

Arguments

ch is the serial channel and can be either COMM1 or COMM2.

Return Value

The function returns TRUE when the buffer is full and FALSE otherwise.

Notes/Warnings

If you specify an incorrect channel number, the function returns TRUE to prevent you from calling CommPutChar() thinking that data can be sent to the serial port.

Example

```
void Task (void *pdata)
{
    INT8U  err;
    char   *s;


        .
    for (;;) {
        .
        if (CommIsFull(COMM1) == FALSE) {
            err = CommPutChar(COMM1, '$', 0);
        }
        .
    }
    .
    .
}
```

11

CommPutChar()

UBYTE CommPutChar(INT8U ch, UBYTE ch, INT16U to);
(COMMRTOS.C)

CommPutChar() allows your application to send data to a serial port (one byte at a time). CommPutChar() suspends the calling task if the transmit ring buffer is full. CommPutChar() will resume when a byte is removed from the ring buffer by the transmit ISR.

Arguments

ch is the serial channel and can be either COMM1 or COMM2.

c is the byte that your application sends to the serial port. The byte sent can have any value between 0x00 and 0xFF (i.e., you can send binary data).

to specifies the amount of time (in "clock ticks") that CommPutChar() will wait for the buffer to clear up. If a byte is not transmitted on the serial port within this time, CommPutChar() will return to your application. Your task will wait forever when you specify a timeout of 0.

Return Value

CommPutChar() returns a value representing the outcome of the function call as follows:

COMM_NO_ERR   the byte was placed in the ring buffer and will be sent by the UART if a byte is available from the ring buffer.

COMM_BAD_CH   if you do not specify either COMM1 or COMM2.

COMM_TX_TIMEOUT   indicates that the buffer didn't clear up within the allowed time.

Notes/Warnings

If you configured the serial port to 7 data bits then you will not be able to send binary data.

Example

```
char Message[] = "Hello World!";



void Task (void)
{
    INT8U  err;
    char  *s;


    .
    for (;;) {
        .
        s    = &Message[0];
        err  = COMM_NO_ERR;
        while (*s && err == COMM_NO_ERR) {
            err = CommPutChar(COMM1, *s++, 0);
        }
        .
    }
    .
    .
}
```

11.08 Configuration

Configuration of the communications driver is very simple because all you have to do is change a few #defines to accomodate your environment.

COMM_PC.H (or CFG.H):

COMM1_BASE and COMM2_BASE are the base port address for the PC's COM1 and COM2. In most cases, you will not have to change these.

COMM_MAX_RX sets the number of bytes that the UART buffers internally. For the NS16550 UART, you should set this constant to 16 because the NS16550 can be receiving a byte while another byte is waiting to be processed by the CPU.

COMMBGND.H, COMMRTOS.H (or CFG.H):

COMM_RX_BUF_SIZE sets the size of the receive ring buffer for both serial ports. The size of the receive buffer can be as large as 65534 bytes.

COMM_TX_BUF_SIZE sets the size of the transmit ring buffer for both serial ports. As with the receive ring buffer, the size can be as large as 65534 bytes.

11.09 How to use the COMM_PC and the COMMBGND Module

If you write a foreground/background application you will need to use the COMM_PC (assuming you are using a PC) and the COMMBGND modules. The first thing you need to do is to configure the module by setting the value of the #defines described in section 11.08. Next, you will need to call functions to initialize the modules and the serial port(s) that you are planning on using. For example, if you are using the PC's COM1, you would need to have the following code:

```
void main(void)
{
    .
    .
    .
    CommInit();                    /* Initialize COMMBGND        */
    .
    .
    .
    CommCfgPort(COMM1, 9600, 8, COMM_PARITY_NONE, 1);
    CommSetIntVect(COMM1);    /* Install the interrupt vector  */
    CommRxIntEn(COMM1);       /* Enable receive interrupts     */
    .
    .
    .
}
```

You should note that you don't need to enable transmit interrupts because this is done automatically when you send data on the serial port. When all your initialization is done, your background loop could check for incoming data, as shown.

```c
void main(void)
{
    INT8U c;
    INT8U err;
    .
    .
    /* Initialization code described above ---------------------------*/
    .
    .
    while (1) {                     /* Background loop (infinite loop)      */
        .
        .
        if (!CommIsEmpty(COMM1)) {
            c = CommGetChar(COMM1, &err);
            if (err == COMM_NO_ERR) {
                /* Process received data ------------------------------*/
                .
                .
                CommPutChar(COMM1, ???);  /* Send response              */
            } else {
                /* Process communications error ---------------------*/
            }
        }
    }
}
```

11.10 How to use the COMM_PC and the COMMRTOS Module

If you write an application using a real-time kernel you will need to use the COMM_PC (assuming you are using a PC) and the COMMRTOS modules. Again, the first thing you need to do is to configure the module by setting the value of the #defines described in section 11.08. Your startup code will need to create the task(s) that will be responsible for servicing the serial port(s). You should have one task for each serial port. The following segment of code is used to create the task that will handle COM1. You should consult TEST.C (see Chapter 1) to see what else you need to properly initialize µC/OS-II.

```
#define COMM_TASK_PRIO    20        /* Define the priority of the task      */


OS_STK    CommTaskStk[512];


void main(void)
{
    .
    OSInit()                         /* Initialize the O.S. (uC/OS-II)       */
    .

    .
    OSTaskCreate(CommTask, (void *)0, &CommTaskStk[511], COMM_TASK_PRIO);
    .

    .
    OSStart();
}
```

You should initialize the serial communications code from within the task that will handle the port(s). Using the PC's COM1, you would have the following code:

```
void CommTask(void *pdata)
{
    INT8U c;
    INT8U err;


    CommInit();                           /* Initialize COMMRTOS          */
    CommCfgPort(COMM1, 9600, 8 COMM_PARITY_NONE, 1);
    CommSetIntVect(COMM1);                /* Install the interrupt vector */
    CommRxIntEn(COMM1);                   /* Enable receive interrupts    */
    for (;;) {
        c = CommGetChar(COMM1, 0, &err);
        if (err == COMM_NO_ERR) {
            /* Process received byte ----------------------------------*/
            .

            .
            CommPutChar(COMM1, ..);    /* Send response                */
            .
        } else {
            /* Process communication error ---------------------------*/
        }
        .

        .
    }
}
```

11.11 Bibliography

Campbell, Joe
C Programmer's Guide to Serial Communications (Second Edition)
Sams Publishing, 1993
Indianapolis, Indiana
ISBN 0-672-30286-1

Choiser, John P., Foster, John O.
The XT-AT Handbook
Annabooks, 1993
ISBN 0-929392-00-0

Erdelsky, Philip
"PC Interrupt-Driven Serial I/O"
From the book: *MS-DOS System Programming*
R&D Publications, 1990
ISBN 0-923667-20-2

Halsall, Fred
Data Communications, Computer Networks and Open Systems (Third Edition)
Addison-Wesley, 1992
ISBN 0-201-56506-4

Pippenger, D.E. and Tobaben, E.J.
Linear and Interface Circuits Applications
Volume 2: Line Circuits and Display Drivers
Texas Instruments, 1985
ISBN 0-89512-185-9

Tanenbaum, Andrew S.
Computer Networks (Second Edition)
PTR Prentice-Hall, Inc., 1989
ISBN 0-13-162959-X

11

Listing 11.1 COMM_PC.C

```
/*
*********************************************************************************************
*                           Embedded Systems Building Blocks
*                          Complete and Ready-to-Use Modules in C
*
*                              Asynchronous Serial Communications
*                              IBM-PC Serial I/O Low Level Driver
*
*                          (c) Copyright 1999, Jean J. Labrosse, Weston, FL
*                                     All Rights Reserved
*
* Filename    : COMM_PC.C
* Programmer  : Jean J. Labrosse
*
* Notes       : 1) The code in this file assumes that you are using a National Semiconductor NS16450 (most
*                  PCs do or, an Intel i82C50) serial communications controller.
*
*               2) The functions (actually macros) OS_ENTER_CRITICAL() and OS_EXIT_CRITICAL() are used to
*                  disable and enable interrupts, respectively.  If using the Borland C++ compiler V3.1,
*                  all you need to do is to define these macros as follows:
*
*                      #define OS_ENTER_CRITICAL()  disable()
*                      #define OS_EXIT_CRITICAL()   enable()
*
*               3) You will need to define the following constants:
*                      COMM1_BASE    is the base address of COM1 on your PC (typically 0x03F8)
*                      COMM2_BASE    is the base address of COM2 on your PC (typically 0x02F8)
*                      COMM_MAX_RX   is the number of characters buffered by the UART
*                                      2 for the NS16450
*                                     16 for the NS16550
*
*               4) COMM_BAD_CH, COMM_NO_ERR and COMM_TX_EMPTY,
*                  COMM_NO_PARITY, COMM_ODD_PARITY and COMM_EVEN_PARITY
*                     are all defined in other modules (i.e. COMM1.H, COMM2.H or COMM3.H)
*********************************************************************************************
*/

/*
*********************************************************************************************
*                                        INCLUDES
*********************************************************************************************
*/

#include "includes.h"

/*$PAGE*/
```

Listing 11.1 *(continued)* COMM_PC.C

```
/*
*********************************************************************************************
*                                       CONSTANTS
*********************************************************************************************
*/

#define  BIT0                     0x01
#define  BIT1                     0x02
#define  BIT2                     0x04
#define  BIT3                     0x08
#define  BIT4                     0x10
#define  BIT5                     0x20
#define  BIT6                     0x40
#define  BIT7                     0x80

#define  PIC_INT_REG_PORT         0x0020
#define  PIC_MSK_REG_PORT         0x0021

#define  COMM_UART_RBR              0
#define  COMM_UART_THR              0
#define  COMM_UART_DIV_LO           0
#define  COMM_UART_DIV_HI           1
#define  COMM_UART_IER              1
#define  COMM_UART_IIR              2
#define  COMM_UART_LCR              3
#define  COMM_UART_MCR              4
#define  COMM_UART_LSR              5
#define  COMM_UART_MSR              6
#define  COMM_UART_SCR              7

/*
*********************************************************************************************
*                                  LOCAL GLOBAL VARIABLES
*********************************************************************************************
*/

static  INT16U  Comm1ISROldOffset;
static  INT16U  Comm1ISROldSegment;

static  INT16U  Comm2ISROldOffset;
static  INT16U  Comm2ISROldSegment;

/*$PAGE*/
```

11

Listing 11.1 (continued) `COMM_PC.C`

```
/*
*******************************************************************************************************
*                                         CONFIGURE PORT
*
* Description : This function is used to configure a serial I/O port.  This code is for IBM-PCs and
*               compatibles and assumes a National Semiconductor NS16450.
*
* Arguments   : 'ch'          is the COMM port channel number and can either be:
*                                  COMM1
*                                  COMM2
*               'baud'        is the desired baud rate (anything, standard rates or not)
*               'bits'        defines the number of bits used and can be either 5, 6, 7 or 8.
*               'parity'      specifies the 'parity' to use:
*                                  COMM_PARITY_NONE
*                                  COMM_PARITY_ODD
*                                  COMM_PARITY_EVEN
*               'stops'       defines the number of stop bits used and can be either 1 or 2.
*
* Returns     : COMM_NO_ERR   if the channel has been configured.
*               COMM_BAD_CH   if you have specified an incorrect channel.
*
* Notes       : 1) Refer to the NS16450 Data sheet
*               2) The constant 115200 is based on a 1.8432 MHz crystal oscillator and a 16 x Clock.
*               3) 'lcr' is the Line Control Register and is define as:
*
*                      B7  B6  B5  B4  B3  B2  B1  B0
*                                          ------ #Bits  (00 = 5, 01 = 6, 10 = 7 and 11 = 8)
*                                      --          #Stops (0 = 1 stop, 1 = 2 stops)
*                                  --              Parity enable (1 = parity is enabled)
*                              --                  Even parity when set to 1.
*                          --                      Stick parity (see 16450 data sheet)
*                      --                          Break control (force break when 1)
*                  --                              Divisor access bit (set to 1 to access divisor)
*               4) This function enables Rx interrupts but not Tx interrupts.
*******************************************************************************************************
*/
```

Listing 11.1 (continued) `COMM_PC.C`

```c
INT8U  CommCfgPort (INT8U ch, INT16U baud, INT8U bits, INT8U parity, INT8U stops)
{
    INT16U  div;                                    /* Baud rate divisor                          */
    INT8U   divlo;
    INT8U   divhi;
    INT8U   lcr;                                    /* Line Control Register                      */
    INT16U  base;                                   /* COMM port base address                     */


    switch (ch) {                                   /* Obtain base address of COMM port           */
        case COMM1:
            base = COMM1_BASE;
            break;

        case COMM2:
            base = COMM2_BASE;
            break;

        default:
            return (COMM_BAD_CH);
    }

    div   = (INT16U)(115200L / (INT32U)baud);       /* Compute divisor for desired baud rate      */
    divlo = div & 0x00FF;                           /* Split divisor into LOW and HIGH bytes      */
    divhi = (div >> 8) & 0x00FF;
    lcr   = ((stops - 1) << 2) + (bits - 5);
    switch (parity) {
        case COMM_PARITY_ODD:
            lcr |= 0x08;                            /* Odd  parity                                */
            break;

        case COMM_PARITY_EVEN:
            lcr |= 0x18;                            /* Even parity                                */
            break;
    }
    OS_ENTER_CRITICAL();
    outp(base + COMM_UART_LCR, BIT7);               /* Set divisor access bit                     */
    outp(base + COMM_UART_DIV_LO, divlo);           /* Load divisor                               */
    outp(base + COMM_UART_DIV_HI, divhi);
    outp(base + COMM_UART_LCR, lcr);                /* Set line control register (Bit 8 is 0)     */
    outp(base + COMM_UART_MCR, BIT3 | BIT1 | BIT0); /* Assert DTR and RTS and, allow interrupts   */
    outp(base + COMM_UART_IER, 0x00);               /* Disable both Rx and Tx interrupts          */
    OS_EXIT_CRITICAL();
    CommRxFlush(ch);                                /* Flush the Rx input                         */
    return (COMM_NO_ERR);
}

/*$PAGE*/
```

11

Listing 11.1 (continued) `COMM_PC.C`

```
/*
*********************************************************************************************
*                                    COMM ISR HANDLER
*
* Description : This function processes an interrupt from a COMM port.  The function verifies whether the
*               interrupt comes from a received character, the completion of a transmitted character or
*               both.
* Arguments   : 'ch'    is the COMM port channel number and can either be:
*                          COMM1
*                          COMM2
* Notes       : 'switch' statements are used for expansion.
*********************************************************************************************
*/
void  CommISRHandler (INT8U ch)
{
    INT8U   c;
    INT8U   iir;                                    /* Interrupt Identification Register (IIR)   */
    INT8U   stat;
    INT16U  base;                                   /* COMM port base address                    */
    INT8U   err;
    INT8U   max;                                    /* Max. number of interrupts serviced        */
```

Listing 11.1 (continued)   `COMM_PC.C`

```
switch (ch) {                                      /* Obtain pointer to communications channel  */
    case COMM1:
        base = COMM1_BASE;
        break;

    case COMM2:
        base = COMM2_BASE;
        break;

    default:
        base = COMM1_BASE;
        break;
}

max = COMM_MAX_RX;
iir = (INT8U)inp(base + COMM_UART_IIR) & 0x07;     /* Get contents of IIR                       */
while (iir != 1 && max > 0) {                       /* Process ALL interrupts                    */
    switch (iir) {
        case 0:                                    /* See if we have a Modem Status interrupt   */
            c   = (INT8U)inp(base + COMM_UART_MSR); /* Clear interrupt (do nothing about it!)    */
            break;

        case 2:                                    /* See if we have a Tx interrupt             */
            c = CommGetTxChar(ch, &err);           /* Get next character to send.               */
            if (err == COMM_TX_EMPTY) {            /* Do we have anymore characters to send ?   */
                                                   /* No, Disable Tx interrupts                 */
                stat = (INT8U)inp(base + COMM_UART_IER) & ~BIT1;
                outp(base + COMM_UART_IER, stat);
            } else {
                outp(base + COMM_UART_THR, c);     /* Yes, Send character                       */
            }
            break;

        case 4:                                    /* See if we have an Rx interrupt            */
            c   = (INT8U)inp(base + COMM_UART_RBR); /* Process received character                */
            CommPutRxChar(ch, c);                  /* Insert received character into buffer     */
            break;

        case 6:                                    /* See if we have a Line Status interrupt    */
            c   = (INT8U)inp(base + COMM_UART_LSR); /* Clear interrupt (do nothing about it!)    */
            break;
    }
    iir = (INT8U)inp(base + COMM_UART_IIR) & 0x07;  /* Get contents of IIR                       */
    max--;
}
switch (ch) {
    case COMM1:
    case COMM2:
        outp(PIC_INT_REG_PORT, 0x20);              /* Reset interrupt controller                */
        break;

    default:
        outp(PIC_INT_REG_PORT, 0x20);
        break;
}
}
```

11

Listing 11.1 (continued) COMM_PC.C

```
/*
*********************************************************************************************
*                              RESTORE OLD INTERRUPT VECTOR
*
* Description : This function restores the old interrupt vector for the desired communications channel.
* Arguments   : 'ch'    is the COMM port channel number and can either be:
*                             COMM1
*                             COMM2
* Note(s)     : This function assumes that the 80x86 is running in REAL mode.
*********************************************************************************************
*/

void  CommRclIntVect (INT8U ch)
{
    INT16U  *pvect;


    switch (ch) {
        case COMM1:
            pvect     = (INT16U *)MK_FP(0x0000, 0x0C << 2);     /* Point to proper IVT location    */
            OS_ENTER_CRITICAL();
            *pvect++ = Comm1ISROldOffset;                       /* Restore saved vector            */
            *pvect     = Comm1ISROldSegment;
            OS_EXIT_CRITICAL();
            break;

        case COMM2:
            pvect     = (INT16U *)MK_FP(0x0000, 0x0B << 2);     /* Point to proper IVT location    */
            OS_ENTER_CRITICAL();
            *pvect++ = Comm2ISROldOffset;                       /* Restore saved vector            */
            *pvect     = Comm2ISROldSegment;
            OS_EXIT_CRITICAL();
            break;
    }
}

/*$PAGE*/
```

Listing 11.1 (continued)   COMM_PC.C

```
/*
*********************************************************************************************
*                                    FLUSH RX PORT
*
* Description : This function is called to flush any input characters still in the receiver.  This
*               function is useful when you replace the NS16450 with the more powerful NS16450.
* Arguments   : 'ch'    is the COMM port channel number and can either be:
*                           COMM1
*                           COMM2
*********************************************************************************************
*/

void  CommRxFlush (INT8U ch)
{
    INT8U  ctr;
    INT16U base;


    switch (ch) {
        case COMM1:
             base = COMM1_BASE;
             break;

        case COMM2:
             base = COMM2_BASE;
             break;
    }
    ctr = COMM_MAX_RX;                                 /* Flush Rx input              */
    OS_ENTER_CRITICAL();
    while (ctr-- > 0) {
        inp(base + 0);
    }
    OS_EXIT_CRITICAL();
}

/*$PAGE*/
```

11

Listing 11.1 (continued) COMM_PC.C

```c
/*
*********************************************************************************
*                              DISABLE RX INTERRUPTS
*
* Description : This function disables the Rx interrupt.
* Arguments   : 'ch'    is the COMM port channel number and can either be:
*                       COMM1
*                       COMM2
*********************************************************************************
*/

void  CommRxIntDis (INT8U ch)
{
    INT8U stat;


    switch (ch) {
        case COMM1:
            OS_ENTER_CRITICAL();
                                                    /* Disable Rx interrupts          */
            stat = (INT8U)inp(COMM1_BASE + COMM_UART_IER) & ~BIT0;
            outp(COMM1_BASE + COMM_UART_IER, stat);
            if (stat == 0x00) {                     /* Both Tx & Rx interrupts are disabled ?   */
                                                    /* Yes, disable IRQ4 on the PC    */
                outp(PIC_MSK_REG_PORT, (INT8U)inp(PIC_MSK_REG_PORT) | BIT4);
            }
            OS_EXIT_CRITICAL();
            break;

        case COMM2:
            OS_ENTER_CRITICAL();
                                                    /* Disable Rx interrupts          */
            stat = (INT8U)inp(COMM2_BASE + COMM_UART_IIR) & ~BIT0;
            outp(COMM2_BASE + COMM_UART_IER, stat);
            if (stat == 0x00) {                     /* Both Tx & Rx interrupts are disabled ?   */
                                                    /* Yes, disable IRQ3 on the PC    */
                outp(PIC_MSK_REG_PORT, (INT8U)inp(PIC_MSK_REG_PORT) | BIT3);
            }
            OS_EXIT_CRITICAL();
            break;
    }
}

/*$PAGE*/
```

Listing 11.1 (continued) `COMM_PC.C`

```
/*
*********************************************************************************************
*                                    ENABLE RX INTERRUPTS
*
* Description : This function enables the Rx interrupt.
* Arguments   : 'ch'    is the COMM port channel number and can either be:
*                              COMM1
*                              COMM2
*********************************************************************************************
*/

void  CommRxIntEn (INT8U ch)
{
    INT8U stat;


    switch (ch) {
        case COMM1:
            OS_ENTER_CRITICAL();
                                                     /* Enable Rx interrupts            */
            stat = (INT8U)inp(COMM1_BASE + COMM_UART_IER) | BIT0;
            outp(COMM1_BASE + COMM_UART_IER, stat);
                                                     /* Enable IRQ4 on the PC           */
            outp(PIC_MSK_REG_PORT, (INT8U)inp(PIC_MSK_REG_PORT) & ~BIT4);
            OS_EXIT_CRITICAL();
            break;

        case COMM2:
            OS_ENTER_CRITICAL();
                                                     /* Enable Rx interrupts            */
            stat = (INT8U)inp(COMM2_BASE + COMM_UART_IER) | BIT0;
            outp(COMM2_BASE + COMM_UART_IER, stat);
                                                     /* Enable IRQ3 on the PC           */
            outp(PIC_MSK_REG_PORT, (INT8U)inp(PIC_MSK_REG_PORT) & ~BIT3);
            OS_EXIT_CRITICAL();
            break;
    }
}

/*$PAGE*/
```

11

Listing 11.1 (continued)   COMM_PC.C

```
/*
*********************************************************************************************
*                                    SET INTERRUPT VECTOR
*
* Description : This function installs the interrupt vector for the desired communications channel.
* Arguments   : 'ch'    is the COMM port channel number and can either be:
*                            COMM1
*                            COMM2
* Note(s)     : This function assumes that the 80x86 is running in REAL mode.
*********************************************************************************************
*/

void  CommSetIntVect (INT8U ch)
{
    INT16U   segment;
    INT16U   offset;
    INT16U  *pvect;


    switch (ch) {
        case COMM1:
            pvect          = (INT16U *)MK_FP(0x0000, 0x0C << 2);  /* Point to proper IVT location     */
            OS_ENTER_CRITICAL();
            Comm1ISROldOffset  = *pvect++;                        /* Save current vector              */
            Comm1ISROldSegment = *pvect;
            pvect--;
            *pvect++           = FP_OFF(Comm1ISR);                /* Set new vector                   */
            *pvect             = FP_SEG(Comm1ISR);
            OS_EXIT_CRITICAL();
            break;

        case COMM2:
            pvect          = (INT16U *)MK_FP(0x0000, 0x0B << 2);  /* Point to proper IVT location     */
            OS_ENTER_CRITICAL();
            Comm2ISROldOffset  = *pvect++;                        /* Save current vector              */
            Comm2ISROldSegment = *pvect;
            pvect--;
            *pvect++           = FP_OFF(Comm2ISR);                /* Set new vector                   */
            *pvect             = FP_SEG(Comm2ISR);
            OS_EXIT_CRITICAL();
            break;
    }
}

/*$PAGE*/
```

Listing 11.1 *(continued)* COMM_PC.C

```
/*
*********************************************************************************************************
*                                        DISABLE TX INTERRUPTS
*
* Description : This function disables the character transmission.
* Arguments   : 'ch'    is the COMM port channel number and can either be:
*                           COMM1
*                           COMM2
*********************************************************************************************************
*/

void  CommTxIntDis (INT8U ch)
{
    INT8U   stat;
    INT8U   cmd;


    switch (ch) {
        case COMM1:
            OS_ENTER_CRITICAL();
                                                        /* Disable Tx interrupts                    */
            stat = (INT8U)inp(COMM1_BASE + COMM_UART_IER) & ~BIT1;
            outp(COMM1_BASE + COMM_UART_IER, stat);
            if (stat == 0x00) {                         /* Both Tx & Rx interrupts are disabled ?   */
                cmd = (INT8U)inp(PIC_MSK_REG_PORT) | BIT4;
                outp(PIC_MSK_REG_PORT, cmd);            /* Yes, disable IRQ4 on the PC              */
            }
            OS_EXIT_CRITICAL();
            break;

        case COMM2:
            OS_ENTER_CRITICAL();
                                                        /* Disable Tx interrupts                    */
            stat = (INT8U)inp(COMM2_BASE + COMM_UART_IER) & ~BIT1;
            outp(COMM2_BASE + COMM_UART_IER, stat);
            if (stat == 0x00) {                         /* Both Tx & Rx interrupts are disabled ?   */
                cmd = (INT8U)inp(PIC_MSK_REG_PORT) | BIT3;
                outp(PIC_MSK_REG_PORT, cmd);            /* Yes, disable IRQ3 on the PC              */
            }
            OS_EXIT_CRITICAL();
            break;
    }
}
/*$PAGE*/
```

11

Listing 11.1 (continued) COMM_PC.C

```c
/*
*********************************************************************************************
*                                  ENABLE TX INTERRUPTS
*
* Description : This function enables transmission of characters.  Transmission of characters is
*               interrupt driven.  If you are using a multi-drop driver, the code must enable the driver
*               for transmission.
* Arguments   : 'ch'   is the COMM port channel number and can either be:
*                         COMM1
*                         COMM2
*********************************************************************************************
*/

void  CommTxIntEn (INT8U ch)
{
    INT8U   stat;
    INT8U   cmd;


    switch (ch) {
        case COMM1:
             OS_ENTER_CRITICAL();
             stat = (INT8U)inp(COMM1_BASE + COMM_UART_IER) | BIT1;     /* Enable Tx inter-
rupts            */
             outp(COMM1_BASE + COMM_UART_IER, stat);
             cmd  = (INT8U)inp(PIC_MSK_REG_PORT) & ~BIT4;
            outp(PIC_MSK_REG_PORT, cmd);                             /* Enable IRQ4 on the PC
*/
             OS_EXIT_CRITICAL();
             break;

        case COMM2:
             OS_ENTER_CRITICAL();
             stat = (INT8U)inp(COMM2_BASE + COMM_UART_IER) | BIT1;     /* Enable Tx inter-
rupts            */
             outp(COMM2_BASE + COMM_UART_IER, stat);
             cmd  = (INT8U)inp(PIC_MSK_REG_PORT) & ~BIT3;
            outp(PIC_MSK_REG_PORT, cmd);                             /* Enable IRQ3 on the PC
*/
             OS_EXIT_CRITICAL();
             break;
    }
}
```

Listing 11.2   COMM_PC.H

```
/*
*********************************************************************************************
*                              Embedded Systems Building Blocks
*                             Complete and Ready-to-Use Modules in C
*
*                              Asynchronous Serial Communications
*                              IBM-PC Serial I/O Low Level Driver
*
*                          (c) Copyright 1999, Jean J. Labrosse, Weston, FL
*                                       All Rights Reserved
*
* Filename   : COMM_PC.H
* Programmer : Jean J. Labrosse
*
*********************************************************************************************
*/


/*
*********************************************************************************************
*                                 CONFIGURATION CONSTANTS
*********************************************************************************************
*/

#ifndef  CFG_H

#define  COMM1_BASE        0x03F8             /* Base address of PC's COM1                 */
#define  COMM2_BASE        0x02F8             /* Base address of PC's COM2                 */

#define  COMM_MAX_RX       2                  /* NS16450 has 2 byte buffer                 */

#endif


/*
*********************************************************************************************
*                                 FUNCTION PROTOTYPES
*********************************************************************************************
*/

void     Comm1ISR(void);
void     Comm2ISR(void);
INT8U    CommCfgPort(INT8U ch, INT16U baud, INT8U bits, INT8U parity, INT8U stops);
void     CommISRHandler(INT8U ch);
void     CommRxFlush(INT8U ch);
void     CommRxIntDis(INT8U ch);
void     CommRxIntEn(INT8U ch);
void     CommTxIntDis(INT8U ch);
void     CommTxIntEn(INT8U ch);
void     CommRclIntVect(INT8U ch);
void     CommSetIntVect(INT8U ch);
```

11

Listing 11.3   COMM_PCA.ASM

```
;***********************************************************************************************
;                              Embedded Systems Building Blocks
;                              Complete and Ready-to-Use Modules in C
;
;                              Asynchronous Serial Communications
;                              IBM-PC Serial I/O Low Level Driver
;
;                         (c) Copyright 1999, Jean J. Labrosse, Weston, FL
;                                       All Rights Reserved
;
; Filename   : COMM_PCA.ASM
; Programmer : Jean J. Labrosse
; Notes      : If you are not using uC/OS-II you will need to DELETE the increments of OSIntNesting and
;              the calls to OSIntExit().
;***********************************************************************************************

            PUBLIC  _Comm1ISR
            PUBLIC  _Comm2ISR

            EXTRN   _OSIntExit:FAR
            EXTRN   _CommISRHandler:FAR

            EXTRN   _OSIntNesting:BYTE

.MODEL      LARGE
.CODE
.186

;/*$PAGE*/
```

Listing 11.3 (continued) COMM_PCA.ASM

```
;*********************************************************************************************
;                                    HANDLE COM1 ISR
;*********************************************************************************************
;
_Comm1ISR  PROC   FAR
;
           PUSHA                                 ; Save interrupted task's context
           PUSH   ES
           PUSH   DS
;
           MOV    AX, DGROUP                     ; Reload DS with DGROUP
           MOV    DS, AX
;
;          NOTE: Comment OUT the next line (i.e. INC _OSIntNesting) if you don't use uC/OS-II.
           INC    BYTE PTR _OSIntNesting         ; Notify uC/OS-II of ISR
;
           PUSH   1                              ; Indicate COMM1
           CALL   FAR PTR _CommISRHandler        ; Process COMM interrupt
           ADD    SP,2
;
;          NOTE: Comment OUT the next line (i.e. CALL _OSIntExit) if you don't use uC/OS-II.
           CALL   FAR PTR _OSIntExit             ; Notify OS of end of ISR
;
           POP    DS                             ; Restore interrupted task's context
           POP    ES
           POPA
;
           IRET                                  ; Return to interrupted task
;
_Comm1ISR  ENDP
;
;/*$PAGE*/
```

11

Listing 11.3 (continued) COMM_PCA.ASM

```
;********************************************************************************
;                                   HANDLE COM2 ISR
;********************************************************************************
;
_Comm2ISR    PROC    FAR
;
             PUSHA                                  ; Save interrupted task's context
             PUSH  ES
             PUSH  DS
;
             MOV   AX, DGROUP                       ; Reload DS with DGROUP
             MOV   DS, AX
;
;            NOTE: Comment OUT the next line (i.e. INC _OSIntNesting) if you don't use uC/OS-II.
             INC   BYTE PTR _OSIntNesting           ; Notify uC/OS-II of ISR
;
             PUSH  2                                ; Indicate COMM2
             CALL  FAR PTR _CommISRHandler          ; Process COMM interrupt
             ADD   SP,2
;
;            NOTE: Comment OUT the next line (i.e. CALL _OSIntExit) if you don't use uC/OS-II.
             CALL  FAR PTR _OSIntExit               ; Notify OS of end of ISR
;
             POP   DS                               ; Restore interrupted task's context
             POP   ES
             POPA
;
             IRET                                   ; Return to interrupted task
;
_Comm2ISR    ENDP
;
             END
```

Listing 11.4   COMMBGND.C

```
/*
********************************************************************************************************
*                                   Embedded Systems Building Blocks
*                                   Complete and Ready-to-Use Modules in C
*
*                                   Asynchronous Serial Communications
*                                          Buffered Serial I/O
*                                      (Foreground/Background Systems)
*
*                            (c) Copyright 1999, Jean J. Labrosse, Weston, FL
*                                        All Rights Reserved
*
* Filename    : COMMBGND.C
* Programmer  : Jean J. Labrosse
*
* Notes       : The functions (actually macros) OS_ENTER_CRITICAL() and OS_EXIT_CRITICAL() are used to
*               disable and enable interrupts, respectively.  If using the Borland C++ compiler V3.1,
*               all you need to do is to define these macros as follows:
*
*                       #define OS_ENTER_CRITICAL()  disable()
*                       #define OS_EXIT_CRITICAL()   enable()
********************************************************************************************************
*/

/*
********************************************************************************************************
*                                             INCLUDES
********************************************************************************************************
*/

#include "includes.h"

/*$PAGE*/
```

11

Listing 11.4 *(continued)* COMMBGND.C

```
/*
*************************************************************************************************
*                                          CONSTANTS
*************************************************************************************************
*/

/*
*************************************************************************************************
*                                          DATA TYPES
*************************************************************************************************
*/

typedef struct {
    INT16U  RingBufRxCtr;                    /* Number of characters in the Rx ring buffer       */
    INT8U  *RingBufRxInPtr;                  /* Pointer to where next character will be inserted  */
    INT8U  *RingBufRxOutPtr;                 /* Pointer from where next character will be extracted */
    INT8U   RingBufRx[COMM_RX_BUF_SIZE];     /* Ring buffer character storage (Rx)                */
    INT16U  RingBufTxCtr;                    /* Number of characters in the Tx ring buffer       */
    INT8U  *RingBufTxInPtr;                  /* Pointer to where next character will be inserted  */
    INT8U  *RingBufTxOutPtr;                 /* Pointer from where next character will be extracted */
    INT8U   RingBufTx[COMM_TX_BUF_SIZE];     /* Ring buffer character storage (Tx)                */
} COMM_RING_BUF;

/*
*************************************************************************************************
*                                          GLOBAL VARIABLES
*************************************************************************************************
*/

COMM_RING_BUF  Comm1Buf;
COMM_RING_BUF  Comm2Buf;

/*$PAGE*/
```

Listing 11.4 (continued) COMMBGND.C

```
/*
*********************************************************************************************
*                                REMOVE CHARACTER FROM RING BUFFER
*
*
* Description : This function is called by your application to obtain a character from the communications
*               channel.
* Arguments   : 'ch'   is the COMM port channel number and can either be:
*                           COMM1
*                           COMM2
*               'err'  is a pointer to where an error code will be placed:
*                           *err is set to COMM_NO_ERR   if a character is available
*                           *err is set to COMM_RX_EMPTY if the Rx buffer is empty
*                           *err is set to COMM_BAD_CH   if you have specified an invalid channel
* Returns     : The character in the buffer (or NUL if the buffer is empty)
*********************************************************************************************
*/

INT8U  CommGetChar (INT8U ch, INT8U *err)
{
    INT8U           c;
    COMM_RING_BUF  *pbuf;


    switch (ch) {                                      /* Obtain pointer to communications channel */
        case COMM1:
             pbuf = &Comm1Buf;
             break;

        case COMM2:
             pbuf = &Comm2Buf;
             break;

        default:
             *err = COMM_BAD_CH;
             return (NUL);
    }
    OS_ENTER_CRITICAL();
    if (pbuf->RingBufRxCtr > 0) {                       /* See if buffer is empty                  */
        pbuf->RingBufRxCtr--;                           /* No, decrement character count          */
        c = *pbuf->RingBufRxOutPtr++;                   /* Get character from buffer              */
        if (pbuf->RingBufRxOutPtr == &pbuf->RingBufRx[COMM_RX_BUF_SIZE]) {    /* Wrap OUT pointer  */
            pbuf->RingBufRxOutPtr = &pbuf->RingBufRx[0];
        }
        OS_EXIT_CRITICAL();
        *err = COMM_NO_ERR;
        return (c);
    } else {
        OS_EXIT_CRITICAL();
        *err = COMM_RX_EMPTY;
        c    = NUL;                                     /* Buffer is empty, return NUL            */
        return (c);
    }
}

/*$PAGE*/
```

11

Listing 11.4 (continued)   COMMBGND.C

```
/*
*********************************************************************************************
*                              GET TX CHARACTER FROM RING BUFFER
*
*
* Description : This function is called by the Tx ISR to extract the next character from the Tx buffer.
*               The function returns FALSE if the buffer is empty after the character is extracted from
*               the buffer.  This is done to signal the Tx ISR to disable interrupts because this is the
*               last character to send.
* Arguments   : 'ch'   is the COMM port channel number and can either be:
*                          COMM1
*                          COMM2
*               'err'  is a pointer to where an error code will be deposited:
*                          *err is set to COMM_NO_ERR       if at least one character was left in the
*                                                           buffer.
*                          *err is set to COMM_TX_EMPTY     if the Tx buffer is empty.
*                          *err is set to COMM_BAD_CH       if you have specified an incorrect channel
* Returns     : The next character in the Tx buffer or NUL if the buffer is empty.
*********************************************************************************************
*/

INT8U  CommGetTxChar (INT8U ch, INT8U *err)
{
    INT8U          c;
    COMM_RING_BUF *pbuf;


    switch (ch) {                                       /* Obtain pointer to communications channel */
        case COMM1:
             pbuf = &Comm1Buf;
             break;

        case COMM2:
             pbuf = &Comm2Buf;
             break;

        default:
             *err = COMM_BAD_CH;
             return (NUL);
    }
    if (pbuf->RingBufTxCtr > 0) {                        /* See if buffer is empty                   */
        pbuf->RingBufTxCtr--;                            /* No, decrement character count           */
        c = *pbuf->RingBufTxOutPtr++;                    /* Get character from buffer               */
        if (pbuf->RingBufTxOutPtr == &pbuf->RingBufTx[COMM_TX_BUF_SIZE]) {    /* Wrap OUT pointer    */
            pbuf->RingBufTxOutPtr = &pbuf->RingBufTx[0];
        }
        *err = COMM_NO_ERR;
        return (c);                                      /* Characters are still available          */
    } else {
        *err = COMM_TX_EMPTY;
        return (NUL);                                    /* Buffer is empty                         */
    }
}

/*$PAGE*/
```

Listing 11.4 (continued) `COMMBGND.C`

```
/*
*********************************************************************************************
*                              INITIALIZE COMMUNICATIONS MODULE
*
*
* Description : This function is called by your application to initialize the communications module.  You
*               must call this function before calling any other functions.
* Arguments   : none
*********************************************************************************************
*/

void  CommInit (void)
{
    COMM_RING_BUF *pbuf;


    pbuf                 = &Comm1Buf;                   /* Initialize the ring buffer for COMM1     */
    pbuf->RingBufRxCtr    = 0;
    pbuf->RingBufRxInPtr  = &pbuf->RingBufRx[0];
    pbuf->RingBufRxOutPtr = &pbuf->RingBufRx[0];
    pbuf->RingBufTxCtr    = 0;
    pbuf->RingBufTxInPtr  = &pbuf->RingBufTx[0];
    pbuf->RingBufTxOutPtr = &pbuf->RingBufTx[0];

    pbuf                 = &Comm2Buf;                   /* Initialize the ring buffer for COMM2     */
    pbuf->RingBufRxCtr    = 0;
    pbuf->RingBufRxInPtr  = &pbuf->RingBufRx[0];
    pbuf->RingBufRxOutPtr = &pbuf->RingBufRx[0];
    pbuf->RingBufTxCtr    = 0;
    pbuf->RingBufTxInPtr  = &pbuf->RingBufTx[0];
    pbuf->RingBufTxOutPtr = &pbuf->RingBufTx[0];
}

/*$PAGE*/
```

11

Listing 11.4 (continued) COMMBGND.C

```
/*
*********************************************************************************************
*                           SEE IF RX CHARACTER BUFFER IS EMPTY
*
*
* Description : This function is called by your application to see if any character is available from the
*               communications channel.  If at least one character is available, the function returns
*               FALSE otherwise, the function returns TRUE.
* Arguments   : 'ch'    is the COMM port channel number and can either be:
*                            COMM1
*                            COMM2
* Returns     : TRUE    if the buffer IS empty.
*               FALSE   if the buffer IS NOT empty or you have specified an incorrect channel
*********************************************************************************************
*/

BOOLEAN  CommIsEmpty (INT8U ch)
{
    BOOLEAN          empty;
    COMM_RING_BUF *pbuf;


    switch (ch) {                                       /* Obtain pointer to communications channel */
        case COMM1:
            pbuf = &Comm1Buf;
            break;

        case COMM2:
            pbuf = &Comm2Buf;
            break;

        default:
            return (TRUE);
    }
    OS_ENTER_CRITICAL();
    if (pbuf->RingBufRxCtr > 0) {                       /* See if buffer is empty                  */
        empty = FALSE;                                  /* Buffer is NOT empty                     */
    } else {
        empty = TRUE;                                   /* Buffer is empty                         */
    }
    OS_EXIT_CRITICAL();
    return (empty);
}

/*$PAGE*/
```

Listing 11.4 (continued) COMMBGND.C

```
/*
*********************************************************************************************************
*                               SEE IF TX CHARACTER BUFFER IS FULL
*
*
* Description : This function is called by your application to see if any more characters can be placed
*               in the Tx buffer.  In other words, this function check to see if the Tx buffer is full.
*               If the buffer is full, the function returns TRUE otherwise, the function returns FALSE.
* Arguments   : 'ch'    is the COMM port channel number and can either be:
*                           COMM1
*                           COMM2
* Returns     : TRUE    if the buffer IS full.
*               FALSE   if the buffer IS NOT full or you have specified an incorrect channel
*********************************************************************************************************
*/

BOOLEAN  CommIsFull (INT8U ch)
{
    BOOLEAN        full;
    COMM_RING_BUF *pbuf;


    switch (ch) {                                      /* Obtain pointer to communications channel */
        case COMM1:
             pbuf = &Comm1Buf;
             break;

        case COMM2:
             pbuf = &Comm2Buf;
             break;

        default:
             return (TRUE);
    }
    OS_ENTER_CRITICAL();
    if (pbuf->RingBufTxCtr < COMM_TX_BUF_SIZE) {       /* See if buffer is full                     */
        full = FALSE;                                  /* Buffer is NOT full                        */
    } else {
        full = TRUE;                                   /* Buffer is full                           */
    }
    OS_EXIT_CRITICAL();
    return (full);
}

/*$PAGE*/
```

11

Listing 11.4 (continued)   COMMBGND.C

```
/*
*********************************************************************************************
*                                    OUTPUT CHARACTER
*
*
* Description : This function is called by your application to send a character on the communications
*               channel.  The character to send is first inserted into the Tx buffer and will be sent by
*               the Tx ISR.  If this is the first character placed into the buffer, the Tx ISR will be
*               enabled.  If the Tx buffer is full, the character will not be sent (i.e. it will be lost)
* Arguments  : 'ch'    is the COMM port channel number and can either be:
*                           COMM1
*                           COMM2
*              'c'      is the character to send.
* Returns    : COMM_NO_ERR    if the function was successful (the buffer was not full)
*              COMM_TX_FULL  if the buffer was full
*              COMM_BAD_CH   if you have specified an incorrect channel
*********************************************************************************************
*/

INT8U  CommPutChar (INT8U ch, INT8U c)
{
    COMM_RING_BUF *pbuf;


    switch (ch) {                                       /* Obtain pointer to communications channel */
        case COMM1:
            pbuf = &Comm1Buf;
            break;

        case COMM2:
            pbuf = &Comm2Buf;
            break;

        default:
            return (COMM_BAD_CH);
    }
    OS_ENTER_CRITICAL();
    if (pbuf->RingBufTxCtr < COMM_TX_BUF_SIZE) {        /* See if buffer is full            */
        pbuf->RingBufTxCtr++;                           /* No, increment character count    */
        *pbuf->RingBufTxInPtr++ = c;                    /* Put character into buffer        */
        if (pbuf->RingBufTxInPtr == &pbuf->RingBufTx[COMM_TX_BUF_SIZE]) { /* Wrap IN pointer    */
            pbuf->RingBufTxInPtr = &pbuf->RingBufTx[0];
        }
        if (pbuf->RingBufTxCtr == 1) {                  /* See if this is the first character    */
            CommTxIntEn(ch);                            /* Yes, Enable Tx interrupts        */
            OS_EXIT_CRITICAL();
        } else {
            OS_EXIT_CRITICAL();
        }
        return (COMM_NO_ERR);
    } else {
        OS_EXIT_CRITICAL();
        return (COMM_TX_FULL);
    }
}

/*$PAGE*/
```

Listing 11.4 (continued) COMMBGND.C

```
/*
*********************************************************************************************
*                              INSERT CHARACTER INTO RING BUFFER
*
*
* Description : This function is called by the Rx ISR to insert a character into the receive ring buffer.
* Arguments   : 'ch'   is the COMM port channel number and can either be:
*                          COMM1
*                          COMM2
*                'c'    is the character to insert into the ring buffer.  If the buffer is full, the
*                       character will not be inserted, it will be lost.
*********************************************************************************************
*/

void  CommPutRxChar (INT8U ch, INT8U c)
{
    COMM_RING_BUF *pbuf;


    switch (ch) {                                       /* Obtain pointer to communications channel */
        case COMM1:
            pbuf = &Comm1Buf;
            break;

        case COMM2:
            pbuf = &Comm2Buf;
            break;

        default:
            return;
    }
    if (pbuf->RingBufRxCtr < COMM_RX_BUF_SIZE) {        /* See if buffer is full                     */
        pbuf->RingBufRxCtr++;                           /* No, increment character count            */
        *pbuf->RingBufRxInPtr++ = c;                    /* Put character into buffer                */
        if (pbuf->RingBufRxInPtr == &pbuf->RingBufRx[COMM_RX_BUF_SIZE]) { /* Wrap IN pointer        */
            pbuf->RingBufRxInPtr = &pbuf->RingBufRx[0];
        }
    }
}
```

11

Listing 11.5   COMMBGND.H

```
/*
********************************************************************************************
*                              Embedded Systems Building Blocks
*                             Complete and Ready-to-Use Modules in C
*
*                              Asynchronous Serial Communications
*                                    Buffered Serial I/O
*                                 (Foreground/Background Systems)
*
*                          (c) Copyright 1999, Jean J. Labrosse, Weston, FL
*                                      All Rights Reserved
*
* Filename   : COMMBGND.H
* Programmer : Jean J. Labrosse
********************************************************************************************
*/

/*
********************************************************************************************
*                                  CONFIGURATION CONSTANTS
********************************************************************************************
*/

#ifndef  CFG_H

#define   COMM_RX_BUF_SIZE    128               /* Number of characters in Rx ring buffer        */
#define   COMM_TX_BUF_SIZE    128               /* Number of characters in Tx ring buffer        */

#endif

/*
********************************************************************************************
*                                      CONSTANTS
********************************************************************************************
*/

#ifndef  NUL
#define  NUL            0x00
#endif

#define  COMM1          1
#define  COMM2          2

                                                /* ERROR CODES                                    */
#define  COMM_NO_ERR    0                       /* Function call was successful                   */
#define  COMM_BAD_CH    1                       /* Invalid communications port channel           */
#define  COMM_RX_EMPTY  2                       /* Rx buffer is empty, no character available    */
#define  COMM_TX_FULL   3                       /* Tx buffer is full, could not deposit character */
#define  COMM_TX_EMPTY  4                       /* If the Tx buffer is empty.                     */


#ifdef   COMM_GLOBALS
#define  COMM_EXT
#else
#define  COMM_EXT  extern
#endif
/*$PAGE*/
```

Listing 11.5 (continued) COMMBGND.H

```
/*
*********************************************************************************************
*                                    FUNCTION PROTOTYPES
*********************************************************************************************
*/

INT8U    CommGetChar(INT8U ch, INT8U *err);
INT8U    CommGetTxChar(INT8U ch, INT8U *err);
void     CommInit(void);
BOOLEAN  CommIsEmpty(INT8U ch);
BOOLEAN  CommIsFull(INT8U ch);
INT8U    CommPutChar(INT8U ch, INT8U c);
void     CommPutRxChar(INT8U ch, INT8U c);
```

11

Listing 11.6 COMMRTOS.C

```
/*
*********************************************************************************************************
*                                  Embedded Systems Building Blocks
*                                 Complete and Ready-to-Use Modules in C
*
*                                  Asynchronous Serial Communications
*                                          Buffered Serial I/O
*                                                (RTOS)
*
*                          (c) Copyright 1999, Jean J. Labrosse, Weston, FL
*                                        All Rights Reserved
*
* Filename    : COMMRTOS.C
* Programmer  : Jean J. Labrosse
*********************************************************************************************************
*/

/*
*********************************************************************************************************
*                                             INCLUDES
*********************************************************************************************************
*/

#include "includes.h"

/*
*********************************************************************************************************
*                                            DATA TYPES
*********************************************************************************************************
*/

typedef struct {
    INT16U    RingBufRxCtr;                 /* Number of characters in the Rx ring buffer            */
    OS_EVENT  *RingBufRxSem;                /* Pointer to Rx semaphore                               */
    INT8U     *RingBufRxInPtr;              /* Pointer to where next character will be inserted      */
    INT8U     *RingBufRxOutPtr;             /* Pointer from where next character will be extracted   */
    INT8U     RingBufRx[COMM_RX_BUF_SIZE];  /* Ring buffer character storage (Rx)                    */
    INT16U    RingBufTxCtr;                 /* Number of characters in the Tx ring buffer            */
    OS_EVENT  *RingBufTxSem;                /* Pointer to Tx semaphore                               */
    INT8U     *RingBufTxInPtr;              /* Pointer to where next character will be inserted      */
    INT8U     *RingBufTxOutPtr;             /* Pointer from where next character will be extracted   */
    INT8U     RingBufTx[COMM_TX_BUF_SIZE];  /* Ring buffer character storage (Tx)                    */
} COMM_RING_BUF;

/*
*********************************************************************************************************
*                                          GLOBAL VARIABLES
*********************************************************************************************************
*/

COMM_RING_BUF  Comm1Buf;
COMM_RING_BUF  Comm2Buf;

/*$PAGE*/
```

Listing 11.6 (continued) COMMRTOS.C

```
/*
*********************************************************************************************
*                              REMOVE CHARACTER FROM RING BUFFER
*
*
* Description : This function is called by your application to obtain a character from the communications
*               channel.  The function will wait for a character to be received on the serial channel or
*               until the function times out.
* Arguments   : 'ch'    is the COMM port channel number and can either be:
*                           COMM1
*                           COMM2
*               'to'    is the amount of time (in clock ticks) that the calling function is willing to
*                       wait for a character to arrive.  If you specify a timeout of 0, the function will
*                       wait forever for a character to arrive.
*               'err'   is a pointer to where an error code will be placed:
*                           *err is set to COMM_NO_ERR     if a character has been received
*                           *err is set to COMM_RX_TIMEOUT if a timeout occurred
*                           *err is set to COMM_BAD_CH     if you specify an invalid channel number
* Returns     : The character in the buffer (or NUL if a timeout occurred)
*********************************************************************************************
*/

INT8U  CommGetChar (INT8U ch, INT16U to, INT8U *err)
{
    INT8U           c;
    INT8U           oserr;
    COMM_RING_BUF *pbuf;


    switch (ch) {                                     /* Obtain pointer to communications channel */
        case COMM1:
             pbuf = &Comm1Buf;
             break;

        case COMM2:
             pbuf = &Comm2Buf;
             break;

        default:
             *err = COMM_BAD_CH;
             return (NUL);
    }
    OSSemPend(pbuf->RingBufRxSem, to, &oserr);         /* Wait for character to arrive             */
    if (oserr == OS_TIMEOUT) {                         /* See if characters received within timeout*/
        *err = COMM_RX_TIMEOUT;                        /* No, return error code                    */
        return (NUL);
    } else {
        OS_ENTER_CRITICAL();
        pbuf->RingBufRxCtr--;                          /* Yes, decrement character count           */
        c = *pbuf->RingBufRxOutPtr++;                  /* Get character from buffer                */
        if (pbuf->RingBufRxOutPtr == &pbuf->RingBufRx[COMM_RX_BUF_SIZE]) {    /* Wrap OUT pointer  */
            pbuf->RingBufRxOutPtr = &pbuf->RingBufRx[0];
        }
        OS_EXIT_CRITICAL();
        *err = COMM_NO_ERR;
        return (c);
    }
}

/*$PAGE*/
```

Listing 11.6 (continued) COMMRTOS.C

```
/*
*********************************************************************************************
*                            GET TX CHARACTER FROM RING BUFFER
*
*
* Description : This function is called by the Tx ISR to extract the next character from the Tx buffer.
*               The function returns FALSE if the buffer is empty after the character is extracted from
*               the buffer.  This is done to signal the Tx ISR to disable interrupts because this is the
*               last character to send.
* Arguments   : 'ch'   is the COMM port channel number and can either be:
*                          COMM1
*                          COMM2
*               'err'   is a pointer to where an error code will be deposited:
*                          *err is set to COMM_NO_ERR       if at least one character was available
*                                                           from the buffer.
*                          *err is set to COMM_TX_EMPTY     if the Tx buffer is empty.
*                          *err is set to COMM_BAD_CH       if you have specified an incorrect channel
* Returns     : The next character in the Tx buffer or NUL if the buffer is empty.
*********************************************************************************************
*/

INT8U  CommGetTxChar (INT8U ch, INT8U *err)
{
    INT8U          c;
    COMM_RING_BUF *pbuf;


    switch (ch) {                                        /* Obtain pointer to communications channel */
        case COMM1:
            pbuf = &Comm1Buf;
            break;

        case COMM2:
            pbuf = &Comm2Buf;
            break;

        default:
            *err = COMM_BAD_CH;
            return (NUL);
    }
    if (pbuf->RingBufTxCtr > 0) {                         /* See if buffer is empty                   */
        pbuf->RingBufTxCtr--;                            /* No, decrement character count            */
        c = *pbuf->RingBufTxOutPtr++;                    /* Get character from buffer                */
        if (pbuf->RingBufTxOutPtr == &pbuf->RingBufTx[COMM_TX_BUF_SIZE]) {     /* Wrap OUT pointer    */
            pbuf->RingBufTxOutPtr = &pbuf->RingBufTx[0];
        }
        OSSemPost(pbuf->RingBufTxSem);                   /* Indicate that character will be sent     */
        *err = COMM_NO_ERR;
        return (c);                                      /* Characters are still available           */
    } else {
        *err = COMM_TX_EMPTY;
        return (NUL);                                    /* Buffer is empty                          */
    }
}

/*$PAGE*/
```

Listing 11.6 (continued) COMMRTOS.C

```
/*
*********************************************************************************************
*                                INITIALIZE COMMUNICATIONS MODULE
*
*
* Description : This function is called by your application to initialize the communications module.  You
*               must call this function before calling any other functions.
* Arguments   : none
*********************************************************************************************
*/

void  CommInit (void)
{
    COMM_RING_BUF *pbuf;


    pbuf                  = &Comm1Buf;                    /* Initialize the ring buffer for COMM1    */
    pbuf->RingBufRxCtr    = 0;
    pbuf->RingBufRxInPtr  = &pbuf->RingBufRx[0];
    pbuf->RingBufRxOutPtr = &pbuf->RingBufRx[0];
    pbuf->RingBufRxSem    = OSSemCreate(0);
    pbuf->RingBufTxCtr    = 0;
    pbuf->RingBufTxInPtr  = &pbuf->RingBufTx[0];
    pbuf->RingBufTxOutPtr = &pbuf->RingBufTx[0];
    pbuf->RingBufTxSem    = OSSemCreate(COMM_TX_BUF_SIZE);

    pbuf                  = &Comm2Buf;                    /* Initialize the ring buffer for COMM2    */
    pbuf->RingBufRxCtr    = 0;
    pbuf->RingBufRxInPtr  = &pbuf->RingBufRx[0];
    pbuf->RingBufRxOutPtr = &pbuf->RingBufRx[0];
    pbuf->RingBufRxSem    = OSSemCreate(0);
    pbuf->RingBufTxCtr    = 0;
    pbuf->RingBufTxInPtr  = &pbuf->RingBufTx[0];
    pbuf->RingBufTxOutPtr = &pbuf->RingBufTx[0];
    pbuf->RingBufTxSem    = OSSemCreate(COMM_TX_BUF_SIZE);
}

/*$PAGE*/
```

11

Listing 11.6 (continued)   COMMRTOS.C

```
/*
*********************************************************************************************
*                             SEE IF RX CHARACTER BUFFER IS EMPTY
*
*
* Description : This function is called by your application to see if any character is available from the
*               communications channel.  If at least one character is available, the function returns
*               FALSE otherwise, the function returns TRUE.
* Arguments   : 'ch'    is the COMM port channel number and can either be:
*                           COMM1
*                           COMM2
* Returns     : TRUE    if the buffer IS empty.
*               FALSE   if the buffer IS NOT empty or you have specified an incorrect channel.
*********************************************************************************************
*/

BOOLEAN  CommIsEmpty (INT8U ch)
{
    BOOLEAN          empty;
    COMM_RING_BUF *pbuf;


    switch (ch) {                                    /* Obtain pointer to communications channel */
        case COMM1:
             pbuf = &Comm1Buf;
             break;

        case COMM2:
             pbuf = &Comm2Buf;
             break;

        default:
             return (TRUE);
    }
    OS_ENTER_CRITICAL();
    if (pbuf->RingBufRxCtr > 0) {                     /* See if buffer is empty              */
        empty = FALSE;                               /* Buffer is NOT empty                 */
    } else {
        empty = TRUE;                                /* Buffer is empty                    */
    }
    OS_EXIT_CRITICAL();
    return (empty);
}

/*$PAGE*/
```

Listing 11.6 (continued) COMMRTOS.C

```
/*
******************************************************************************************************
*                               SEE IF TX CHARACTER BUFFER IS FULL
*
*
* Description : This function is called by your application to see if any more characters can be placed
*               in the Tx buffer.  In other words, this function check to see if the Tx buffer is full.
*               If the buffer is full, the function returns TRUE otherwise, the function returns FALSE.
* Arguments   : 'ch'    is the COMM port channel number and can either be:
*                         COMM1
*                         COMM2
* Returns     : TRUE    if the buffer IS full.
*               FALSE   if the buffer IS NOT full or you have specified an incorrect channel.
******************************************************************************************************
*/

BOOLEAN  CommIsFull (INT8U ch)
{
    BOOLEAN         full;
    COMM_RING_BUF *pbuf;


    switch (ch) {                                        /* Obtain pointer to communications channel */
        case COMM1:
             pbuf = &Comm1Buf;
             break;

        case COMM2:
             pbuf = &Comm2Buf;
             break;

        default:
             return (TRUE);
    }
    OS_ENTER_CRITICAL();
    if (pbuf->RingBufTxCtr < COMM_TX_BUF_SIZE) {         /* See if buffer is full                     */
        full = FALSE;                                    /* Buffer is NOT full                        */
    } else {
        full = TRUE;                                     /* Buffer is full                           */
    }
    OS_EXIT_CRITICAL();
    return (full);
}

/*$PAGE*/
```

Listing 11.6 (continued)   COMMRTOS.C

```
/*
*********************************************************************************************
*                                   OUTPUT CHARACTER
*
*
* Description : This function is called by your application to send a character on the communications
*               channel.  The function will wait for the buffer to empty out if the buffer is full.
*               The function returns to your application if the buffer doesn't empty within the specified
*               timeout.  A timeout value of 0 means that the calling function will wait forever for the
*               buffer to empty out.  The character to send is first inserted into the Tx buffer and will
*               be sent by the Tx ISR.  If this is the first character placed into the buffer, the Tx ISR
*               will be enabled.
* Arguments   : 'ch'   is the COMM port channel number and can either be:
*                         COMM1
*                         COMM2
*               'c'    is the character to send.
*               'to'   is the timeout (in clock ticks) to wait in case the buffer is full.  If you
*                      specify a timeout of 0, the function will wait forever for the buffer to empty.
* Returns     : COMM_NO_ERR      if the character was placed in the Tx buffer
*               COMM_TX_TIMEOUT  if the buffer didn't empty within the specified timeout period
*               COMM_BAD_CH      if you specify an invalid channel number
*********************************************************************************************
*/

INT8U  CommPutChar (INT8U ch, INT8U c, INT16U to)
{
    INT8U          oserr;
    COMM_RING_BUF *pbuf;


    switch (ch) {                                       /* Obtain pointer to communications channel */
        case COMM1:
             pbuf = &Comm1Buf;
             break;

        case COMM2:
             pbuf = &Comm2Buf;
             break;

        default:
             return (COMM_BAD_CH);
    }
    OSSemPend(pbuf->RingBufTxSem, to, &oserr);           /* Wait for space in Tx buffer             */
    if (oserr == OS_TIMEOUT) {
        return (COMM_TX_TIMEOUT);                        /* Timed out, return error code            */
    }
    OS_ENTER_CRITICAL();
    pbuf->RingBufTxCtr++;                                /* No, increment character count           */
    *pbuf->RingBufTxInPtr++ = c;                         /* Put character into buffer               */
    if (pbuf->RingBufTxInPtr == &pbuf->RingBufTx[COMM_TX_BUF_SIZE]) {    /* Wrap IN pointer          */
        pbuf->RingBufTxInPtr = &pbuf->RingBufTx[0];
    }
    if (pbuf->RingBufTxCtr == 1) {                       /* See if this is the first character       */
        CommTxIntEn(ch);                                 /* Yes, Enable Tx interrupts                */
    }
    OS_EXIT_CRITICAL();
    return (COMM_NO_ERR);
}

/*$PAGE*/
```

Listing 11.6 (continued) COMMRTOS.C

```
/*
*********************************************************************************************
*                               INSERT CHARACTER INTO RING BUFFER
*
*
* Description : This function is called by the Rx ISR to insert a character into the receive ring buffer.
* Arguments   : 'ch'   is the COMM port channel number and can either be:
*                          COMM1
*                          COMM2
*                'c'    is the character to insert into the ring buffer.  If the buffer is full, the
*                       character will not be inserted, it will be lost.
*********************************************************************************************
*/

void  CommPutRxChar (INT8U ch, INT8U c)
{
    COMM_RING_BUF *pbuf;


    switch (ch) {                                       /* Obtain pointer to communications channel */
        case COMM1:
             pbuf = &Comm1Buf;
             break;

        case COMM2:
             pbuf = &Comm2Buf;
             break;

        default:
             return;
    }
    if (pbuf->RingBufRxCtr < COMM_RX_BUF_SIZE) {        /* See if buffer is full                     */
        pbuf->RingBufRxCtr++;                           /* No, increment character count            */
        *pbuf->RingBufRxInPtr++ = c;                    /* Put character into buffer                */
        if (pbuf->RingBufRxInPtr == &pbuf->RingBufRx[COMM_RX_BUF_SIZE]) { /* Wrap IN pointer         */
            pbuf->RingBufRxInPtr = &pbuf->RingBufRx[0];
        }
        OSSemPost(pbuf->RingBufRxSem);                  /* Indicate that character was received     */
    }
}
```

11

Listing 11.7 COMMRTOS.H

```
/*
**************************************************************************************************
*                            Embedded Systems Building Blocks
*                            Complete and Ready-to-Use Modules in C
*
*                            Asynchronous Serial Communications
*                                  Buffered Serial I/O
*                                       (RTOS)
*
*                        (c) Copyright 1999, Jean J. Labrosse, Weston, FL
*                                   All Rights Reserved
*
* Filename   : COMMRTOS.H
* Programmer : Jean J. Labrosse
**************************************************************************************************
*/

/*
**************************************************************************************************
*                                CONFIGURATION CONSTANTS
**************************************************************************************************
*/

#ifndef  CFG_H

#define  COMM_RX_BUF_SIZE      64            /* Number of characters in Rx ring buffer       */
#define  COMM_TX_BUF_SIZE      64            /* Number of characters in Tx ring buffer       */

#endif

/*
**************************************************************************************************
*                                      CONSTANTS
**************************************************************************************************
*/

#ifndef  NUL
#define  NUL               0x00
#endif

#define  COMM1                 1
#define  COMM2                 2

                                             /* ERROR CODES                                  */
#define  COMM_NO_ERR           0             /* Function call was successful                 */
#define  COMM_BAD_CH           1             /* Invalid communications port channel          */
#define  COMM_RX_EMPTY         2             /* Rx buffer is empty, no character available   */
#define  COMM_TX_FULL          3             /* Tx buffer is full, could not deposit character */
#define  COMM_TX_EMPTY         4             /* If the Tx buffer is empty.                   */
#define  COMM_RX_TIMEOUT       5             /* If a timeout occurred while waiting for a character*/
#define  COMM_TX_TIMEOUT       6             /* If a timeout occurred while waiting to send a char.*/


#define  COMM_PARITY_NONE      0             /* Defines for setting parity                   */
#define  COMM_PARITY_ODD       1
#define  COMM_PARITY_EVEN      2
```

Listing 11.7 (continued) COMMRTOS.H

```
#ifdef    COMM_GLOBALS
#define   COMM_EXT
#else
#define   COMM_EXT   extern
#endif
/*$PAGE*/
/*
*********************************************************************************************************
*                                         FUNCTION PROTOTYPES
*********************************************************************************************************
*/

INT8U     CommGetChar(INT8U ch, INT16U to, INT8U *err);
INT8U     CommGetTxChar(INT8U ch, INT8U *err);
void      CommInit(void);
BOOLEAN   CommIsEmpty(INT8U ch);
BOOLEAN   CommIsFull(INT8U ch);
INT8U     CommPutChar(INT8U ch, INT8U c, INT16U to);
void      CommPutRxChar(INT8U ch, INT8U c);
```

11

Chapter 12

PC Services

The code in this book was tested on a PC. It was convenient to create a number of services (i.e., functions) to access some of the capabilities of a PC. These services are invoked from the test code and are encapsulated in a file called PC.C. Because industrial PCs are so popular as embedded system platforms, the functions provided in this chapter could be of some use to you. These services assume that you are running under DOS or a DOS box under Windows 95/98 or NT. You should note that under Windows 95/98 or NT, you have an emulated DOS and not an actual one (i.e., a Virtual x86 session). The behavior of some functions may be altered because of this.

The files PC.C (Listing 12.3) and PC.H (Listing 12.4) are found in the \SOFT-WARE\BLOCKS\PC\BC45 directory. Unlike the first edition of ESBB, I decided to encapsulate these functions (as they should have been) to avoid defining them in the example code and also, to allow you to easily adapt the code to a different compiler. PC.C basically contains three types of services: character based display, elapsed time measurement, and miscellaneous. All functions start with the prefix PC_.

12.00 Character Based Display

PC.C provides services to display ASCII (and special) characters on a PC's VGA display. In normal mode (i.e., character mode), a PC's display can hold up to 2000 characters organized as 25 rows (i.e., y) by 80 columns (i.e., x) as shown in Figure 12.1. Please disregard the aspect ratio of the figure. The actual aspect ratio of a monitor is generally 4×3. Video memory on a PC is memory mapped and, on a VGA monitor, video memory starts at absolute memory location 0x000B8000 (or using a segment:offset notation, B800:0000).

Figure 12.1   80 x 25 characters on a VGA monitor.

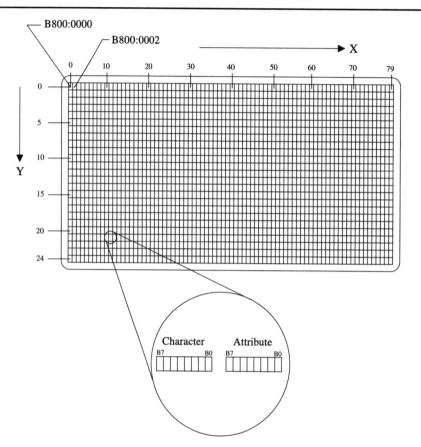

Each displayable *character* requires two bytes to display. The first byte (lowest memory location) is the character that you want to display while the second byte (next memory location) is an attribute that determines the foreground/background color combination of the character. The foreground color is specified in the lower 4 bits of the attribute while the background color appears in bits 4 to 6. Finally, the most-significant bit determines whether the character will blink (when 1) or not (when 0). The character and attribute bytes are shown in Figure 12.2.

Figure 12.2   Character and attribute bytes on a VGA monitor.

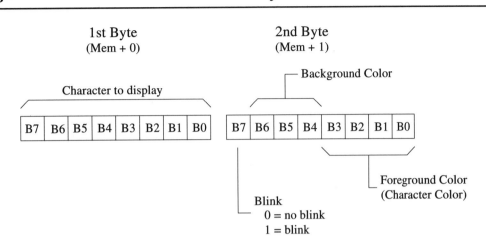

Table 12.1 shows the possible colors that can be obtained from the PC's VGA character mode.

You will note that you can only have 8 possible background colors but a choice of 16 foreground colors. PC.H contains #defines which allow you to select the proper combination of foreground and background colors. These #defines are shown in Table 12.1. For example, to obtain a non-blinking WHITE character on a BLACK background, you would simply add DISP_FGND_WHITE and DISP_BGND_BLACK (FGND means foreground and BGND is background). This corresponds to a HEX value of 0x07 which happens to be the default video attribute of a displayable character on a PC. You should note that because DISP_BGND_BLACK has a value of 0x00, you don't actually need to specify it and thus, the attribute for the same WHITE character could just as well have been specified as DISP_FGND_WHITE. You should use the #define constants instead of the HEX values to make your code more readable.

The display functions in PC.C are used to write ASCII (and special) characters anywhere on the screen using *x* and *y* coordinates. The coordinate system of the display is shown in Figure 12.1. You should note that position 0,0 is located at the upper left corner as opposed to the bottom left corner as you may have expected. This makes the computation of the location of each character to display easier to determine. The address in video memory for any character on the screen is given by:

$$\text{Address of Character} = \texttt{0x000B8000 + Y * 160 + X * 2}$$

The address of the attribute byte is at the next memory location or:

$$\text{Address of Attribute} = \texttt{0x000B8000 + Y * 160 + X * 2 + 1}$$

The display functions provided in PC.C perform direct writes to video RAM even though BIOS (Basic Input Output System) services in most PCs can do the same thing but in a portable fashion. I chose to write directly to video memory for performance reasons.

PC.C contains the following five functions which are further described in the interface section of this chapter.

12

`PC_DispChar()`	To display a single ASCII character anywhere on the screen
`PC_DispClrCol()`	To clear a single column
`PC_DispClrRow()`	To clear a single row (or line)
`PC_DispClrScr()`	To clear the screen
`PC_DispStr()`	To display an ASCII string anywhere on the screen

Table 12.1   Attribute byte values.

Blink			Background Color			Foreground Color		
(B7)			(B6 B5 B4)			(B3 B2 B1 B0)		
Blink?	#define	HEX	Color	#define	HEX	Color	#define	HEX
No		0x00	Black	`DISP_BGND_BLACK`	0x00	Black	`DISP_FGND_BLACK`	0x00
Yes	`DISP_BLINK`	0x80	Blue	`DISP_BGND_BLUE`	0x10	Blue	`DISP_FGND_BLUE`	0x01
			Green	`DISP_BGND_GREEN`	0x20	Green	`DISP_FGND_GREEN`	0x02
			Cyan	`DISP_BGND_CYAN`	0x30	Cyan	`DISP_FGND_CYAN`	0x03
			Red	`DISP_BGND_RED`	0x40	Red	`DISP_FGND_RED`	0x04
			Purple	`DISP_BGND_PURPLE`	0x50	Purple	`DISP_FGND_PURPLE`	0x05
			Brown	`DISP_BGND_BROWN`	0x60	Brown	`DISP_FGND_BROWN`	0x06
			Light Gray	`DISP_BGND_LIGHT_GRAY`	0x70	Light Gray	`DISP_FGND_LIGHT_GRAY`	0x07
						Dark Gray	`DISP_FGND_DARK_GRAY`	0x08
						Light Blue	`DISP_FGND_LIGHT_BLUE`	0x09
						Light Green	`DISP_FGND_LIGHT_GREEN`	0x0A
						Light Cyan	`DISP_FGND_LIGHT_CYAN`	0x0B
						Light Red	`DISP_FGND_LIGHT_RED`	0x0C
						Light Purple	`DISP_FGND_LIGHT_PURPLE`	0x0D
						Yellow	`DISP_FGND_YELLOW`	0x0E
						White	`DISP_FGND_WHITE`	0x0F

12.01 Saving and Restoring DOS's Context

The current DOS environment is saved by calling `PC_DOSSaveReturn()` (see Listing 12.1) and would be called by `main()` to:

1. Setup μC/OS-II's context switch vector,

2. Setup the tick ISR vector,

3. Save DOS's context so that we can return back to DOS when we need to terminate execution of a μC/OS-II based application.

A lot happens in `PC_DOSSaveReturn()` so you may need to look at the code in Listing 12.1 to follow along. `PC_DOSSaveReturn()` starts by setting the flag `PC_ExitFlag` to FALSE [L12.1(1)] indicating that we are not returning to DOS. Then, `PC_DOSSaveReturn()` initializes `OSTickDOSCtr` to 8 [L12.1(2)] because this variable will be decremented in `OSTickISR()`. A value of 0 would have caused this value to wrap around to 255 when decremented by `OSTickISR()`. `PC_DOSSaveReturn()` then saves DOS's tick handler in a free vector table [L12.1(3)–(4)] entry so it can be called by µC/OS-II's tick handler (this is called *chaining* the vectors). Next, `PC_DOSSaveReturn()` calls `setjmp()` [L12.1(5)], which captures the state of the processor (i.e., the contents of all important registers) into a structure called `PC_JumpBuf`. Capturing the processor's context will allow us to return to `PC_DOSSaveReturn()` and execute the code immediately following the call to `setjmp()`. Because `PC_ExitFlag` was initialized to FALSE [L12.1(1)], `PC_DOSSaveReturn()` skips the code in the `if` statement [i.e., L12.1(6)–(9)] and returns to the caller (i.e., `main()`).

When you want to return to DOS, all you have to do is call `PC_DOSReturn()` (see Listing 12.2) which sets `PC_ExitFlag` to TRUE [L12.2(1)] and execute a `longjmp()` [L12.2(2)]. This brings the processor back in `PC_DOSSaveReturn()` (just after the call to `setjmp()`) [L12.1(5)]. This time, however, `PC_ExitFlag` is TRUE and the code following the if statement is executed. `PC_DOSSaveReturn()` changes the tick rate back to 18.2 Hz [L12.1(6)], restores the PC's tick ISR handler [L12.1(7)], clears the screen [L12.1(8)], and returns to the DOS prompt through the `exit(0)` function [L12.1(9)].

Listing 12.1   Saving the DOS environment.

```
void PC_DOSSaveReturn (void)
{
    PC_ExitFlag  = FALSE;                                     (1)
    OSTickDOSCtr =     8;                                     (2)
    PC_TickISR   = PC_VectGet(VECT_TICK);                     (3)

    OS_ENTER_CRITICAL();
    PC_VectSet(VECT_DOS_CHAIN, PC_TickISR);                   (4)
    OS_EXIT_CRITICAL();

    setjmp(PC_JumpBuf);                                       (5)
    if (PC_ExitFlag == TRUE) {
        OS_ENTER_CRITICAL();
        PC_SetTickRate(18);                                   (6)
        PC_VectSet(VECT_TICK, PC_TickISR);                    (7)
        OS_EXIT_CRITICAL();
        PC_DispClrScr(DISP_FGND_WHITE + DISP_BGND_BLACK);     (8)
        exit(0);                                              (9)
    }
}
```

12

Listing 12.2   *Setting up to return to DOS.*

```
void PC_DOSReturn (void)
{
    PC_ExitFlag = TRUE;                                      (1)
    longjmp(PC_JumpBuf, 1);                                  (2)
}
```

12.02 Elapsed Time Measurement

The elapsed time measurement functions are used to determine how much time a function takes to execute. Time measurement is performed by using the PC's 82C54 timer #2. You make time measurement by wrapping the code to measure by the two functions PC_ElapsedStart() and PC_ElapsedStop(). However, before you can use these two functions, you need to call the function PC_ElapsedInit(). PC_ElapsedInit() basically computes the overhead associated with the other two functions. This way, the execution time (in microseconds) returned by PC_ElapsedStop() consist exclusively of the code you are measuring. Note that none of these functions are reentrant and thus, you must be careful that you do not invoke them from multiple tasks at the same time.

12.03 Miscellaneous

PC_GetDateTime() is a function that obtains the PC's current date and time, and formats this information into an ASCII string. The format is:

```
"YYYY-MM-DD  HH:MM:SS"
```

and you will need at least 21 characters (including the NUL character) to hold this string. You should note that there are 2 spaces between the date and the time which explains why you need 21 characters instead of 20. PC_GetDateTime() uses the Borland C/C++ library functions gettime() and getdate() which should have their equivalent on other DOS compilers.

PC_GetKey() is a function that checks to see if a key was pressed and if so, obtains that key, and returns it to the caller. PC_GetKey() uses the Borland C/C++ library functions kbhit() and getch() which again, have their equivalent on other DOS compilers.

PC_SetTickRate() allows you to change the tick rate for µC/OS-II by specifying the desired frequency. Under DOS, a tick occurs 18.20648 times per second or, every 54.925 mS. This is because the 82C54 chip used didn't get its counter initialized and the default value of 65535 takes effect. Had the chip been initialized with a divide by 59659, the tick rate would have been a very nice 20.000 Hz! I decided to change the tick rate to something more 'exciting' and thus, decided to use about 200 Hz (actually 199.9966). The code found in OS_CPU_A.OBJ calls the DOS tick handler one time out of 11. This is done to ensure that some of the housekeeping needed in DOS is maintained. You would not need to do this if you were to set the tick rate to 20 Hz. Before returning to DOS, PC_SetTickRate() is called by specifying 18 as the desired frequency. PC_SetTickRate() will know that you actually mean 18.2 Hz and will correctly set the 82C54.

The last two functions in `PC.C` are used to get and set an interrupt vector. `PC_VectGet()` and `PC_VectSet()` should be compiler independent as long as the compiler support the macros `MK_FP()` (make far pointer), `FP_OFF()` (get the offset portion of a far pointer) and, `FP_SEG()` (get the segment of a far pointer).

12.04 Interface Functions

This section provides a reference section for the PC services.

12

PC_DispChar()

void PC_DispChar(INT8U x, INT8U y, INT8U c, INT8U color);

PC_DispChar() allows you to display a single ASCII (or special) character anywhere on the display.

Arguments

x and **y** specifies the coordinates (col, row) where the character will appear. rows (i.e., lines) are numbered from 0 to DISP_MAX_Y − 1, and columns are numbered from 0 to DISP_MAX_X − 1 (see Listing 12.3, PC.C).

c is the character to display. You can specify any ASCII characters and special characters if c has a value higher than 128. You can see what characters (i.e., symbols) will be displayed based on the value of c by running the test code provided in this book as follows:

> C:\SOFTWARE\BLOCKS\SAMPLE\TEST > **TEST display**

color specifies the contents of the attribute byte and thus the color combination of the character to be displayed. You can add one DISP_FGND_??? (see Listing 12.4, PC.H) and one DISP_BGND_??? (see Listing 12.4, PC.H) to obtain the desired color combination.

Return Value

None

Notes/Warnings

None

Example

```
void Task (void *pdata)
{
    .
    .
    for (;;) {

        PC_DispChar(0, 0, '$', DISP_FGND_WHITE);
        .

    }
}
```

PC_DispClrCol()

```
void PC_DispClrCol(INT8U x, INT8U color);
```

PC_DispClrCol() allows you to clear the contents of a column (all 25 characters).

Arguments

x specifies which column will be cleared. Columns are numbered from 0 to DISP_MAX_X − 1 (see Listing 12.3, PC.C).

color specifies the contents of the attribute byte. Because the character used to clear a column is the space character (i.e., ' '), only the background color will appear. You can thus specify any of the DISP_BGND_??? colors.

Return Value

None

Notes/Warnings

None

Example

```
void Task (void *pdata)
{
    .
    .
    for (;;) {
        .
        PC_DispClrCol(0, DISP_BGND_BLACK);
        .
        .
    }
}
```

12

PC_DispClrRow()

void PC_DispClrRow(INT8U y, INT8U color);

PC_DispClrRow() allows you to clear the contents of a row (all 80 characters).

Arguments

y specifies which row (i.e., line) will be cleared. Rows are numbered from 0 to DISP_MAX_Y − 1 (see Listing 12.3, PC.C).

color specifies the contents of the attribute byte. Because the character used to clear a row is the space character (i.e., ' '), only the background color will appear. You can thus specify any of the DISP_BGND_??? colors.

Return Value

None

Notes/Warnings

None

Example

```
void Task (void *pdata)
{
    .
    .
    for (;;) {
        .
        PC_DispClrRow(10, DISP_BGND_BLACK);
        .
        .
    }
}
```

PC_DispClrScr()

void PC_DispClrScr(INT8U color);

PC_DispClrScr() allows you to clear the entire display.

Arguments

color specifies the contents of the attribute byte. Because the character used to clear the screen is the space character (i.e., ' '), only the background color will appear. You can thus specify any of the DISP_BGND_??? colors.

Return Value

None

Notes/Warnings

You should use DISP_FGND_WHITE instead of DISP_BGND_BLACK because you don't want to leave the attribute field with black on black.

Example

```
void Task (void *pdata)
{
    .
    .

    PC_DispClrScr(DISP_FGND_WHITE);
    for (;;) {
        .
        .
        .
    }
}
```

12

PC_DispStr()

void PC_DispStr(INT8U x, INT8U y, INT8U *s, INT8U color);

PC_DispStr() allows you to display an ASCII string. In fact, you could display an array containing any of 255 characters as long as the array itself is NUL terminated.

Arguments

x and **y** specifies the coordinates (col, row) where the first character will appear. rows (i.e., lines) are numbered from 0 to DISP_MAX_Y − 1, and columns are numbered from 0 to DISP_MAX_X − 1 (see Listing 12.3, PC.C).

s is a pointer to the array of characters to display. The array *must* be NUL terminated. Note that you can display any characters from 0x01 to 0xFF. You can see what characters (i.e., symbols) will be displayed based on the value of c by running the test code provided in this book as follows:

    C:\SOFTWARE\BLOCKS\SAMPLE\TEST > **TEST display**

color specifies the contents of the attribute byte and thus the color combination of the characters to be displayed. You can add one DISP_FGND_??? (see Listing 12.4, PC.H) and one DISP_BGND_??? (see Listing 12.4, PC.H) to obtain the desired color combination.

Return Value

None

Notes/Warnings

All the characters of the string or array will be displayed with the same color attributes.

Example #1

The code below displays the current value of a global variable called Temperature. The color used depends on whether the temperature is below 100 (white), below 200 (yellow) or if it exceeds 200 (blinking white on a red background).

```
FP32 Temperature;



void Task (void *pdata)
{
    char s[20];


        .

        .

    PC_DispStr(0, 0, "Temperature:", DISP_FGND_YELLOW + DISP_BGND_BLUE);
```

```
   for (;;) {
      sprintf(s, "%6.1f", Temperature);
      if (Temperature < 100.0) {
         color = DISP_FGND_WHITE;
      } else if (Temperature < 200.0) {
         color = DISP_FGND_YELLOW;
      } else {
         color = DISP_FGND_WHITE + DISP_BGND_RED + DISP_BLINK;
      PC_DispStr(13, 0, s, color);
         .
         .
         .
   }
}
```

Example #2

The code below displays a square box 10 characters wide by 7 characters high in the center of the screen.

```
INT8U  Box[7][11] = {
    {0xDA, 0xC4, 0xC4, 0xC4, 0xC4, 0xC4, 0xC4, 0xC4, 0xC4, 0xBF, 0x00},
    {0xB3, 0x20, 0x20, 0x20, 0x20, 0x20, 0x20, 0x20, 0x20, 0xB3, 0x00},
    {0xB3, 0x20, 0x20, 0x20, 0x20, 0x20, 0x20, 0x20, 0x20, 0xB3, 0x00},
    {0xB3, 0x20, 0x20, 0x20, 0x20, 0x20, 0x20, 0x20, 0x20, 0xB3, 0x00},
    {0xB3, 0x20, 0x20, 0x20, 0x20, 0x20, 0x20, 0x20, 0x20, 0xB3, 0x00},
    {0xB3, 0x20, 0x20, 0x20, 0x20, 0x20, 0x20, 0x20, 0x20, 0xB3, 0x00},
    {0xC0, 0xC4, 0xC4, 0xC4, 0xC4, 0xC4, 0xC4, 0xC4, 0xC4, 0xD9, 0x00}
};


void Task (void *pdata)
{
    INT8U i;


        .

        .

    for (i = 0; i < 7; i++) {
        PC_DispStr(35, i + 9, Box[i], DISP_FGND_WHITE);
    }
    for (;;) {

        .

        .

    }
}
```

12

PC_DOSReturn()

void PC_DOSReturn(void);

PC_DOSReturn() allows your application to return back to DOS. It is assumed that you have previously called PC_DOSSaveReturn() in order to save the processor's important registers in order to properly return to DOS. See section 12.01 for a description on how to use this function.

Arguments

None

Return Value

None

Notes/Warnings

You *must* have called PC_DOSSaveReturn() prior to calling PC_DOSReturn().

Example

```
void Task (void *pdata)
{
    INT16U key;

    .
    .

    for (;;) {

        .

        .

        if (PC_GetKey(&key) == TRUE) {
            if (key == 0x1B) {
                PC_DOSReturn();                    /* Return to DOS */
            }
        }

        .

        .

    }
}
```

PC_DOSSaveReturn()

void PC_DOSSaveReturn(void);

PC_DOSSaveReturn() allows your application to save the processor's important registers in order to properly return to DOS before you actually start multitasking with μC/OS-II. You would normally call this function from main() as shown in the example code provided below.

Arguments

None

Return Value

None

Notes/Warnings

You *must* call this function prior to setting μC/OS-II's context switch vector (as shown below).

Example

```
void  main (void)
{
    OSInit();                       /* Initialize uC/OS-II          */
    .
    PC_DOSSaveReturn();             /* Save DOS's environment       */
    .
    PC_VectSet(uCOS, OSCtxSw);  /* uC/OS-II's context switch vector */
    OSTaskCreate(…);
    .
    .
    OSStart();                      /* Start multitasking           */
}
```

12

PC_ElapsedInit()

void PC_ElapsedInit(void);

PC_ElapsedInit() is invoked to compute the overhead associated with the PC_ElapsedStart() and PC_ElapsedStop() calls. This allows PC_ElapsedStop() to return return the execution time (in microseconds) of the code you are trying to measure.

Arguments

None

Return Value

None

Notes/Warnings

You *must* call this function prior to calling either PC_ElapsedStart() and PC_ElapsedStop().

Example

```
void  main (void)
{
    OSInit();               /* Initialize uC/OS-II           */
    .
    .
    PC_ElapsedInit();       /* Compute overhead of elapse meas. */
    .
    .
    OSStart();              /* Start multitasking            */
}
```

PC_ElapsedStart()

`void PC_ElapsedStart(void);`

`PC_ElapsedStart()` is used in conjunction with `PC_ElapsedStop()` to measure the execution time of some of your application code.

Arguments

None

Return Value

None

Notes/Warnings

You *must* call `PC_ElapsedInit()` before you use either `PC_ElapsedStart()` and `PC_ElapsedStop()`.

This function is non-reentrant and cannot be called by multiple tasks without proper protection mechanisms (i.e., semaphores, locking the scheduler, etc.).

The execution time of your code must be less than 54.93 milliseconds in order for the elapsed time measurement functions to work properly.

12

Example

```
void  main (void)
{
    OSInit();                    /* Initialize uC/OS-II              */
    .
    .
    PC_ElapsedInit();            /* Compute overhead of elapse meas. */
    .
    .
    OSStart();                   /* Start multitasking               */
}




void Task (void *pdata)
{
    INT16U time_us;


    .
    .
    for (;;) {
        .

        .
        PC_ElapsedStart();
        /* Code you want to measure the execution time */
        time_us = PC_ElaspedStop();
        .
        .
    }
}
```

PC_ElapsedStop()

INT16U PC_ElapsedStop(void);

PC_ElapsedStop() is used in conjunction with PC_ElapsedStart() to measure the execution time of some of your application code.

Arguments

None

Return Value

The execution time of your code that was wrapped between PC_ElapsedStart() and PC_ElapsedStop(). The execution time is returned in microseconds.

Notes/Warnings

You *must* call PC_ElapsedInit() before you use either PC_ElapsedStart() and PC_ElapsedStop().

This function is non-reentrant and cannot be called by multiple tasks without proper protection mechanisms (i.e., semaphores, locking the scheduler, etc.).

The execution time of your code must be less than 54.93 milliseconds in order for the elapsed time measurement functions to work properly.

Example

See PC_ElapsedStart() on page 511.

12

PC_GetDateTime()

void PC_GetDateTime(char *s);

PC_GetDateTime() is used to obtain the current date and time from the PC's real-time clock chip and return this information in an ASCII string that can hold at least 19 characters.

Arguments

s is a pointer to the storage area where the ASCII string will be deposited. The format of the ASCII string is:

```
"YYYY-MM-DD  HH:MM:SS"
```

and requires 21 bytes of storage (note that there is 2 spaces between the date and the time).

Return Value

None

Notes/Warnings

None

Example

```
void Task (void *pdata)
{
    char s[80];

        .

        .

    for (;;) {

          .

          .

        PC_GetDateTime(&s[0]);
        PC_DispStr(0, 24, s, DISP_FGND_WHITE);

          .

          .

    }
}
```

PC_GetKey()

BOOLEAN PC_GetDateTime(INT16S *key);

PC_GetKey() is used to see if a key was pressed at the PC's keyboard and if so, obtain the value of the key pressed. You would normally invoke this function every so often (i.e., poll the keyboard) to see if a key was pressed. Note that the PC actually obtains key presses through an ISR and buffers key presses. Up to 10 keys are buffered by the PC.

Arguments

key is a pointer to where the key value will be stored. If no key has been pressed, the value will contain 0x00.

Return Value

TRUE is a key was pressed and FALSE otherwise.

Notes/Warnings

None

Example

```
void Task (void *pdata)
{
    INT16S    key;
    BOOLEAN   avail;

    .

    .

    for (;;) {

        .

        .

        avail = PC_GetKey(&key);
        if (avail == TRUE) {
            /* Process key pressed */
        }

        .

        .

    }
}
```

12

PC_SetTickRate()

void PC_SetTickRate(INT16U freq);

PC_SetTickRate() is used to change the PC's tick rate from the standard 18.20648 Hz to something faster. A tick rate of 200 Hz is a multiple of 18.20648 Hz (the multiple is 11).

Arguments

freq is the desired frequency of the ticker.

Return Value

None

Notes/Warnings

You can only make the ticker faster than 18.20648 Hz.
The higher the frequency, the more overhead you will impose on the CPU.
You will have to change OSTickISR() in order to account for the increased rate (see *MicroC/OS-II, The Real-Time Kernel*, R&D Books, ISBN 0-87930-543-6).

```
void  Task (void *pdata)
{
    .
    .
    OS_ENTER_CRITICAL();
    PC_VectSet(0x08, OSTickISR);
    PC_SetTickRate(400);          /* Reprogram PC's tick rate to 400 Hz */
    OS_EXIT_CRITICAL();
    .
    .
    for (;;) {
        .
        .
    }
}
```

PC_VectGet()

void *PC_VectGet(INT8U vect);

PC_VectGet() is used to obtain the address of the interrupt handler specified by the interrupt vector number. An 80x86 processor supports up to 256 interrupt/exception handlers.

Arguments

vect is the interrupt vector number, a number between 0 and 255.

Return Value

The address of the current interrupt/exception handler for the specified interrupt vector number.

Notes/Warnings

Vector number 0 corresponds to the RESET handler.

It is assumed that the 80x86 code is compiled using the 'large model' option and thus all pointers returned are 'far pointers'.

It is assumed that the 80x86 is running in 'real mode'.

Example

```
void Task (void *pdata)
{
    void (*p_tick_isr)(void);


    .
    .

    p_tick_isr = PC_VectGet(0x08);  /* Get tick handler address */
    .

    .

    for (;;) {

        .
        .

    }
}
```

12

PC_VectSet()

```
void PC_VectSet(INT8U vect, void *(pisr)(void));
```

PC_VectSet() is used to set the contents of an interrupt vector table location. An 80x86 processor supports up to 256 interrupt/exception handlers.

Arguments

vect is the interrupt vector number, a number between 0 and 255.

pisr is the address of the interrupt/exception handler.

Return Value

None

Notes/Warnings

You should be careful when setting interrupt vectors. Some interrupt vectors are used by the operating system (DOS and/or µC/OS-II).

It is assumed that the 80x86 code is compiled using the 'large model' option and thus all pointers returned are 'far pointers'.

If your interrupt handler works in conjunction with µC/OS-II, it must follow the rules imposed by µC/OS-II (see page 91 of *MicroC/OS-II, The Real-Time Kernel*, ISBN 0-87930-543-6).

Example

```
void  InterruptHandler (void)
{
}




void  Task (void *pdata)
{
    .

    .

    PC_VectSet(64, InterruptHandler);

    .

    .

    for (;;) {

        .

        .

    }
}
```

12.05 Bibliography

Chappell, Geoff
DOS Internals
Reading, Massachusetts
Addison-Wesley, 1994
ISBN 0-201-60835-9

Tischer, Michael
PC Intern, System Programming, 5th Edition
Grand Rapids, Michigan
Abacus, 1995
ISBN 1-55755-282-7

Villani, Pat
FreeDOS Kernel, An MS-DOS Emulator for Platform Independence & Embedded Systems Development
Lawrence, Kansas
R&D Books, 1996
ISBN 0-87930-436-7

12

Listing 12.3 PC.C

```
/*
********************************************************************************************
*                                    PC SUPPORT FUNCTIONS
*
*                       (c) Copyright 1992-1999, Jean J. Labrosse, Weston, FL
*                                    All Rights Reserved
*
* File : PC.C
* By   : Jean J. Labrosse
********************************************************************************************
*/

#include "includes.h"

/*
********************************************************************************************
*                                        CONSTANTS
********************************************************************************************
*/
#define  DISP_BASE                  0xB800      /* Base segment of display (0xB800=VGA, 0xB000=Mono) */
#define  DISP_MAX_X                     80      /* Maximum number of columns                         */
#define  DISP_MAX_Y                     25      /* Maximum number of rows                            */

#define  TICK_T0_8254_CWR            0x43      /* 8254 PIT Control Word Register address.           */
#define  TICK_T0_8254_CTR0           0x40      /* 8254 PIT Timer 0 Register address.                */
#define  TICK_T0_8254_CTR1           0x41      /* 8254 PIT Timer 1 Register address.                */
#define  TICK_T0_8254_CTR2           0x42      /* 8254 PIT Timer 2 Register address.                */

#define  TICK_T0_8254_CTR0_MODE3     0x36      /* 8254 PIT Binary Mode 3 for Counter 0 control word. */
#define  TICK_T0_8254_CTR2_MODE0     0xB0      /* 8254 PIT Binary Mode 0 for Counter 2 control word. */
#define  TICK_T0_8254_CTR2_LATCH     0x80      /* 8254 PIT Latch command control word               */

#define  VECT_TICK                   0x08      /* Vector number for 82C54 timer tick                */
#define  VECT_DOS_CHAIN              0x81      /* Vector number used to chain DOS                   */

/*
********************************************************************************************
*                                  LOCAL GLOBAL VARIABLES
********************************************************************************************
*/

static INT16U    PC_ElapsedOverhead;
static jmp_buf   PC_JumpBuf;
static BOOLEAN   PC_ExitFlag;
void             (*PC_TickISR)(void);

/*$PAGE*/
```

Listing 12.3 (continued) PC.C

```
/*
*********************************************************************************************************
*                         DISPLAY A SINGLE CHARACTER AT 'X' & 'Y' COORDINATE
*
* Description : This function writes a single character anywhere on the PC's screen.  This function
*               writes directly to video RAM instead of using the BIOS for speed reasons.  It assumed
*               that the video adapter is VGA compatible.  Video RAM starts at absolute address
*               0x000B8000.  Each character on the screen is composed of two bytes: the ASCII character
*               to appear on the screen followed by a video attribute.  An attribute of 0x07 displays
*               the character in WHITE with a black background.
*
* Arguments   : x      corresponds to the desired column on the screen.  Valid columns numbers are from
*                      0 to 79.  Column 0 corresponds to the leftmost column.
*               y      corresponds to the desired row on the screen.  Valid row numbers are from 0 to 24.
*                      Line 0 corresponds to the topmost row.
*               c      Is the ASCII character to display.  You can also specify a character with a
*                      numeric value higher than 128.  In this case, special character based graphics
*                      will be displayed.
*               color  specifies the foreground/background color to use (see PC.H for available choices)
*                      and whether the character will blink or not.
*
* Returns     : None
*********************************************************************************************************
*/
void PC_DispChar (INT8U x, INT8U y, INT8U c, INT8U color)
{
    INT8U  far *pscr;
    INT16U     offset;


    offset   = (INT16U)y * DISP_MAX_X * 2 + (INT16U)x * 2;  /* Calculate position on the screen       */
    pscr     = (INT8U far *)MK_FP(DISP_BASE, offset);
    *pscr++  = c;                                           /* Put character in video RAM             */
    *pscr    = color;                                       /* Put video attribute in video RAM       */
}
/*$PAGE*/
```

12

Listing 12.3 (continued) PC.C

```
/*
*********************************************************************************************
*                                    CLEAR A COLUMN
*
* Description : This function clears one of the 80 columns on the PC's screen by directly accessing video
*               RAM instead of using the BIOS.  It assumed that the video adapter is VGA compatible.
*               Video RAM starts at absolute address 0x000B8000.  Each character on the screen is
*               composed of two bytes: the ASCII character to appear on the screen followed by a video
*               attribute.  An attribute of 0x07 displays the character in WHITE with a black background.
*
* Arguments   : x           corresponds to the desired column to clear.  Valid column numbers are from
*                           0 to 79.  Column 0 corresponds to the leftmost column.
*
*               color       specifies the foreground/background color combination to use
*                           (see PC.H for available choices)
*
* Returns     : None
*********************************************************************************************
*/
void PC_DispClrCol (INT8U x, INT8U color)
{
    INT8U far *pscr;
    INT8U     i;


    pscr = (INT8U far *)MK_FP(DISP_BASE, (INT16U)x * 2);
    for (i = 0; i < DISP_MAX_Y; i++) {
        *pscr++ = ' ';                      /* Put ' ' character in video RAM           */
        *pscr   = color;                    /* Put video attribute in video RAM         */
        pscr    = pscr + DISP_MAX_X * 2;     /* Position on next row                     */
    }
}
/*$PAGE*/
```

Listing 12.3 (continued)   PC.C

```
/*
*********************************************************************************************************
*                                          CLEAR A ROW
*
* Description : This function clears one of the 25 lines on the PC's screen by directly accessing video
*               RAM instead of using the BIOS.  It assumed that the video adapter is VGA compatible.
*               Video RAM starts at absolute address 0x000B8000.  Each character on the screen is
*               composed of two bytes: the ASCII character to appear on the screen followed by a video
*               attribute.  An attribute of 0x07 displays the character in WHITE with a black background.
*
* Arguments   : y             corresponds to the desired row to clear.  Valid row numbers are from
*                             0 to 24.  Row 0 corresponds to the topmost line.
*
*               color         specifies the foreground/background color combination to use
*                             (see PC.H for available choices)
*
* Returns     : None
*********************************************************************************************************
*/
void PC_DispClrRow (INT8U y, INT8U color)
{
    INT8U far *pscr;
    INT8U      i;


    pscr = (INT8U far *)MK_FP(DISP_BASE, (INT16U)y * DISP_MAX_X * 2);
    for (i = 0; i < DISP_MAX_X; i++) {
        *pscr++ = ' ';                          /* Put ' ' character in video RAM             */
        *pscr++ = color;                        /* Put video attribute in video RAM           */
    }
}
/*$PAGE*/
```

12

Listing 12.3 (continued) PC.C

```
/*
*********************************************************************************************************
*                                           CLEAR SCREEN
*
* Description : This function clears the PC's screen by directly accessing video RAM instead of using
*               the BIOS.  It assumed that the video adapter is VGA compatible.  Video RAM starts at
*               absolute address 0x000B8000.  Each character on the screen is composed of two bytes:
*               the ASCII character to appear on the screen followed by a video attribute.  An attribute
*               of 0x07 displays the character in WHITE with a black background.
*
* Arguments   : color   specifies the foreground/background color combination to use
*                       (see PC.H for available choices)
*
* Returns     : None
*********************************************************************************************************
*/
void PC_DispClrScr (INT8U color)
{
    INT8U   far *pscr;
    INT16U      i;


    pscr = (INT8U far *)MK_FP(DISP_BASE, 0x0000);
    for (i = 0; i < (DISP_MAX_X * DISP_MAX_Y); i++) { /* PC display has 80 columns and 25 lines      */
        *pscr++ = ' ';                               /* Put ' ' character in video RAM              */
        *pscr++ = color;                             /* Put video attribute in video RAM            */
    }
}
/*$PAGE*/
```

Listing 12.3 (continued)   PC.C

```
/*
*********************************************************************************************
*                          DISPLAY A STRING  AT 'X' & 'Y' COORDINATE
*
* Description : This function writes an ASCII string anywhere on the PC's screen.  This function writes
*               directly to video RAM instead of using the BIOS for speed reasons.  It assumed that the
*               video adapter is VGA compatible.  Video RAM starts at absolute address 0x000B8000.  Each
*               character on the screen is composed of two bytes: the ASCII character to appear on the
*               screen followed by a video attribute.  An attribute of 0x07 displays the character in
*               WHITE with a black background.
*
* Arguments   : x       corresponds to the desired column on the screen.  Valid columns numbers are from
*                       0 to 79.  Column 0 corresponds to the leftmost column.
*               y       corresponds to the desired row on the screen.  Valid row numbers are from 0 to 24.
*                       Line 0 corresponds to the topmost row.
*               s       Is the ASCII string to display.  You can also specify a string containing
*                       characters with numeric values higher than 128.  In this case, special character
*                       based graphics will be displayed.
*               color   specifies the foreground/background color to use (see PC.H for available choices)
*                       and whether the characters will blink or not.
*
* Returns     : None
*********************************************************************************************
*/
void PC_DispStr (INT8U x, INT8U y, INT8U *s, INT8U color)
{
    INT8U   far *pscr;
    INT16U      offset;


    offset = (INT16U)y * DISP_MAX_X * 2 + (INT16U)x * 2;   /* Calculate position of 1st character   */
    pscr   = (INT8U far *)MK_FP(DISP_BASE, offset);
    while (*s) {
        *pscr++ = *s++;                                     /* Put character in video RAM            */
        *pscr++ = color;                                   /* Put video attribute in video RAM      */
    }
}
/*$PAGE*/
```

12

Listing 12.3 (continued) PC.C

```
/*
*********************************************************************************************
*                                  RETURN TO DOS
*
* Description : This functions returns control back to DOS by doing a 'long jump' back to the saved
*               location stored in 'PC_JumpBuf'.  The saved location was established by the function
*               'PC_DOSSaveReturn()'.  After execution of the long jump, execution will resume at the
*               line following the 'set jump' back in 'PC_DOSSaveReturn()'.  Setting the flag
*               'PC_ExitFlag' to TRUE ensures that the 'if' statement in 'PC_DOSSaveReturn()' executes.
*
* Arguments   : None
*
* Returns     : None
*********************************************************************************************
*/
void PC_DOSReturn (void)
{
    PC_ExitFlag = TRUE;                                 /* Indicate we are returning to DOS        */
    longjmp(PC_JumpBuf, 1);                             /* Jump back to saved environment          */
}
/*$PAGE*/
/*
*********************************************************************************************
*                              SAVE DOS RETURN LOCATION
*
* Description : This function saves the location of where we are in DOS so that it can be recovered.
*               This allows us to abort multitasking under uC/OS-II and return back to DOS as if we had
*               never left.  When this function is called by 'main()', it sets 'PC_ExitFlag' to FALSE
*               so that we don't take the 'if' branch.  Instead, the CPU registers are saved in the
*               long jump buffer 'PC_JumpBuf' and we simply return to the caller.  If a 'long jump' is
*               performed using the jump buffer then, execution would resume at the 'if' statement and
*               this time, if 'PC_ExitFlag' is set to TRUE then we would execute the 'if' statements and
*               restore the DOS environment.
*
* Arguments   : None
*
* Returns     : None
*********************************************************************************************
*/
void PC_DOSSaveReturn (void)
{
    PC_ExitFlag  = FALSE;                               /* Indicate that we are not exiting yet!   */
    OSTickDOSCtr =    1;                                /* Initialize the DOS tick counter         */
    PC_TickISR   = PC_VectGet(VECT_TICK);              /* Get MS-DOS's tick vector                */

    OS_ENTER_CRITICAL();
    PC_VectSet(VECT_DOS_CHAIN, PC_TickISR);            /* Store MS-DOS's tick to chain            */
    OS_EXIT_CRITICAL();

    setjmp(PC_JumpBuf);                                /* Capture where we are in DOS             */
    if (PC_ExitFlag == TRUE) {                         /* See if we are exiting back to DOS       */
        OS_ENTER_CRITICAL();
        PC_SetTickRate(18);                            /* Restore tick rate to 18.2 Hz            */
        PC_VectSet(VECT_TICK, PC_TickISR);            /* Restore DOS's tick vector               */
        OS_EXIT_CRITICAL();
        PC_DispClrScr(DISP_FGND_WHITE + DISP_BGND_BLACK); /* Clear the display                    */
        exit(0);                                       /* Return to DOS                           */
    }
}
/*$PAGE*/
```

Listing 12.3 (continued) PC.C

```
/*
*********************************************************************************************
*                                 ELAPSED TIME INITIALIZATION
*
* Description : This function initialize the elapsed time module by determining how long the START and
*               STOP functions take to execute.  In other words, this function calibrates this module
*               to account for the processing time of the START and STOP functions.
*
* Arguments   : None.
*
* Returns     : None.
*********************************************************************************************
*/
void PC_ElapsedInit(void)
{
    PC_ElapsedOverhead = 0;
    PC_ElapsedStart();
    PC_ElapsedOverhead = PC_ElapsedStop();
}
/*$PAGE*/


/*
*********************************************************************************************
*                                 INITIALIZE PC'S TIMER #2
*
* Description : This function initialize the PC's Timer #2 to be used to measure the time between events.
*               Timer #2 will be running when the function returns.
*
* Arguments   : None.
*
* Returns     : None.
*********************************************************************************************
*/
void PC_ElapsedStart(void)
{
    INT8U  data;


    OS_ENTER_CRITICAL();
    data  = (INT8U)inp(0x61);                                  /* Disable timer #2                   */
    data &= 0xFE;
    outp(0x61, data);
    outp(TICK_T0_8254_CWR,  TICK_T0_8254_CTR2_MODE0);          /* Program timer #2 for Mode 0        */
    outp(TICK_T0_8254_CTR2, 0xFF);
    outp(TICK_T0_8254_CTR2, 0xFF);
    data |= 0x01;                                              /* Start the timer                    */
    outp(0x61, data);
    OS_EXIT_CRITICAL();
}
/*$PAGE*/
```

12

Listing 12.3 (continued) PC.C

```
/*
*********************************************************************************************
*                          STOP THE PC'S TIMER #2 AND GET ELAPSED TIME
*
* Description : This function stops the PC's Timer #2, obtains the elapsed counts from when it was
*               started and converts the elapsed counts to micro-seconds.
*
* Arguments   : None.
*
* Returns     : The number of micro-seconds since the timer was last started.
*
* Notes       : - The returned time accounts for the processing time of the START and STOP functions.
*               - 54926 represents 54926S-16 or, 0.838097 which is used to convert timer counts to
*                 micro-seconds.  The clock source for the PC's timer #2 is 1.19318 MHz (or 0.838097 uS)
*********************************************************************************************
*/
INT16U PC_ElapsedStop(void)
{
    INT8U   data;
    INT8U   low;
    INT8U   high;
    INT16U  cnts;


    OS_ENTER_CRITICAL();
    data  = (INT8U)inp(0x61);                             /* Disable the timer                 */
    data &= 0xFE;
    outp(0x61, data);
    outp(TICK_TO_8254_CWR, TICK_TO_8254_CTR2_LATCH);      /* Latch the timer value             */
    low  = inp(TICK_TO_8254_CTR2);
    high = inp(TICK_TO_8254_CTR2);
    cnts = (INT16U)0xFFFF - (((INT16U)high << 8) + (INT16U)low); /* Compute time it took for operation */
    OS_EXIT_CRITICAL();
    return ((INT16U)((ULONG)cnts * 54926L >> 16) - PC_ElapsedOverhead);
}
/*$PAGE*/
```

Listing 12.3 (continued) PC.C

```
/*
*********************************************************************************************
*                                GET THE CURRENT DATE AND TIME
*
* Description: This function obtains the current date and time from the PC.
*
* Arguments  : s      is a pointer to where the ASCII string of the current date and time will be stored.
*                     You must allocate at least 21 bytes (includes the NUL) of storage in the return
*                     string.  The date and time will be formatted as follows:
*
*                        "YYYY-MM-DD  HH:MM:SS"
*
* Returns    : none
*********************************************************************************************
*/
void PC_GetDateTime (char *s)
{
    struct time now;
    struct date today;


    gettime(&now);
    getdate(&today);
    sprintf(s, "%04d-%02d-%02d  %02d:%02d:%02d",
                today.da_year,
                today.da_mon,
                today.da_day,
                now.ti_hour,
                now.ti_min,
                now.ti_sec);
}
/*$PAGE*/
```

12

Listing 12.3 (continued) PC.C

```
/*
*********************************************************************************************
*                             CHECK AND GET KEYBOARD KEY
*
* Description: This function checks to see if a key has been pressed at the keyboard and returns TRUE if
*              so.  Also, if a key is pressed, the key is read and copied where the argument is pointing
*              to.
*
* Arguments  : c    is a pointer to where the read key will be stored.
*
* Returns    : TRUE  if a key was pressed
*              FALSE otherwise
*********************************************************************************************
*/
BOOLEAN PC_GetKey (INT16S *c)
{
    if (kbhit()) {                              /* See if a key has been pressed            */
        *c = (INT16S)getch();                   /* Get key pressed                          */
        return (TRUE);
    } else {
        *c = 0x00;                              /* No key pressed                           */
        return (FALSE);
    }
}
/*$PAGE*/
```

Listing 12.3 (continued) PC.C

```
/*
**********************************************************************************************
*                              SET THE PC'S TICK FREQUENCY
*
* Description: This function is called to change the tick rate of a PC.
*
* Arguments  : freq     is the desired frequency of the ticker (in Hz)
*
* Returns    : none
*
* Notes      : 1) The magic number 2386360 is actually twice the input frequency of the 8254 chip which
*                 is always 1.193180 MHz.
*              2) The equation computes the counts needed to load into the 8254.  The strange equation
*                 is actually used to round the number using integer arithmetic.  This is equivalent to
*                 the floating point equation:
*
*                          1193180.0 Hz
*                count = ------------ + 0.5
*                              freq
**********************************************************************************************
*/
void PC_SetTickRate (INT16U freq)
{
    INT16U  count;


    if (freq == 18) {                              /* See if we need to restore the DOS frequency      */
        count = 0;
    } else if (freq > 0) {
                                                   /* Compute 8254 counts for desired frequency and ... */
                                                   /* ... round to nearest count                        */
        count = (INT16U)(((INT32U)2386360L / freq + 1) >> 1);
    } else {
        count = 0;
    }
    OS_ENTER_CRITICAL();
    outp(TICK_T0_8254_CWR,  TICK_T0_8254_CTR0_MODE3); /* Load the 8254 with desired frequency           */
    outp(TICK_T0_8254_CTR0, count & 0xFF);            /* Low  byte                                      */
    outp(TICK_T0_8254_CTR0, (count >> 8) & 0xFF);     /* High byte                                      */
    OS_EXIT_CRITICAL();
}
/*$PAGE*/
```

Listing 12.3 (continued) PC.C

```c
/*
*********************************************************************************************
*                               OBTAIN INTERRUPT VECTOR
*
* Description: This function reads the pointer stored at the specified vector.
*
* Arguments  : vect  is the desired interrupt vector number, a number between 0 and 255.
*
* Returns    : The address of the Interrupt handler stored at the desired vector location.
*********************************************************************************************
*/
void *PC_VectGet (INT8U vect)
{
    INT16U    *pvect;
    INT16U    off;
    INT16U    seg;


    pvect = (INT16U *)MK_FP(0x0000, vect * 4);       /* Point into IVT at desired vector location   */
    OS_ENTER_CRITICAL();
    off   = *pvect++;                                /* Obtain the vector's OFFSET                  */
    seg   = *pvect;                                  /* Obtain the vector's SEGMENT                 */
    OS_EXIT_CRITICAL();
    return (MK_FP(seg, off));
}


/*
*********************************************************************************************
*                               INSTALL INTERRUPT VECTOR
*
* Description: This function sets an interrupt vector in the interrupt vector table.
*
* Arguments  : vect  is the desired interrupt vector number, a number between 0 and 255.
*              isr   is a pointer to a function to execute when the interrupt or exception occurs.
*
* Returns    : none
*********************************************************************************************
*/
void PC_VectSet (INT8U vect, void (*isr)(void))
{
    INT16U  *pvect;


    pvect    = (INT16U *)MK_FP(0x0000, vect * 4);    /* Point into IVT at desired vector location    */
    OS_ENTER_CRITICAL();
    *pvect++ = (INT16U)FP_OFF(isr);                  /* Store ISR offset                            */
    *pvect   = (INT16U)FP_SEG(isr);                  /* Store ISR segment                           */
    OS_EXIT_CRITICAL();
}
```

Listing 12.4   PC.H

```
/*
*********************************************************************************************
*                                 PC SUPPORT FUNCTIONS
*
*                     (c) Copyright 1992-1999, Jean J. Labrosse, Weston, FL
*                                      All Rights Reserved
*
* File : PC.H
* By   : Jean J. Labrosse
*********************************************************************************************
*/


/*
*********************************************************************************************
*                                      CONSTANTS
*                            COLOR ATTRIBUTES FOR VGA MONITOR
*
* Description: These #defines are used in the PC_Disp???() functions.  The 'color' argument in these
*              function MUST specify a 'foreground' color, a 'background' and whether the display will
*              blink or not.  If you don't specify a background color, BLACK is assumed.  You would
*              specify a color combination as follows:
*
*              PC_DispChar(0, 0, 'A', DISP_FGND_WHITE + DISP_BGND_BLUE + DISP_BLINK);
*
*              To have the ASCII character 'A' blink with a white letter on a blue background.
*********************************************************************************************
*/
#define DISP_FGND_BLACK          0x00
#define DISP_FGND_BLUE           0x01
#define DISP_FGND_GREEN          0x02
#define DISP_FGND_CYAN           0x03
#define DISP_FGND_RED            0x04
#define DISP_FGND_PURPLE         0x05
#define DISP_FGND_BROWN          0x06
#define DISP_FGND_LIGHT_GRAY     0x07
#define DISP_FGND_DARK_GRAY      0x08
#define DISP_FGND_LIGHT_BLUE     0x09
#define DISP_FGND_LIGHT_GREEN    0x0A
#define DISP_FGND_LIGHT_CYAN     0x0B
#define DISP_FGND_LIGHT_RED      0x0C
#define DISP_FGND_LIGHT_PURPLE   0x0D
#define DISP_FGND_YELLOW         0x0E
#define DISP_FGND_WHITE          0x0F

#define DISP_BGND_BLACK          0x00
#define DISP_BGND_BLUE           0x10
#define DISP_BGND_GREEN          0x20
#define DISP_BGND_CYAN           0x30
#define DISP_BGND_RED            0x40
#define DISP_BGND_PURPLE         0x50
#define DISP_BGND_BROWN          0x60
#define DISP_BGND_LIGHT_GRAY     0x70

#define DISP_BLINK               0x80
```

12

Listing 12.4 *(continued)* `PC.H`

```
/*
*************************************************************************************************
*                                    FUNCTION PROTOTYPES
*************************************************************************************************
*/

void    PC_DispChar(INT8U x, INT8U y, INT8U c, INT8U color);
void    PC_DispClrCol(INT8U x, INT8U bgnd_color);
void    PC_DispClrRow(INT8U y, INT8U bgnd_color);
void    PC_DispClrScr(INT8U bgnd_color);
void    PC_DispStr(INT8U x, INT8U y, INT8U *s, INT8U color);

void    PC_DOSReturn(void);
void    PC_DOSSaveReturn(void);

void    PC_ElapsedInit(void);
void    PC_ElapsedStart(void);
INT16U  PC_ElapsedStop(void);

void    PC_GetDateTime(char *s);
BOOLEAN PC_GetKey(INT16S *c);

void    PC_SetTickRate(INT16U freq);

void   *PC_VectGet(INT8U vect);
void    PC_VectSet(INT8U vect, void (*isr)(void));
```

Appendix A

µC/OS-II, The Real-Time Kernel

µC/OS-II is a portable, ROM-able, preemptive, real-time, multitasking kernel that can manage up to 63 tasks. µC/OS-II is comparable in performance to many commercially available kernels. µC/OS-II was written in C with microprocessor-specific code written in assembly language. Assembly language was kept to a minimum so that µC/OS-II can easily be ported to other target microprocessors.

Most modules presented in this book assume that services are provided by a real-time multitasking kernel. Because of this, I have provided, in object form, a scaled down version of *µC/OS-II, The Real-Time Kernel v2.00* that will allow you to test all of the code in this book. In other words, only the features needed to run the examples are provided.

The complete source code (along with a port for the Intel 80x86, large model) for µC/OS-II is available in my book: *MicroC/OS-II, The Real-Time Kernel* (ISBN 0-87930-543-6), also published by R&D Books (See the ad in the back of this book.) The source code for µC/OS-II is available on a floppy diskette (MS-DOS format) which is included with the book. Along with providing the source code for µC/OS-II, the book describes the internals, explains how the kernel works, and allows you to port µC/OS-II to other microprocessors (if needed). You can also obtain port to many processors through the official µC/OS and µC/OS-II web site at www.uCOS-II.com. µC/OS-II provides the following features:

- create and manage up to 63 tasks,
- create and manage a large number of semaphores,
- delay tasks for an integral number of ticks or a user-specified amount of time in hours, minutes, seconds, and milliseconds,
- lock/unlock the scheduler,
- service interrupts,
- allows you to change the priority of tasks,
- lets you delete tasks,
- allows tasks to suspend and resume other tasks,
- manages a large number of message mailboxes and queues for intertask communications,

- provides fixed-sized memory block management,
- manages a 32-bit system clock.

Even though *Embedded Systems Building Blocks, Second Edition* assumes the presence of μC/OS-II, you can easily adapt the code in this book to any other real-time kernel as long as the kernel provides the same services (most other kernels do). If you do not have a real-time kernel, you can easily modify some of the code to work in a *Foreground/Background* environment.

The version of μC/OS-II in this book is provided in object form for the Intel 80x86 Large Model and has been compiled with the Borland's C++ v4.51. The compiler was instructed to generate code for any Intel 80x86 which has hardware floating-point support. You can thus use the code on any PC having either an Intel 80486, Pentium, Pentium-II, Pentium-III and processors from AMD which have floating-point hardware.

I configured μC/OS-II to limit the number of tasks to 15 and the number of semaphores to 10. You will not be able to invoke either the queue or memory management feature of μC/OS-II because they have been disabled in OS_CFG.H.

The object code for μC/OS-II is found in the \SOFTWARE\BLOCKS\SAMPLE\OBJ directory in these files:

uCOS_II.OBJ	μC/OS-II (compiled from the C source)
OS_CPU_C.OBJ	80x86 microprocessor specifics, large model with hardware floating-point support(compiled from the C source)
OS_CPU_A.OBJ	80x86 microprocessor specifics (assembled from the ASM source)

You will need to link these files with your application if you are planning on using this version of μC/OS-II.

When you use μC/OS-II, you will need to include the following header files in your source code:

- OS_CPU.H which is found in \SOFTWARE\uCOS-II\Ix86L-FP\BC45\SOURCE
- UCOS_II.H which is found in \SOFTWARE\uCOS-II\SOURCE.

You should note that OS_CPU.H must be listed first. Also, you cannot change any of the #defines that are provided in these files. If you do, your application may not work properly. The only way to change the #defines is to obtain the full source code for μC/OS-II (see forementioned ad).

I included a μC/OS-II mini-reference section which contains only the functions used in this book.

OSInit()

void OSInit(void);

OSInit() is used to initialize µC/OS-II. OSInit() must be called prior to calling OSStart() which will actually start multitasking.

Arguments

None

Return Value

None

Notes/Warnings

OSInit() must be called before OSStart().

Example

```
void main (void)
{
    .
    .
    OSInit();        /* Initialize uC/OS-II */
    .
    .
    OSStart();       /* Start Multitasking  */
}
```

A

OSSemCreate()

OS_EVENT *OSSemCreate(WORD value);

OSSemCreate() is used to create and initialize a semaphore. A semaphore is used to:

1. Allow a task to synchronize with either an ISR or a task
2. Gain exclusive access to a resource
3. Signal the occurrence of an event

Arguments

value is the initial value of the semaphore. The initial value of the semaphore is allowed to be between 0 and 65535.

Return Value

A pointer to the event control block allocated to the semaphore. If no event control block is available, OSSemCreate() will return a NULL pointer.

Notes/Warnings

Semaphores must be created before they are used.

Example

```
OS_EVENT *DispSem;


void main(void)
{
    .
    .
    OSInit();                          /* Initialize µC/OS-II         */
    .
    .
    DispSem = OSSemCreate(1);          /* Create Display Semaphore     */
    .
    .
    OSStart();                         /* Start Multitasking           */
}
```

OSSemPend()

```
void OSSemPend(OS_EVENT *pevent, INT16U timeout, INT8U *err);
```

OSSemPend() is used when a task desires to get exclusive access to a resource, synchronize its activities with an ISR, a task, or until an event occurs. If a task calls OSSemPend() and the value of the semaphore is greater than 0, then OSSemPend() will decrement the semaphore and return to its caller. However, if the value of the semaphore is equal to zero, OSSemPend() places the calling task in the waiting list for the semaphore. The task will thus wait until a task or an ISR signals the semaphore or, the specified timeout expires. If the semaphore is signaled before the timeout expires, µC/OS-II will resume the highest priority task that is waiting for the semaphore. A pended task that has been suspended with OSTaskSuspend() can obtain the semaphore. The task will, however, remain suspended until the task is resumed by calling OSTaskResume().

Arguments

pevent is a pointer to the semaphore. This pointer is returned to your application when the semaphore is created (see OSSemCreate() on page 538).

timeout is used to allow the task to resume execution if a message is not received from the mailbox within the specified number of clock ticks. A timeout value of 0 indicates that the task desires to wait forever for the message. The maximum timeout is 65535 clock ticks. The timeout value is not synchronized with the clock tick. The timeout count starts being decremented on the next clock tick which could potentially occur immediately.

err is a pointer to a variable which will be used to hold an error code. OSSemPend() sets *err to either:

1. OS_NO_ERR, the semaphore is available
2. OS_TIMEOUT, the semaphore was not signaled within the specified timeout
3. OS_ERR_PEND_ISR, you called this function from an ISR and µC/OS-II would have to suspend the ISR. In general, you should not call OSMboxPend(). µC/OS-II checks for this situation in case you do anyway.

Return Value

None

Notes/Warnings

Semaphores must be created before they are used.

Example

```
OS_EVENT *DispSem;


void DispTask(void *pdata)
{
    INT8U  err;


    pdata = pdata;
    for (;;) {
        .
        .
        OSSemPend(DispSem, 0, &err);
        .                /* The only way this task continues is if … */
        .                /* … the semaphore is signaled!            */
    }
}
```

OSSemPost()

```
INT8U OSSemPost(OS_EVENT *pevent);
```

A semaphore is signaled by calling OSSemPost(). If the semaphore value is greater than or equal to zero, the semaphore is incremented and OSSemPost() returns to its caller. If tasks are waiting for the semaphore to be signaled then, OSSemPost() removes the highest priority task pending (waiting) for the semaphore from the waiting list and makes this task ready to run. The scheduler is then called to determine if the awakened task is now the highest priority task ready to run.

Arguments

pevent is a pointer to the semaphore. This pointer is returned to your application when the semaphore is created (see OSSemCreate() on page 538).

Return Value

OSSemPost() returns one of these two error codes:

1. OS_NO_ERR, if the semaphore was successfully signaled
2. OS_SEM_OVF, if the semaphore count overflowed

Notes/Warnings

Semaphores must be created before they are used.

A

Example

```
OS_EVENT *DispSem;


void TaskX(void *pdata)
{
    INT8U   err;


    pdata = pdata;
    for (;;) {
        .
        .
        err = OSSemPost(DispSem);
        if (err == OS_NO_ERR) {
            .                         /* Semaphore signaled      */
            .
        } else {
            .                         /* Semaphore has overflowed */
            .
        }
        .
        .
    }
}
```

OSStart()

void OSStart(void);

OSStart() is used to start multitasking under µC/OS-II.

Arguments

None

Return Value

None

Notes/Warnings

OSInit() must be called prior to calling OSStart(). OSStart() should only called once by your application code. If you do call OSStart() more than once, OSStart() will not do anything on the second and subsequent calls.

Example

```
void main(void)
{
        .                   /* User Code              */
        .
   OSInit();                /* Initialize µC/OS-II  */
        .                   /* User Code              */
        .
   OSStart();               /* Start Multitasking   */
}
```

A

OSStatInit()

`void OSStatInit(void);`

`OSStatInit()` is used to have µC/OS-II determine the maximum value that a 32-bit counter can reach when no other task is executing. This function must be called when there is only one task created in your application and, when multitasking has started. In other words, this function must be called from the first, and only created task.

Arguments

None

Return Value

None

Notes/Warnings

None

Example

```
void FirstAndOnlyTask (void *pdata)
{
   .
   .
   .
   OSStatInit();       /* Compute CPU capacity with no task running */
   .
   OSTaskCreate(...);   /* Create the other tasks                    */
   OSTaskCreate(...);
   .
   for (;;) {
      .
      .
   }
}
```

OSTaskCreate()

`INT8U OSTaskCreate(void (*task)(void *pd), void *pdata, OS_STK *ptos, INT8U prio);`

`OSTaskCreate()` allows an application to create a task so it can be managed by µC/OS-II. Tasks can either be created prior to the start of multitasking or by a running task. A task cannot be created by an ISR. A task *must* be written as an infinite loop as shown in the example below and, *must not* return.

    `OSTaskCreate()` is used for backward compatibility with µC/OS and when the added features of `OSTaskCreateExt()` are not needed.

    Depending on how the stack frame was built, your task will either have interrupts enabled or disabled. You will need to check with the processor specific code for details.

Arguments

task is a pointer to the task's code.

pdata is a pointer to an optional data area which can be used to pass parameters to the task when it is created. Where the task is concerned, it thinks it was invoked and passed the argument pdata as follows:

```
void Task (void *pdata)
{
        .                       /* Do something with 'pdata'              */
    for (;;) {                  /* Task body, always an infinite loop.    */
            .

            .
        /* Must call one of the following services:                       */
        /*      OSMboxPend()                                              */
        /*      OSQPend()                                                 */
        /*      OSSemPend()                                               */
        /*      OSTimeDly()                                               */
        /*      OSTimeDlyHMSM()                                           */
        /*      OSTaskSuspend()       (Suspend self)                      */
        /*      OSTaskDel()           (Delete  self)                      */
            .

            .
    }
}
```

ptos is a pointer to the task's top of stack. The stack is used to store local variables, function parameters, return addresses, and CPU registers during an interrupt. The size of the stack is determined by the task's requirements and, the anticipated interrupt nesting. Determining the size of the stack involves knowing how many bytes are required for storage of local variables for the task itself, all nested functions, as well as requirements for interrupts (accounting for nesting). If the configuration constant `OS_STK_GROWTH` is set to 1, the stack is assumed to grow downward (i.e., from high memory to low

A

memory). `ptos` will thus need to point to the highest *valid* memory location on the stack. If `OS_STK_GROWTH` is set to 0, the stack is assumed to grow in the opposite direction (i.e., from low memory to high memory).

prio is the task priority. A unique priority number must be assigned to each task and the lower the number, the higher the priority.

Return Value

`OSTaskCreate()` returns one of the following error codes:

1. `OS_NO_ERR`, if the function was successful
2. `OS_PRIO_EXIST`, if the requested priority already exist

Notes/Warnings

The stack *must* be declared with the `OS_STK` type.

A task *must* always invoke one of the services provided by µC/OS-II to either wait for time to expire, suspend the task or, wait an event to occur (wait on a mailbox, queue, or semaphore). This will allow other tasks to gain control of the CPU.

You should not use task priorities 0, 1, 2, 3 and `OS_LOWEST_PRIO-3`, `OS_LOWEST_PRIO-2`, `OS_LOWEST_PRIO-1` and `OS_LOWEST_PRIO` because they are reserved for µC/OS-II's use. This thus leaves you with up to 56 application tasks.

Example

This examples shows that the argument that `Task1()` will receive is not used and thus, the pointer `pdata` is set to `NULL`. Note that I assumed that the stack grows from high memory to low memory because I passed the address of the highest valid memory location of the stack `Task1Stk[]`. If the stack grows in the opposite direction for the processor you are using, you will need to pass `Task1Stk[0]` as the task's top-of-stack.

```
OS_STK  *Task1Stk[1024];
INT8U    Task1Data;



void main(void)
{
   INT8U err;


      .
   OSInit();                    /* Initialize µC/OS-II          */
      .
   OSTaskCreate(Task1,
                (void *)&Task1Data,
                &Task1Stk[1023],
                25);
      .
   OSStart();                   /* Start Multitasking           */
}



void Task1(void *pdata)
{
   pdata = pdata;
   for (;;) {
         .                      /* Task code                    */

         .
   }
}
```

A

OSTaskCreateExt()

```
INT8U OSTaskCreateExt(void (*task)(void *pd), void *pdata, OS_STK *ptos, INT8U prio,
       INT16U id, OS_STK *pbos, INT32U stk_size, void *pext, INT16U opt);
```

OSTaskCreateExt() allows an application to create a task so it can be managed by µC/OS-II. This function serves the same purpose as OSTaskCreate() except that it allows you to specify additional information about your task to µC/OS-II. Tasks can either be created prior to the start of multitasking or by a running task. A task cannot be created by an ISR. A task *must* be written as an infinite loop as shown in the example code below and, *must not* return. Depending on how the stack frame was built, your task will either have interrupts enabled or disabled. You will need to check with the processor specific code for details. You should note that the first four arguments are exactly the same as the ones for OSTaskCreate(). This was done to simplify the migration to this new, and more powerful function.

Arguments

task is a pointer to the task's code.

pdata is a pointer to an optional data area which can be used to pass parameters to the task when it is created. Where the task is concerned, it thinks it was invoked and passed the argument pdata as follows:

```
void Task (void *pdata)
{
        .                      /* Do something with 'pdata'               */
    for (;;) {                 /* Task body, always an infinite loop.     */
        .
        .

    /* Must call one of the following services:                           */
        /*      OSMboxPend()                                              */
        /*      OSQPend()                                                 */
        /*      OSSemPend()                                               */
        /*      OSTimeDly()                                               */
        /*      OSTimeDlyHMSM()                                           */
        /*      OSTaskSuspend()       (Suspend self)                      */
        /*      OSTaskDel()           (Delete  self)                      */
        .
        .

    }
}
```

ptos is a pointer to the task's top of stack. The stack is used to store local variables, function parameters, return addresses, and CPU registers during an interrupt. The size of this stack is determined by the task's requirements, and the anticipated interrupt nesting. Determining the size of the stack involves knowing how many bytes are required for storage of local variables for the task itself, all nested functions, as well as requirements for interrupts (accounting for nesting). If the configuration constant OS_STK_GROWTH is set to 1, the stack is assumed to grow downward (i.e., from high memory to low memory). ptos will thus need to point to the highest *valid* memory location on the stack. If OS_STK_GROWTH is set to 0, the stack is assumed to grow in the opposite direction (i.e., from low memory to high memory).

prio is the task priority. A unique priority number must be assigned to each task and the lower the number, the higher the priority (i.e., the importance) of the task.

id is the task's ID number. At this time, the ID is not currently used in any other function and has simply been added in OSTaskCreateExt() for future expansion. You should set the id to the same value as the task's priority.

pbos is a pointer to the task's bottom of stack. If the configuration constant OS_STK_GROWTH is set to 1, the stack is assumed to grow downward (i.e., from high memory to low memory) and thus, pbos must point to the lowest valid stack location. If OS_STK_GROWTH is set to 0, the stack is assumed to grow in the opposite direction (i.e., from low memory to high memory) and thus, pbos must point to the highest valid stack location. pbos is used by the stack checking function OSTaskStkChk().

stk_size is used to specify the size of the task's stack (in number of elements). If OS_STK is set to INT8U, then stk_size corresponds to the number of bytes available on the stack. If OS_STK is set to INT16U, then stk_size contains the number of 16-bit entries available on the stack. Finally, if OS_STK is set to INT32U, then stk_size contains the number of 32-bit entries available on the stack.

pext is a pointer to a user supplied memory location (typically a data structure) which is used as a TCB extension. For example, this user memory can hold the contents of floating-point registers during a context switch, the time each task takes to execute, the number of times the task is switched-in, etc.

opt contains task specific options. The lower 8 bits are reserved by µC/OS-II but you can use the upper 8 bits for application specific options. Each option consist of a bit. The option is selected when the bit is set. The current version of µC/OS-II supports the following options:

- OS_TASK_OPT_STK_CHK specifies whether stack checking is allowed for the task.
- OS_TASK_OPT_STK_CLR specifies whether the stack needs to be cleared.
- OS_TASK_OPT_SAVE_FP specifies whether floating-point registers will be saved.

You should refer to uCOS_II.H for other options, i.e., OS_TASK_OPT_???.

Return Value

OSTaskCreateExt() returns one of the following error codes:

1. OS_NO_ERR, if the function was successful
2. OS_PRIO_EXIST, if the requested priority already exist

Notes/Warnings

The stack *must* be declared with the OS_STK type.

A task *must* always invoke one of the services provided by µC/OS-II to either wait for time to expire, suspend the task or, wait an event to occur (wait on a mailbox, queue, or semaphore). This will allow other tasks to gain control of the CPU.

A

You should not use task priorities 0, 1, 2, 3 and OS_LOWEST_PRIO-3, OS_LOWEST_PRIO-2, OS_LOWEST_PRIO-1 and OS_LOWEST_PRIO because they are reserved for µC/OS-II's use. This thus leaves you with up to 56 application tasks.

Example

The task control block is extended (1) using a 'user defined' data structure called TASK_USER_DATA (2) which, in this case, contains the name of the task as well as other fields. The task name is initialized with the strcpy() standard library function (3). Note that stack checking has been enabled (4) for this task and thus, you are allowed to call OSTaskStkChk(). Also, we assume here that the stack grown downward (5) on the processor used (i.e., OS_STK_GROWTH is set to 1). TOS stands for 'Top-Of-Stack' and BOS stands for 'Bottom-Of-Stack'.

```
typedef struct {                    /* (2) User defined data structure */
  char      TaskName[20];
  INT16U    TaskCtr;
  INT16U    TaskExecTime;
  INT32U    TaskTotExecTime;
} TASK_USER_DATA;


OS_STK            *TaskStk[1024];
TASK_USER_DATA    TaskUserData;



void main(void)
{
   INT8U err;


   .
   OSInit();                                /* Initialize µC/OS-II       */
   .
   strcpy(TaskUserData.TaskName, "MyTaskName");  /* (3) Name of task     */
   err = OSTaskCreateExt(Task,
            (void *)0,
            &TaskStk[1023],             /* (5) Stack grows down (TOS) */
            10,
            10,
            &TaskStk[0],                /* (5) Stack grows down (BOS) */
            1024,
            (void *)&TaskUserData,      /* (1) TCB Extension          */
            OS_TASK_OPT_STK_CHK);       /* (4) Stack checking enabled */

   .
   OSStart();                               /* Start Multitasking        */
}
```

```
void Task(void *pdata)
{
   pdata = pdata;          /* Avoid compiler warning          */
   for (;;) {
         .                 /* Task code                       */

         .
   }
}
```

A

OSTimeDly()

void OSTimeDly(INT16U ticks);

OSTimeDly() allows a task to delay itself for a number of clock ticks. Rescheduling always occurs when the number of clock ticks is greater than zero. Valid delays range from 0 to 65535 ticks. A delay of 0 means that the task will not be delayed and OSTimeDly() will return immediately to the caller. The actual delay time depends on the tick rate (see OS_TICKS_PER_SEC in the configuration file OS_CFG.H).

Arguments

ticks is the number of clock ticks to delay the current task.

Return Value

None

Notes/Warnings

Note that calling this function with a delay of 0 results in no delay and thus the function returns immediately to the caller. To ensure that a task delays for the specified number of ticks, you should consider using a delay value that is one tick higher. For example, to delay a task for at least 10 ticks, you should specify a value of 11.

Example

```
void TaskX(void *pdata)
{
   for (;;) {
      .
      .
      OSTimeDly(10);        /* Delay task for 10 clock ticks */
      .
      .
   }
}
```

OSTimeDlyHMSM()

```
void OSTimeDlyHMSM(INT8U hours, INT8U minutes, INT8U seconds, INT8U milli);
```

OSTimeDlyHMSM() allows a task to delay itself for a user-specified amount of time specified in hours, minutes, seconds, and milliseconds. This is a more convenient and natural format than ticks. Rescheduling always occurs when at least one of the parameters is non-zero.

Arguments

hours is the number of hours that the task will be delayed. The valid range of values is from 0 to 255 hours.

minutes is the number of minutes that the task will be delayed. The valid range of values is from 0 to 59.

seconds is the number of seconds that the task will be delayed. The valid range of values is from 0 to 59.

milli is the number of milliseconds that the task will be delayed. The valid range of values is from 0 to 999. Note that the resolution of this argument is in multiples of the tick rate. For instance, if the tick rate is set to 10 mS then a delay of 5 mS would result in no delay. The delay is actually rounded to the nearest tick. Thus, a delay of 15 mS would actually result in a delay of 20 mS.

Return Value

OSTimeDlyHMSM() returns one of the following error codes:

1. OS_NO_ERR, if you specified valid arguments and the call was successful.
2. OS_TIME_INVALID_MINUTES, if the minutes argument is greater than 59.
3. OS_TIME_INVALID_SECONDS, if the seconds argument is greater than 59.
4. OS_TIME_INVALID_MS, if the milliseconds argument is greater than 999.
5. OS_TIME_ZERO_DLY, if all four arguments are 0.

Notes/Warnings

Note that calling this function with a delay of 0 hours, 0 minutes, 0 seconds, and 0 milliseconds results in no delay and thus the function returns immediately to the caller. Also, if the total delay time ends up being larger than 65535 clock ticks then, you will not be able to abort the delay and resume the task by calling OSTimeDlyResume().

A

Example

```
void TaskX(void *pdata)
{
   for (;;) {
      .
      .

      OSTimeDlyHMSM(0, 0, 1, 0);  /* Delay task for 1 second */
      .
      .
   }
}
```

OSVersion()

`INT16U OSVersion(void);`

OSVersion() is used to obtain the current version of μC/OS-II.

Arguments

None

Return Value

The version is returned as x.yy multiplied by 100. In other words, version 2.00 is returned as 200.

Notes/Warnings

None

Example

```
void TaskX(void *pdata)
{
    INT16U os_version;


    for (;;) {

        .
        .

        os_version = OSVersion();   /* Obtain uC/OS-II's version   */

        .
        .

    }
}
```

A

OS_ENTER_CRITICAL() and
OS_EXIT_CRITICAL()

OS_ENTER_CRITICAL() and OS_EXIT_CRITICAL() are macros which are used to disable and enable the processor's interrupts, respectively.

Arguments

None

Return Value

None

Notes/Warnings

These macros must be used in pair.

Example

```
INT32U Val;


void TaskX(void *pdata)
{
    for (;;) {
        .
        .
        OS_ENTER_CRITICAL();    /* Disable interrupts    */
        .
                                /* Access critical code  */
        .
        OS_EXIT_CRITICAL();     /* Enable  interrupts    */ .
        .
        .
    }
}
```

Listing A.1   OS_CPU.H

```
/*
*********************************************************************************************
*                                       uC/OS-II
*                                   The Real-Time Kernel
*
*                      (c) Copyright 1992-1999, Jean J. Labrosse, Weston, FL
*                                   All Rights Reserved
*
*                                 80x86/80x88 Specific code
*                                    LARGE MEMORY MODEL
*
*                                   Borland C/C++ V4.51
*
* File        : OS_CPU.H
* By          : Jean J. Labrosse
* Port Version : V1.00
*********************************************************************************************
*/

#ifdef  OS_CPU_GLOBALS
#define OS_CPU_EXT
#else
#define OS_CPU_EXT   extern
#endif

/*
*********************************************************************************************
*                                       DATA TYPES
*                                   (Compiler Specific)
*********************************************************************************************
*/

typedef unsigned char  BOOLEAN;
typedef unsigned char  INT8U;          /* Unsigned  8 bit quantity                        */
typedef signed   char  INT8S;          /* Signed    8 bit quantity                        */
typedef unsigned int   INT16U;         /* Unsigned 16 bit quantity                        */
typedef signed   int   INT16S;         /* Signed   16 bit quantity                        */
typedef unsigned long  INT32U;         /* Unsigned 32 bit quantity                        */
typedef signed   long  INT32S;         /* Signed   32 bit quantity                        */
typedef float          FP32;           /* Single precision floating point                 */
typedef double         FP64;           /* Double precision floating point                 */

typedef unsigned int   OS_STK;         /* Each stack entry is 16-bit wide                 */

#define BYTE           INT8S           /* Define data types for backward compatibility ...*/
#define UBYTE          INT8U           /* ... to uC/OS V1.xx.  Not actually needed for ...*/
#define WORD           INT16S          /* ... uC/OS-II.                                   */
#define UWORD          INT16U
#define LONG           INT32S
#define ULONG          INT32U
```

A

Listing A.1 (continued)   OS_CPU.H

```
/*
*********************************************************************************************
*                            Intel 80x86 (Real-Mode, Large Model)
*
* Method #1:  Disable/Enable interrupts using simple instructions.  After critical section, interrupts
*             will be enabled even if they were disabled before entering the critical section.  You MUST
*             change the constant in OS_CPU_A.ASM, function OSIntCtxSw() from 10 to 8.
*
* Method #2:  Disable/Enable interrupts by preserving the state of interrupts.  In other words, if
*             interrupts were disabled before entering the critical section, they will be disabled when
*             leaving the critical section.  You MUST change the constant in OS_CPU_A.ASM, function
*             OSIntCtxSw() from 8 to 10.
*********************************************************************************************
*/
#define  OS_CRITICAL_METHOD    2

#if      OS_CRITICAL_METHOD == 1
#define  OS_ENTER_CRITICAL()   asm  CLI                    /* Disable interrupts                   */
#define  OS_EXIT_CRITICAL()    asm  STI                    /* Enable  interrupts                   */
#endif

#if      OS_CRITICAL_METHOD == 2
#define  OS_ENTER_CRITICAL()   asm {PUSHF; CLI}            /* Disable interrupts                   */
#define  OS_EXIT_CRITICAL()    asm  POPF                   /* Enable  interrupts                   */
#endif


/*
*********************************************************************************************
*                         Intel 80x86 (Real-Mode, Large Model) Miscellaneous
*********************************************************************************************
*/

#define  OS_STK_GROWTH         1                           /* Stack grows from HIGH to LOW memory on 80x86  */

#define  uCOS                  0x80                        /* Interrupt vector # used for context switch    */

#define  OS_TASK_SW()          asm  INT   uCOS

/*
*********************************************************************************************
*                                      GLOBAL VARIABLES
*********************************************************************************************
*/

OS_CPU_EXT  INT8U  OSTickDOSCtr;      /* Counter used to invoke DOS's tick handler every 'n' ticks   */

/*
*********************************************************************************************
*                                     FUNCTION PROTOTYPES
*********************************************************************************************
*/

void  OSFPInit(void);
void  OSFPRestore(void *pblk);
void  OSFPSave(void *pblk);
```

Listing A.2   *uCOS_II.H*

```
/*
*********************************************************************************************************
*                                            uC/OS-II
*                                       The Real-Time Kernel
*
*                       (c) Copyright 1999, Jean J. Labrosse, Weston, FL
*                                       All Rights Reserved
*
* File : uCOS_II.H
* By   : Jean J. Labrosse
*********************************************************************************************************
*/

/*$PAGE*/
```

A

Listing A.2 *(continued)*   uCOS_II.H

```
/*
********************************************************************************************
*                                      MISCELLANEOUS
********************************************************************************************
*/

#define   OS_VERSION              200     /* Version of uC/OS-II (Vx.yy multiplied by 100)      */

#ifdef   OS_GLOBALS
#define   OS_EXT
#else
#define   OS_EXT   extern
#endif

#define   OS_PRIO_SELF            0xFF    /* Indicate SELF priority                             */

#if      OS_TASK_STAT_EN
#define   OS_N_SYS_TASKS          2                           /* Number of system tasks         */
#else
#define   OS_N_SYS_TASKS          1
#endif

#define   OS_STAT_PRIO           (OS_LOWEST_PRIO - 1)    /* Statistic task priority             */
#define   OS_IDLE_PRIO           (OS_LOWEST_PRIO)        /* IDLE      task priority             */

#define   OS_EVENT_TBL_SIZE   ((OS_LOWEST_PRIO) / 8 + 1)   /* Size of event table               */
#define   OS_RDY_TBL_SIZE     ((OS_LOWEST_PRIO) / 8 + 1)   /* Size of ready table               */

#define   OS_TASK_IDLE_ID         65535   /* I.D. numbers for Idle and Stat tasks              */
#define   OS_TASK_STAT_ID         65534

                                          /* TASK STATUS (Bit definition for OSTCBStat)         */
#define   OS_STAT_RDY             0x00    /* Ready to run                                       */
#define   OS_STAT_SEM             0x01    /* Pending on semaphore                               */
#define   OS_STAT_MBOX            0x02    /* Pending on mailbox                                 */
#define   OS_STAT_Q               0x04    /* Pending on queue                                   */
#define   OS_STAT_SUSPEND         0x08    /* Task is suspended                                  */

#define   OS_EVENT_TYPE_MBOX       1
#define   OS_EVENT_TYPE_Q          2
#define   OS_EVENT_TYPE_SEM        3

                                          /* TASK OPTIONS (see OSTaskCreateExt())               */
#define   OS_TASK_OPT_STK_CHK  0x0001    /* Enable stack checking for the task                 */
#define   OS_TASK_OPT_STK_CLR  0x0002    /* Clear the stack when the task is create            */
#define   OS_TASK_OPT_SAVE_FP  0x0004    /* Save the contents of any floating-point registers  */

#ifndef   FALSE
#define   FALSE                   0
#endif

#ifndef   TRUE
#define   TRUE                    1
#endif
```

Listing A.2 (continued)   uCOS_II.H

```
/*
*********************************************************************************
*                                   ERROR CODES
*********************************************************************************
*/

#define OS_NO_ERR                 0
#define OS_ERR_EVENT_TYPE         1
#define OS_ERR_PEND_ISR           2

#define OS_TIMEOUT               10
#define OS_TASK_NOT_EXIST        11

#define OS_MBOX_FULL             20

#define OS_Q_FULL                30

#define OS_PRIO_EXIST            40
#define OS_PRIO_ERR              41
#define OS_PRIO_INVALID          42

#define OS_SEM_OVF               50

#define OS_TASK_DEL_ERR          60
#define OS_TASK_DEL_IDLE         61
#define OS_TASK_DEL_REQ          62
#define OS_TASK_DEL_ISR          63

#define OS_NO_MORE_TCB           70

#define OS_TIME_NOT_DLY          80
#define OS_TIME_INVALID_MINUTES  81
#define OS_TIME_INVALID_SECONDS  82
#define OS_TIME_INVALID_MILLI    83
#define OS_TIME_ZERO_DLY         84

#define OS_TASK_SUSPEND_PRIO     90
#define OS_TASK_SUSPEND_IDLE     91

#define OS_TASK_RESUME_PRIO     100
#define OS_TASK_NOT_SUSPENDED   101

#define OS_MEM_INVALID_PART     110
#define OS_MEM_INVALID_BLKS     111
#define OS_MEM_INVALID_SIZE     112
#define OS_MEM_NO_FREE_BLKS     113
#define OS_MEM_FULL             114

#define OS_TASK_OPT_ERR         130

/*$PAGE*/
```

A

Listing A.2 (continued) uCOS_II.H

```
/*
*********************************************************************************************************
*                                       EVENT CONTROL BLOCK
*********************************************************************************************************
*/

#if (OS_MAX_EVENTS >= 2)
typedef struct {
    void    *OSEventPtr;                      /* Pointer to message or queue structure                 */
    INT8U   OSEventTbl[OS_EVENT_TBL_SIZE];    /* List of tasks waiting for event to occur              */
    INT16U  OSEventCnt;                       /* Count of used when event is a semaphore               */
    INT8U   OSEventType;                      /* OS_EVENT_TYPE_MBOX, OS_EVENT_TYPE_Q or OS_EVENT_TYPE_SEM */
    INT8U   OSEventGrp;                       /* Group corresponding to tasks waiting for event to occur */
} OS_EVENT;
#endif


/*$PAGE*/
/*
*********************************************************************************************************
*                                       MESSAGE MAILBOX DATA
*********************************************************************************************************
*/

#if OS_MBOX_EN
typedef struct {
    void    *OSMsg;                           /* Pointer to message in mailbox                         */
    INT8U   OSEventTbl[OS_EVENT_TBL_SIZE];    /* List of tasks waiting for event to occur              */
    INT8U   OSEventGrp;                       /* Group corresponding to tasks waiting for event to occur */
} OS_MBOX_DATA;
#endif


/*
*********************************************************************************************************
*                                  MEMORY PARTITION DATA STRUCTURES
*********************************************************************************************************
*/

#if OS_MEM_EN && (OS_MAX_MEM_PART >= 2)
typedef struct {                             /* MEMORY CONTROL BLOCK                                   */
    void    *OSMemAddr;                       /* Pointer to beginning of memory partition              */
    void    *OSMemFreeList;                   /* Pointer to list of free memory blocks                 */
    INT32U  OSMemBlkSize;                     /* Size (in bytes) of each block of memory               */
    INT32U  OSMemNBlks;                       /* Total number of blocks in this partition              */
    INT32U  OSMemNFree;                       /* Number of memory blocks remaining in this partition   */
} OS_MEM;


typedef struct {
    void    *OSAddr;                          /* Pointer to the beginning address of the memory partition */
    void    *OSFreeList;                      /* Pointer to the beginning of the free list of memory blocks */
    INT32U  OSBlkSize;                        /* Size (in bytes) of each memory block                  */
    INT32U  OSNBlks;                          /* Total number of blocks in the partition               */
    INT32U  OSNFree;                          /* Number of memory blocks free                          */
    INT32U  OSNUsed;                          /* Number of memory blocks used                          */
} OS_MEM_DATA;
#endif


/*$PAGE*/
```

Listing A.2 (continued)   uCOS_II.H

```
/*
*********************************************************************************
*                                MESSAGE QUEUE DATA
*********************************************************************************
*/

#if OS_Q_EN
typedef struct {
    void    *OSMsg;                              /* Pointer to next message to be extracted from queue   */
    INT16U  OSNMsgs;                            /* Number of messages in message queue                   */
    INT16U  OSQSize;                            /* Size of message queue                                 */
    INT8U   OSEventTbl[OS_EVENT_TBL_SIZE];      /* List of tasks waiting for event to occur              */
    INT8U   OSEventGrp;                         /* Group corresponding to tasks waiting for event to occur */
} OS_Q_DATA;
#endif


/*
*********************************************************************************
*                                SEMAPHORE DATA
*********************************************************************************
*/

#if OS_SEM_EN
typedef struct {
    INT16U  OSCnt;                              /* Semaphore count                                       */
    INT8U   OSEventTbl[OS_EVENT_TBL_SIZE];      /* List of tasks waiting for event to occur              */
    INT8U   OSEventGrp;                         /* Group corresponding to tasks waiting for event to occur */
} OS_SEM_DATA;
#endif


/*
*********************************************************************************
*                                TASK STACK DATA
*********************************************************************************
*/

#if OS_TASK_CREATE_EXT_EN
typedef struct {
    INT32U  OSFree;                             /* Number of free bytes on the stack                     */
    INT32U  OSUsed;                             /* Number of bytes used on the stack                     */
} OS_STK_DATA;
#endif

/*$PAGE*/
```

A

Listing A.2 (continued)   uCOS_II.H

```
/*
*********************************************************************************************************
*                                           TASK CONTROL BLOCK
*********************************************************************************************************
*/

typedef struct os_tcb {
    OS_STK          *OSTCBStkPtr;           /* Pointer to current top of stack                        */

#if OS_TASK_CREATE_EXT_EN
    void            *OSTCBExtPtr;           /* Pointer to user definable data for TCB extension       */
    OS_STK          *OSTCBStkBottom;        /* Pointer to bottom of stack                             */
    INT32U          OSTCBStkSize;           /* Size of task stack (in number of stack elements)       */
    INT16U          OSTCBOpt;               /* Task options as passed by OSTaskCreateExt()            */
    INT16U          OSTCBId;                /* Task ID (0..65535)                                     */
#endif

    struct os_tcb *OSTCBNext;               /* Pointer to next     TCB in the TCB list                */
    struct os_tcb *OSTCBPrev;               /* Pointer to previous TCB in the TCB list                */

#if (OS_Q_EN && (OS_MAX_QS >= 2)) || OS_MBOX_EN || OS_SEM_EN
    OS_EVENT        *OSTCBEventPtr;         /* Pointer to event control block                         */
#endif

#if (OS_Q_EN && (OS_MAX_QS >= 2)) || OS_MBOX_EN
    void            *OSTCBMsg;              /* Message received from OSMboxPost() or OSQPost()         */
#endif

    INT16U          OSTCBDly;               /* Nbr ticks to delay task or, timeout waiting for event   */
    INT8U           OSTCBStat;              /* Task status                                            */
    INT8U           OSTCBPrio;              /* Task priority (0 == highest, 63 == lowest)             */

    INT8U           OSTCBX;                 /* Bit position in group  corresponding to task priority (0..7) */
    INT8U           OSTCBY;                 /* Index into ready table corresponding to task priority   */
    INT8U           OSTCBBitX;              /* Bit mask to access bit position in ready table          */
    INT8U           OSTCBBitY;              /* Bit mask to access bit position in ready group          */

#if OS_TASK_DEL_EN
    BOOLEAN         OSTCBDelReq;            /* Indicates whether a task needs to delete itself         */
#endif
} OS_TCB;

/*$PAGE*/
```

Listing A.2 *(continued)*   uCOS_II.H

```
/*
*********************************************************************************************
*                                   GLOBAL VARIABLES
*********************************************************************************************
*/

OS_EXT  INT32U       OSCtxSwCtr;                 /* Counter of number of context switches            */

#if     (OS_MAX_EVENTS >= 2)
OS_EXT  OS_EVENT     *OSEventFreeList;           /* Pointer to list of free EVENT control blocks     */
OS_EXT  OS_EVENT     OSEventTbl[OS_MAX_EVENTS];/* Table of EVENT control blocks                     */
#endif

OS_EXT  INT32U       OSIdleCtr;                  /* Idle counter                                     */

#if     OS_TASK_STAT_EN
OS_EXT  INT8S        OSCPUUsage;                 /* Percentage of CPU used                           */
OS_EXT  INT32U       OSIdleCtrMax;               /* Maximum value that idle counter can take in 1 sec. */
OS_EXT  INT32U       OSIdleCtrRun;               /* Value reached by idle counter at run time in 1 sec. */
OS_EXT  BOOLEAN      OSStatRdy;                  /* Flag indicating that the statistic task is ready */
#endif

OS_EXT  INT8U        OSIntNesting;               /* Interrupt nesting level                          */

OS_EXT  INT8U        OSLockNesting;              /* Multitasking lock nesting level                  */

OS_EXT  INT8U        OSPrioCur;                  /* Priority of current task                         */
OS_EXT  INT8U        OSPrioHighRdy;              /* Priority of highest priority task                */

OS_EXT  INT8U        OSRdyGrp;                       /* Ready list group                             */
OS_EXT  INT8U        OSRdyTbl[OS_RDY_TBL_SIZE];      /* Table of tasks which are ready to run        */

OS_EXT  BOOLEAN      OSRunning;                      /* Flag indicating that kernel is running       */

#if     OS_TASK_CREATE_EN  || OS_TASK_CREATE_EXT_EN  || OS_TASK_DEL_EN
OS_EXT  INT8U        OSTaskCtr;                      /* Number of tasks created                      */
#endif

OS_EXT  OS_TCB       *OSTCBCur;                      /* Pointer to currently running TCB             */
OS_EXT  OS_TCB       *OSTCBFreeList;                 /* Pointer to list of free TCBs                 */
OS_EXT  OS_TCB       *OSTCBHighRdy;                  /* Pointer to highest priority TCB ready to run */
OS_EXT  OS_TCB       *OSTCBList;                     /* Pointer to doubly linked list of TCBs        */
OS_EXT  OS_TCB       *OSTCBPrioTbl[OS_LOWEST_PRIO + 1];/* Table of pointers to created TCBs          */

OS_EXT  INT32U       OSTime;                     /* Current value of system time (in ticks)          */


extern  INT8U const  OSMapTbl[8];                /* Priority->Bit Mask lookup table                  */
extern  INT8U const  OSUnMapTbl[256];            /* Priority->Index   lookup table                   */

/*$PAGE*/
```

A

Listing A.2 (continued)   **uCOS_II.H**

```
/*
*********************************************************************************************************
*                                      FUNCTION PROTOTYPES
*                                     (Target Independant Functions)
*********************************************************************************************************
*/

/*
*********************************************************************************************************
*                                      MESSAGE MAILBOX MANAGEMENT
*********************************************************************************************************
*/
#if        OS_MBOX_EN
void       *OSMboxAccept(OS_EVENT *pevent);
OS_EVENT   *OSMboxCreate(void *msg);
void       *OSMboxPend(OS_EVENT *pevent, INT16U timeout, INT8U *err);
INT8U       OSMboxPost(OS_EVENT *pevent, void *msg);
INT8U       OSMboxQuery(OS_EVENT *pevent, OS_MBOX_DATA *pdata);
#endif
/*
*********************************************************************************************************
*                                      MEMORY MANAGEMENT
*********************************************************************************************************
*/
#if        OS_MEM_EN && (OS_MAX_MEM_PART >= 2)
OS_MEM     *OSMemCreate(void *addr, INT32U nblks, INT32U blksize, INT8U *err);
void       *OSMemGet(OS_MEM *pmem, INT8U *err);
INT8U       OSMemPut(OS_MEM *pmem, void *pblk);
INT8U       OSMemQuery(OS_MEM *pmem, OS_MEM_DATA *pdata);
#endif
/*
*********************************************************************************************************
*                                      MESSAGE QUEUE MANAGEMENT
*********************************************************************************************************
*/
#if        OS_Q_EN && (OS_MAX_QS >= 2)
void       *OSQAccept(OS_EVENT *pevent);
OS_EVENT   *OSQCreate(void **start, INT16U size);
INT8U       OSQFlush(OS_EVENT *pevent);
void       *OSQPend(OS_EVENT *pevent, INT16U timeout, INT8U *err);
INT8U       OSQPost(OS_EVENT *pevent, void *msg);
INT8U       OSQPostFront(OS_EVENT *pevent, void *msg);
INT8U       OSQQuery(OS_EVENT *pevent, OS_Q_DATA *pdata);
#endif
/*$PAGE*/
```

Listing A.2 (continued) **uCOS_II.H**

```
/*
*********************************************************************************
*                               SEMAPHORE MANAGEMENT
*********************************************************************************
*/
#if         OS_SEM_EN
INT16U      OSSemAccept(OS_EVENT *pevent);
OS_EVENT    *OSSemCreate(INT16U value);
void        OSSemPend(OS_EVENT *pevent, INT16U timeout, INT8U *err);
INT8U       OSSemPost(OS_EVENT *pevent);
INT8U       OSSemQuery(OS_EVENT *pevent, OS_SEM_DATA *pdata);
#endif
/*
*********************************************************************************
*                               TASK MANAGEMENT
*********************************************************************************
*/
#if         OS_TASK_CHANGE_PRIO_EN
INT8U       OSTaskChangePrio(INT8U oldprio, INT8U newprio);
#endif

INT8U       OSTaskCreate(void (*task)(void *pd), void *pdata, OS_STK *ptos, INT8U prio);

#if         OS_TASK_CREATE_EXT_EN
INT8U       OSTaskCreateExt(void  (*task)(void *pd),
                            void   *pdata,
                            OS_STK *ptos,
                            INT8U  prio,
                            INT16U id,
                            OS_STK *pbos,
                            INT32U stk_size,
                            void   *pext,
                            INT16U opt);
#endif

#if         OS_TASK_DEL_EN
INT8U       OSTaskDel(INT8U prio);
INT8U       OSTaskDelReq(INT8U prio);
#endif

#if         OS_TASK_SUSPEND_EN
INT8U       OSTaskResume(INT8U prio);
INT8U       OSTaskSuspend(INT8U prio);
#endif

#if         OS_TASK_CREATE_EXT_EN
INT8U       OSTaskStkChk(INT8U prio, OS_STK_DATA *pdata);
#endif

INT8U       OSTaskQuery(INT8U prio, OS_TCB *pdata);
```

A

Listing A.2 *(continued)* uCOS_II.H

```
/*
*********************************************************************************************
*                                   TIME MANAGEMENT
*********************************************************************************************
*/
void        OSTimeDly(INT16U ticks);
INT8U       OSTimeDlyHMSM(INT8U hours, INT8U minutes, INT8U seconds, INT16U milli);
INT8U       OSTimeDlyResume(INT8U prio);
INT32U      OSTimeGet(void);
void        OSTimeSet(INT32U ticks);
void        OSTimeTick(void);


/*
*********************************************************************************************
*                                   MISCELLANEOUS
*********************************************************************************************
*/

void        OSInit(void);

void        OSIntEnter(void);
void        OSIntExit(void);

void        OSSchedLock(void);
void        OSSchedUnlock(void);

void        OSStart(void);

void        OSStatInit(void);

INT16U      OSVersion(void);

/*$PAGE*/
```

Listing A.2 (continued)   uCOS_II.H

```
/*
*********************************************************************************************
*                              INTERNAL FUNCTION PROTOTYPES
*                        (Your application MUST NOT call these functions)
*********************************************************************************************
*/

#if        OS_MBOX_EN || OS_Q_EN || OS_SEM_EN
void       OSEventTaskRdy(OS_EVENT *pevent, void *msg, INT8U msk);
void       OSEventTaskWait(OS_EVENT *pevent);
void       OSEventTO(OS_EVENT *pevent);
void       OSEventWaitListInit(OS_EVENT *pevent);
#endif

#if        OS_MEM_EN && (OS_MAX_MEM_PART >= 2)
void       OSMemInit(void);
#endif

#if        OS_Q_EN
void       OSQInit(void);
#endif

void       OSSched(void);

void       OSTaskIdle(void *data);

#if        OS_TASK_STAT_EN
void       OSTaskStat(void *data);
#endif

INT8U      OSTCBInit(INT8U prio, OS_STK *ptos, OS_STK *pbos, INT16U id, INT32U stk_size, void *pext,
INT16U opt);

/*$PAGE*/
```

A

Listing A.2 (continued)   uCOS_II.H

```
/*
*********************************************************************************************************
*                                        FUNCTION PROTOTYPES
*                                      (Target Specific Functions)
*********************************************************************************************************
*/

void        OSCtxSw(void);

void        OSIntCtxSw(void);

void        OSStartHighRdy(void);

void        OSTaskCreateHook(OS_TCB *ptcb);
void        OSTaskDelHook(OS_TCB *ptcb);
void        OSTaskStatHook(void);
OS_STK      *OSTaskStkInit(void (*task)(void *pd), void *pdata, OS_STK *ptos, INT16U opt);
void        OSTaskSwHook(void);

void        OSTickISR(void);

void        OSTimeTickHook(void);
```

Appendix B

Programming Conventions

Conventions should be established early in a project. These conventions are necessary to maintain consistency throughout the project. Adopting conventions increases productivity and simplifies project maintenance. A few years ago, I saw an article in the *Hewlett-Packard Journal* (see Bibliography on page 585) about the processes used by a team of engineers to design the HP54720/10 oscilloscope. One of the aspects of the design consisted of developing a coding convention. "A consistent format made the code much easier to read and understand. At the completion of the project, all of the engineers involved were enthusiastic about using the standard in developing the code". If you are serious about improving your programming skills you should get *Code Complete* by Steve McConnell (see Bibliography on page 585). Steve also highly recommends that you adopt a coding convention before you begin programming. As he says, "It's nearly impossible to change code to match your conventions after the code is written".

In this section I will describe the conventions I have used to develop the software presented in this book.

B.00 Directory Structure

Adopting a consistent directory structure avoids confusion when either more than one programmer is involved in a project, or you are involved in many projects. This section shows the directory structure that I use on a daily basis.

B.00.01 Directory Structure, Products

All software development projects are placed in a \PRODUCTS subdirectory from the root directory. I prefer to create the \PRODUCTS subdirectory because it avoids having a large number of directories in the root directory.

Each project is placed in a subdirectory by itself under the \PRODUCTS directory. Instead of having all files in a project located in a single subdirectory, I like to split project related files in these subdirectories. (There is nothing like looking at a project subdirectory containing dozens of files!). Each product contains a number of subdirectories:

- `\PRODUCTS\project\SOFTWARE`
  This subdirectory contains product specific software. It is assumed that you would use building blocks and thus this directory contains code that is actually specific to the product. The `SOFTWARE` directory further contains subdirectories such as:

  - `\PRODUCTS\project\SOFTWARE\SOURCE`
    This subdirectory contains the actual product specific source code.

  - `\PRODUCTS\project\SOFTWARE\TEST`
    This subdirectory contains the product build instructions (i.e., makefiles, scripts, batch files, etc.) to create a 'test' version of the product to build.

  - `\PRODUCTS\project\SOFTWARE\OBJ`
    This subdirectory contains the compiled and assembled code into relocatable object form of all the files needed to make the product.

  - `\PRODUCTS\project\SOFTWARE\VC`
    This subdirectory contains the version controlled product specific software.

  - `\PRODUCTS\project\SOFTWARE\????`
    You can have additional subdirectories that would contain documentation about the software aspects of your product (`DOC` directory), a directory where you could 'collect' all the source files that make up your product in order to compile and assemble them (`WORK` directory), a directory where you can 'rebuild' any version of a released product (`PROD` directory), and more.

- `\PRODUCTS\project\HARDWARE`
  This subdirectory could contain information about the product's hardware (schematics, PCBs, parts list, wire lists, etc.).

- `\PRODUCTS\project\MECH`
  This subdirectory could contain information about the mechanical aspects of your product (enclosures, injection molds, parts list, etc.).

B.00.02   Directory Structure, Building Blocks

Each building block is placed in a subdirectory by itself under the `\SOFTWARE` directory. The reason the building blocks are placed in a directory from the root is because the building blocks are not supposed to be platform dependent. Below each building block, I have the following subdirectories:

- `\SOFTWARE\building-block\SOURCE`
  This subdirectory contains the source code of the building block.

- `\SOFTWARE\ building-block \DOC`
  This subdirectory contains documentation specific to the building block.

- `\SOFTWARE\ building-block \VC`
  The *VC* (Version Control) subdirectory contains version controlled archive files generated by a version control software package such as the Merant *PVCS Version Manager* (previously called PVCS). This subdirectory contains the revisions and versions of your source code, documentation, and executables. If you are new to version management and configuration building, consult the excellent book by Wayne A. Babich called *Software Configuration Management* or, contact Merant about their excellent software packages.

To remove the frustration of navigating through these subdirectories, I wrote a utility program that allows you to jump to a directory without having to use the DOS change directory command. This utility is called `TO.EXE` and is described in Appendix D.

B.01 C Programming Style

B.01.01 Overview

There are many ways to code a program in C (or any other language). The style you use is just as good as any other as long as you strive to attain the following goals:

- Portability
- Consistency
- Neatness
- Easy maintenance
- Easy understanding
- Simplicity

Whichever style you use, I would emphasize that it should be adopted consistently throughout all your projects. I would further insist that a single style be adopted by all team members in a large project. To this end, I would recommend that a C programming style document be formalized for your organization. Adopting a common coding style reduces code maintenance headaches and costs. Adopting a common style will avoid code rewrites. This section describes the C programming style I use. The main emphasis on the programming style presented here is to make the source code easy to follow and maintain.

I don't like to limit the width of my C source code to 80 characters just because today's monitors only allow you to display 80 characters wide. My limitation is actually how many characters can be printed on an 8.5" × 11" page using compressed mode (17 characters per inch). Using compressed mode, you can accommodate up to 132 characters and have enough room on the left of the page for holes for insertion in a three ring binder. Allowing 132 characters per line prevents having to interleave source code with comments. The code provided in this book uses 105 characters per line. This limitation is imposed by the publisher.

B.01.02 Header

The header of a C source file looks as shown below. Your company name and address can be on the first few lines followed by a title describing the contents of the file. A copyright notice is included to give warning of the proprietary nature of the software.

B

```
/*
*******************************************************************************
*                            Company Name
*                              Address
*
*            (c) Copyright 20xx, Company Name, City, State
*                         All Rights Reserved
*
*
* Filename    :
* Programmer(s):
* Description  :
*******************************************************************************
*/

/*$PAGE*/
```

The name of the file is supplied followed by the name of the programmer(s). The name of the programmer who created the file is given first. The last item in the header is a description of the contents of the file.

I like to dictate when page breaks occur. This is done by inserting the special comment /*$PAGE*/ whenever you want a page break. The file is printed using a utility that I wrote called HPLISTC (see Appendix D). When HPLISTC encounters this comment, it sends a form feed character to the printer.

B.01.03 Revision History

Because of the dynamic nature of software, I always include a section in the source file to describe changes made to the file. You may either maintain version control manually or automate the process by using a version control software package. I prefer to use version control software because it takes care of a number of chores automatically. The version control section contains the different revision levels, date and time and a short description of each of the different revision levels. Revision history should start on a page boundary.

```
/*
*******************************************************************************
*                           REVISION HISTORY
*******************************************************************************
*/

/*$PAGE*/
```

B.01.04 Include Files

The header files needed for your project immediately follow the revision history section. You may either list only the header files required for the module or combine header files in a single header file like I do in a file called INCLUDES.H. I like to use an INCLUDES.H header file because it prevents you from having to remember which header file goes with which source file especially when new modules are added. The only inconvenience is that it takes longer to compile each file.

```
/*
*********************************************************************************
*                              INCLUDE FILES
*********************************************************************************
*/


#include "INCLUDES.H"


/*$PAGE*/
```

B.01.05 Naming Identifiers

C compilers which conform to the ANSI X3J11 standard (most C compilers do by now) allow up to 32 characters for identifier names. Identifiers are variables, structure/union members, functions, macros, #defines, and so on. Descriptive identifiers can be formulated using this 32 character feature and the use of acronyms, abbreviations, and mnemonics (see the Acronym, Abbreviation, and Mnemonic Dictionary, Appendix C). Identifier names should reflect what the identifier is used for. I like to use a hierarchical method when creating an identifier. For instance, the function OSSemPend() indicates that it is part of the operating system (OS), it is a semaphore (Sem) and the operation being performed is to wait (Pend) for the semaphore. This method allows me to group all functions related to semaphores together.

Variable names should be declared on separate lines rather than combining them on a single line. Separate lines make it easy to provide a descriptive comment for each variable.

I use the filename as a prefix for variables that are either local (static) or global to the file. This makes it clear that the variables are being used locally and globally. For example, local and global variables of a file named KEY.C are declared as follows:

```
static   INT16U KeyCharCnt;           /* Number of keys pressed      */
static   char   KeyInBuf[100];        /* Storage buffer to hold chars */
         char   KeyInChar;            /* Character typed             */


/*$PAGE*/
```

Uppercase characters are used to separate words in an identifier. I prefer to use this technique versus making use of the underscore character, (_) because underscores do not add any meaning to names and also use up character spaces.

Global variables (external to the file) can use any name as long as they contain a mixture of uppercase and lowercase characters and are prefixed with the module/file name (i.e., all global keyboard related variable names would be prefixed with the word Key).

Formal arguments to a function and local variables within a function are declared in lowercase. The lowercase makes it obvious that such variables are local to a function; global variables will contain a

B

mixture of upper- and lowercase characters. To make variables readable, you can use the underscore character (i.e., _).

Within functions, certain variable names can be reserved to always have the same meaning. Some examples are given below but others can be used as long as consistency is maintained.

i, j and k	for loop counters.
p1, p2 ... pn	for pointers.
c, c1 ... cn	for characters.
s, s1 ... sn	for strings.
ix, iy and iz	for intermediate integer variables
fx, fy and fz	for intermediate floating point variables

To summarize:

- *formal parameters* in a function declaration should only contain lowercase characters.

- *auto variable* names should only contain lowercase characters.

- *static variables* and *functions* should use the file/module name (or a portion of it) as a prefix and should make use of upper- and lowercase characters.

- *extern variables* and *functions* should use the file/module name (or a portion of it) as a prefix and should make use of upper- and lowercase characters.

B.01.06 Acronyms, Abbreviations, & Mnemonics

When creating names for variables and functions (identifiers), it is often the practice to use acronyms (e.g. OS, ISR, TCB and so on), abbreviations (buf, doc, etc.) and mnemonics (clr, cmp, etc.). The use of acronyms, abbreviations, and mnemonics allows an identifier to be descriptive while requiring fewer characters. Unfortunately, if acronyms, abbreviations, and mnemonics are not used consistently, they may add confusion. To ensure consistency, I have opted to create a list of acronyms, abbreviations, and mnemonics that I use in all my projects. The same acronym, abbreviation, or mnemonic is used throughout, once it is assigned. I call this list the Acronym, Abbreviation, and Mnemonic Dictionary (see Appendix C). As I need more acronyms, abbreviations, or mnemonics, I simply add them to the list.

There might be instances where one list for all products doesn't make sense. For instance, if you are an engineering firm working on a project for different clients and the products that you develop are totally unrelated, then a different list for each project would be more appropriate; the vocabulary for the farming industry is not the same as the vocabulary for the defense industry. I use the rule that if all products are similar, they use the same dictionary.

A common dictionary to a project team will also increase the team's productivity. It is important that consistency be maintained throughout a project, irrespective of the individual programmer(s). Once buf has been agreed to mean "buffer" it should be used by all project members instead of having some individuals use buffer and others use bfr. To further this concept, you should always use buf even if your identifier can accommodate the full name; stick to buf even if you can fully write the word "buffer."

Appendix C provides the acronyms, abbreviations, and mnemonics dictionary that I used for this book. Note that some of the words are the same in both columns. This is done to indicate that there is no acronym, abbreviation, or mnemonic which would better describe the word on the left.

B.01.07 Comments

I find it very difficult to mentally separate code from comments when code and comments are interleaved. Because of this, I never interleave code with comments. Comments are written to the right of the actual C code. When large comments are necessary, they are written in the function description header.

Comments are lined up as shown in the following example. The comment terminators (*/) do not need to be lined up, but for neatness I prefer to do so. It is not necessary to have one comment per line since a comment could apply to a few lines.

```
/*
*********************************************************************************
*                                  atoi()
*
* Description : Function to convert string 's' to an integer.
* Arguments   : ASCII string to convert to integer.
*               (All characters in the string must be decimal digits (0..9))
* Returns     : String converted to an 'int'
*********************************************************************************
*/

int atoi (char *s)
{
    int n;                              /* Partial result of conversion              */


    n = 0;                             /* Initialize result                         */
    while (*s >= '0' && *s <= '9' && *s) {  /* For all valid characters and not end of string */
        n = 10 * n + *s - '0';         /* Convert char to int and add to partial result */
        s++;                           /* Position on next character to convert     */
    }
    return (n);                        /* Return the result of the converted string */
}

/*$PAGE*/
```

B.01.08 #defines

Header files (.H) and C source files (.C) might require that constants and macros be defined. Constants and macros are always written in uppercase with the underscore character used to separate words. Note that hexadecimal numbers are always written with a lowercase x and all uppercase letters for hexadecimal A through F.

B

```
/*
*****************************************************************************
*                            CONSTANTS & MACROS
*****************************************************************************
*/


#define  KEY_FF            0x0F
#define  KEY_CR            0x0D
#define  KEY_BUF_FULL()   (KeyNRd > 0)


/*$PAGE*/
```

B.01.09 Data Types

C allows you to create new data types using the typedef keyword. I declare all data types using upper-case characters, and thus follow the same rule used for constants and macros. There is never a problem confusing constants, macros, and data types because of the context in which they are used. Since different microprocessors have different word length, I like to declare the following data types (assuming Borland C++ V4.51):

```
/*
*****************************************************************************
*                            DATA TYPES
*****************************************************************************
*/


typedef  unsigned char  BOOLEAN;          /* Boolean            */
typedef  unsigned char  INT8U;            /*  8 bit unsigned    */
typedef  char           INT8S;            /*  8 bit signed      */
typedef  unsigned int   INT16U;           /* 16 bit unsigned    */
typedef  int            INT16S;           /* 16 bit signed      */
typedef  unsigned long  INT32U;           /* 32 bit unsigned    */
typedef  long           INT32S;           /* 32 bit signed      */
typedef  float          FP;               /* Floating Point     */


/*$PAGE*/
```

Using these #defines, you will always know the size of each data type.

B.01.010 Local Variables

Some source modules will require that local variables be available. These variables are only needed for the source file (file scope) and should thus be hidden from the other modules. Hiding these variables is accomplished in C by using the `static` keyword. Variables can either be listed in alphabetical order or in functional order.

```
/*
**********************************************************************
*                         LOCAL VARIABLES
**********************************************************************
*/


static  char    KeyBuf[100];
static  INT16S  KeyNRd;


/*$PAGE*/
```

B.01.011 Function Prototypes

This section contains the prototypes (or calling conventions) used by the functions declared in the file. The order in which functions are prototyped should be the order in which the functions are declared in the file. This order allows you to quickly locate the position of a function when the file is printed.

```
/*
**********************************************************************
*                       FUNCTION  PROTOTYPES
**********************************************************************
*/


        void    KeyClrBuf(void);
static  BOOLEAN KeyChkStat(void);
static  INT16S  KeyGetCnt(int ch);


/*$PAGE*/
```

Also note that the `static` keyword, the returned data type, and the function names are all aligned.

B.01.012 Function Declarations

As much as possible, there should only be one function per page when code listings are printed on a printer. A comment block should precede each function. All comment blocks should look as shown below. A description of the function should be given and should include as much information as necessary. If the combination of the comment block and the source code extends past a printed page, a page break should be forced (preferably between the end of the comment block and the start of the function). This allows the function to be on a page by itself and prevents having a page break in the middle of the function. If the function itself is longer than a printed page then it should be broken by a page break comment (`/*$PAGE*/`) in a logical location (i.e., at the end of an `if` statement instead of in the middle of one).

B

More than one small function can be declared on a single page. They should all, however, contain the comment block describing the function. The beginning of a function should start at least two lines after the end of the previous function.

```
/*
*********************************************************************************
*                           CLEAR KEYBOARD BUFFER
*
* Description : Flush keyboard buffer
* Arguments   : none
* Returns     : none
* Notes       : none
*********************************************************************************
*/


void KeyClrBuf (void)
{


}
/*$PAGE*/
```

Functions that are only used within the file should be declared static to hide them from other functions in different files.

By convention, I always call all invocations of the function without a space between the function name and the open parenthesis of the argument list. Because of this, I place a space between the name of the function and the opening parenthesis of the argument list in the function declaration as shown above. This is done so that I can quickly find the function definition using a *grep* utility.

Function names should make use of the filename as a prefix. This prefix makes it easy to locate function declarations in medium to large projects. It also makes it very easy to know where these functions are declared. For example, all functions in a file named KEY.C and functions in a file named VIDEO.C could be declared as follows:

- KEY.C
   KeyGetChar()
   KeyGetLine()
   KeyGetFnctKey()

- VIDEO.C
   VideoGetAttr()
   VideoPutChar()
   VideoPutStr()
   VideoSetAttr()

It's not necessary to use the whole file/module name as a prefix. For example, a file called KEYBOARD.C could have functions starting with Key instead of Keyboard. It is also preferable to use uppercase characters to separate words in a function name instead of using underscores. Again, underscores don't add any meaning to names and they use up character spaces. As mentioned previously, formal parameters and local variables should be in lowercase. This makes it clear that such variables have a scope limited to the function.

Each local variable name *must* be declared on its own line. This allows the programmer to comment each one as needed. Local variables are indented four spaces. The statements for the function are separated from the local variables by three spaces. Declarations of local variables should be physically separated from the statements because they are different.

B.01.013 Indentation

Indentation is important to show the flow of the function. The question is, how many spaces are needed for indentation? One space is obviously not enough while 8 spaces is too much. The compromise I use is four spaces. I also never use TABs, because various printers will interpret TABs differently; and your code may not look as you want. Avoiding TABs does not mean that you can't use the TAB key on your keyboard. A good editor will give you the option to replace TABs with spaces (in this case, 4 spaces).

A space follows the keywords `if`, `for`, `while`, and `do`. The keyword `else` has the privilege of having one before and one after it if curly braces are used. I write `if (condition)` on its own line and the statement(s) to execute on the next following line(s) as follows:

```
if (x < 0)
    z = 25;

if (y > 2) {
    z = 10;
    x = 100;
    p++;
}
```

instead of the following method:

```
if (x < 0) z = 25;
if (y > 2) {z = 10; x = 100; p++;}
```

There are two reasons for this method. The first is that I like to keep the decision portion apart from the execution statement(s). The second reason is consistency with the method I use for `while`, `for`, and `do` statements.

`switch` statements are treated as any other conditional statement. Note that the case statements are lined up with the case label. The important point here is that `switch` statements must be easy to follow. `cases` should also be separated from one another.

```
if (x > 0) {
    y = 10;
    z =  5;
}
```

B

```
if (z < LIM) {
    x = y + z;
    z = 10;
} else {
    x = y - z;
    z = -25;
}
```

```
for (i = 0; i < MAX_ITER; i++) {
    *p2++ = *p1++;
    xx[i] = 0;
}
```

```
while (*p1) {
    *p2++ = *p1++;
    cnt++;
}
```

```
switch (key) {
    case KEY_BS :
        if (cnt > 0) {
            p--;
            cnt--;
        }
        break;

    case KEY_CR :
        *p = NUL;
        break;

    case KEY_LINE_FEED :
        p++;
        break;

    default:
        *p++ = key;
        cnt++;
        break;
}
```

```
do {
    cnt--;
    *p2++ = *p1++;
} while (cnt > 0);
```

B.01.014 Statements & Expressions

All statements and expressions should be made to fit on a single source line. I never use more than one assignment per line such as:

```
x = y = z = 1;
```

Even though this is correct in C, when the variable names get more complicated, the intent might not be as obvious.

The following operators are written with no space around them:

->	Structure pointer operator	p->m
.	Structure member operator	s.m
[]	Array subscripting	a[i]

Parentheses after function names have no space(s) before them. A space should be introduced after each comma to separate each actual argument in a function. Expressions within parentheses are written with no space after the opening parenthesis and no space before the closing parenthesis. Commas and semicolons should have one space after them.

```
strncat(t, s, n);
for (i = 0; i < n; i++)
```

The unary operators are written with no space between them and their operands:

```
!p    ~b    ++i    --j    (long)m    *p    &x    sizeof(k)
```

The binary operators is preceded and followed by one or more spaces, as is the ternary operator:

```
c1 = c2    x + y    i += 2    n > 0 ? n : -n;
```

The keywords if, while, for, switch, and return are followed by one space.

For assignments, numbers are lined up in columns as if you were to add them (assuming you hard-code numbers). The equal signs are also lined up.

```
x        = 100.567;
temp     =  12.700;
var5     =   0.768;
variable =  12;
storage  = &array[0];
```

B

B.01.015 Structures and Unions

Structures are `typedef` since this allows a single name to represent the structure. The structure type is declared using all uppercase characters with underscore characters used to separate words.

```
typedef struct line {          /* Structure that defines a LINE        */
        int  LineStartX;       /* 'X' & 'Y' starting coordinate        */
        int  LineStartY;
        int  LineEndX;         /* 'X' & 'Y' ending   coordinate        */
        int  LineEndY;
        int  LineColor;        /* Color of line to draw                */
} LINE;
```

```
typedef struct point {         /* Structure that defines a POINT       */
        int  PointPosX;        /* 'X' & 'Y' coordinate of point        */
        int  PointPosY;
        int  PointColor;       /* Color of point                       */
} POINT;
```

Structure members start with the same prefix (as shown in the examples above). Member names should start with the name of the structure type (or a portion of it). This makes it clear when pointers are used to reference members of a structure such as:

```
p->LineColor;                  /* We know that 'p' is a pointer to LINE */
```

B.01.016 Reserved Keywords

The following keywords should never be used for identifiers. These keywords are reserved in the C++ language as defined by Bjarne Stroustrup and are thus reserved for future compatibility.

- asm
- class
- delete
- overload
- private
- protected
- public
- friend
- handle
- new
- operator
- template
- this
- virtual

B.02 Bibliography

Babich, Wayne A.
Software Configuration Management
Reading, Massachusetts
Addison-Wesley Publishing Company, 1986
ISBN 0-201-10161-0

Long, David W. and Duff, Christopher P.
A Survey of Processes Used in the Development of Firmware for a
   Multiprocessor Embedded System
Hewlett-Packard Journal, October 1993, p.59-65

McConnell, Steve
Code Complete
Redmond, Washington
Microsoft Press, 1993
ISBN 1-55615-484-4

Merant, Inc.
PVCS Version Manager
735 SW 158th Avenue
Beaverton, OR 97006
(503) 645-1150

Merant, Inc.
PVCS Configuration Builder
735 SW 158th Avenue
Beaverton, OR 97006
(503) 645-1150

B

Appendix C

Acronym, Abbreviation, and Mnemonic Dictionary

Naming functions and variables might seem trivial but good function and variable names are a sign of superior programs. When creating names for variables and functions (identifiers), it is often the practice to use acronyms (e.g., OS, ISR, TCB), abbreviations (buf, doc, etc.), and mnemonics (clr, cmp, etc.). The use of acronyms, abbreviations, and mnemonics allows an identifier to be descriptive while requiring fewer characters. Unfortunately, if acronyms, abbreviations, and mnemonics are not used consistently, they may add confusion. To ensure consistency, I created a list of acronyms, abbreviations, and mnemonics that I use in all my projects. Once assigned, the same acronym, abbreviation, or mnemonic is used consistently. I call this list the Acronym, Abbreviation, and Mnemonic Dictionary. As I need more acronyms, abbreviations, or mnemonics, I simply add them to the list.

Table C.1 shows the acronyms, abbreviations, and mnemonics dictionary that I used for this book. Note that some of the words are the same in both columns. This is done to indicate that there is no acronym, abbreviation, or mnemonic which would better describe the word on the left. A shaded entry in Table C.1 indicates that an acronym, abbreviation, or mnemonic has been used.

You can combine acronyms, abbreviations, and mnemonics to make up a full function or variable name. For example:

1. *Calculate Cursor Position* could be CurCalcPos.
2. *Get Keyboard Buffer* could be KeyBufGet.
3. *Clear Counter Group* could be ClrCtrGrp.
4. *Clear Alarm Status* could be AlmStatClr.

In fact, I prefer to group related items by their names. You may have noticed that all functions and variables within each module in this book start with the acronym, abbreviation, or mnemonic of the module (or file) name. This allows you to quickly know where each function or variable is declared.

587

Table C.1   *Acronyms, abbreviations, and mnemonics dictionary.*

	Description	*Acronym, abbreviation, or mnemonic*
1	Addition	Add
2	Action	Act
3	Analog Input(s)	AI
4	Analog I/O	AIO
5	All	All
6	Alarm	Alm
7	Analog Output(s)	AO
8	Argument(s)	Arg
9	Bar	Bar
10	Bit	Bit
11	Buffer	Buf
12	Bypass	Bypass
13	Calibration	Cal
14	Calculate	Calc
15	Configuration	Cfg
16	Channel	Ch
17	Change	Change
18	Check	Chk
19	Clock	Clk
20	Clear	Clr
21	Clear Screen	Cls
22	Command	Cmd
23	Compare	Cmp
24	Count	Cnt
25	Column	Col
26	Communication	Comm
27	Control	Ctrl
28	Context	Ctx
29	Current	Cur
30	Cursor	Cursor
31	Control Word	CW
32	Date	Date
33	Day	Day
34	Debounce	Debounce
35	Decimal	Dec

Table C.1   ***Acronyms, abbreviations, and mnemonics dictionary.***

	Description	*Acronym, abbreviation, or mnemonic*
36	Decode	Decode
37	Define	Def
38	Delete	Del
39	Detect/Detection	Detect
40	Discrete Input(s)	DI
41	Digit	Dig
42	Discrete I/O	DIO
43	Disable	Dis
44	Display	Disp
45	Division	Div
46	Divisor	Div
47	Division	Div
48	Delay	Dly
49	Discrete Output(s)	DO
50	Day-of-week	DOW
51	Down	Down
52	Dummy	Dummy
53	Edge	Edge
54	Empty	Empty
55	Enable	En
56	Enter	Enter
57	Entries	Entries
58	Error(s)	Err
59	Engineering Units	EU
60	Event(s)	Event
61	Exit	Exit
62	Exponent	Exp
63	Flag	Flag
64	Flush	Flush
65	Function(s)	Fnct
66	Format	Format
67	Fraction	Fract
68	Free	Free
69	Full	Full
70	Gain	Gain

C

Table C.1   Acronyms, abbreviations, and mnemonics dictionary.

	Description	Acronym, abbreviation, or mnemonic
71	Get	Get
72	Group(s)	Grp
73	Handler	Handler
74	Hexadecimal	Hex
75	High	Hi
76	Hit	Hit
77	High Priority Task	HPT
78	Hour(s)	Hr
79	I.D.	Id
80	Idle	Idle
81	Input(s)	In
82	Initialization	Init
83	Initialize	Init
84	Interrupt	Int
85	Invert	Inv
86	Interrupt Service Routine	ISR
87	Index	Ix
88	Key	Key
89	Keyboard	Key
90	Limit	Lim
91	List	List
92	Low	Lo
93	Lower	Lo
94	Lowest	Lo
95	Lock	Lock
96	Low Priority Task	LPT
97	Mantissa	Man
98	Manual	Man
99	Maximum	Max
100	Mailbox	Mbox
101	Minimum	Min
102	Minute(s)	Min
103	Mode	Mode
104	Month	Month
105	Message	Msg

Table C.1   Acronyms, abbreviations, and mnemonics dictionary.

	Description	Acronym, abbreviation, or mnemonic
106	Mask	Msk
107	Multiplication	Mul
108	Multiplex	Mux
109	Number of	N
110	Nesting	Nesting
111	New	New
112	Next	Next
113	Offset	Offset
114	Old	Old
115	Operating System	OS
116	Output	Out
117	Overflow	Ovf
118	Pass	Pass
119	Port	Port
120	Position	Pos
121	Previous	Prev
122	Priority	Prio
123	Printer	Prt
124	Pointer	Ptr
125	Put	Put
126	Queue	Q
127	Raw	Raw
128	Recall	Rcl
129	Read	Rd
130	Ready	Rdy
131	Register	Reg
132	Reset	Reset
133	Resume	Resume
134	Ring	Ring
135	Row	Row
136	Repeat	Rpt
137	Real-Time	RT
138	Running	Running
139	Receive	Rx
140	Scale	Scale

C

Table C.1   Acronyms, abbreviations, and mnemonics dictionary.

	Description	Acronym, abbreviation, or mnemonic
141	Scaling	Scaling
142	Scan	Scan
143	Schedule	Sched
144	Scheduler	Sched
145	Screen	Scr
146	Second(s)	Sec
147	Segment(s)	Seg
148	Select	Sel
149	Semaphore	Sem
150	Set	Set
151	Scale Factor	SF
152	Size	Size
153	Seven-segments	SS
154	Start	Start
155	Statistic(s)	Stat
156	Status	Stat
157	State	State
158	Stack	Stk
159	Stop	Stop
160	String	Str
161	Subtraction	Sub
162	Suspend	Suspend
163	Switch	Sw
164	Synchronize	Sync
165	Task	Task
166	Table	Tbl
167	Threshold	Th
168	Tick	Tick
169	Time	Time
170	Timer	Tmr
171	Trigger	Trig
172	Time-stamp	TS
173	Transmit	Tx
174	Unlock	Unlock
175	Up	Up

Table C.1   ***Acronyms, abbreviations, and mnemonics dictionary.***

	Description	*Acronym, abbreviation, or mnemonic*
176	Update	Update
177	Value	Val
178	Vector	Vect
179	Write	Wr
180	Year	Year
181		
182		
183		
184		
185		
186		
187		
188		
189		
190		
191		
192		
193		
194		
195		
196		
197		
198		
199		

C

HPLISTC and TO

HPLISTC and TO are MS-DOS utilities that are provided in both executable and source form for your convenience.

D.00 HPLISTC

HPLISTC is an MS-DOS utility to print C source files on an HP Laserjet printer. HPLISTC will print your source code in compressed mode; 17 characters per inch (CPI). An 8 1/2" × 11" page (*portrait*) will accommodate up to 132 characters. An 11" × 8 1/2" page (*landscape*) will accommodate up to 175 characters. Once the source code is printed, HPLISTC return the printer to its normal print mode.

The main directory for HPLISTC is C:\SOFTWARE\HPLISTC. HPLISTC is provided in two files: HPLISTC.EXE (see C:\SOFTWARE\HPLISTC\EXE) is the MS-DOS executable and HPLISTC.C (see C:\SOFTWARE\HPLISTC\SOURCE) is the source code.

HPLISTC prints the current date and time, the filename, its extension, and the page number at the top of each page. An optional title can also be printed at the top of each page. As HPLISTC prints the source code, it looks for two special comments: /*$TITLE=*/ or /*$title=*/ and /*$PAGE*/ or /*$page*/.

The /*$TITLE=*/ comment is used to specify the title to be printed on the second line of each page. You can define a new title for each page by using the /*$TITLE=*/ comment. The new title will be printed at the top of the next page. For example:

```
/*$TITLE=Matrix Keyboard Driver*/
```

will set the title for the next page to *Matrix Keyboard Driver*, and this title will be printed on each subsequent page of your source code until the title is changed again.

The /*$PAGE*/ comment is used to force a page break in your source code listing. HPLISTC will not eject the page unless you specifically specify the /*$PAGE*/ comment. If you do not force a page break using the /*$PAGE*/ comment, a short function may be printed on two separate pages if a page break is forced by the printer when it reaches its maximum number of lines per page. The page number on the top of each page actually indicates the number of occurrences of the /*$PAGE*/ comment encountered by LISTC or HPLISTC.

Before each line is printed, HPLISTC prints a line count that can be used for reference purposes. HPLISTC also allows you to print source code in landscape mode. The programs are invoked as follows:

```
HPLISTC filename.ext [L | l] [destination]
```

where filename.ext is the name of the file to print and destination is the destination of the printout. Since HPLISTC sends the output to stdout, the printout can be redirected to a file, a printer (PRN, LPT1, LPT2, etc.), or a COM port (COM1, COM2, etc.) by using the MS-DOS redirector >. By default, HPLISTC outputs to the monitor.

L or l (lowercase L) means to print the file in *landscape* mode, allowing you to print about 175 columns wide!

D.01 *TO*

TO is an MS-DOS utility that allows you to go to a directory without having to type:

```
CD path
```

or

```
CD ..\path
```

TO is probably the MS-DOS utility I use the most because it allows me to move between directories very quickly. At the DOS prompt, you simply type TO followed by the name you associated with a directory and then press Enter as follows:

```
TO name
```

where name is a name you associated with a path. The names and paths are placed in an ASCII file called TO.TBL, which resides in the root directory of the current drive. TO scans TO.TBL for the name you specified on the command line. If the name exists in TO.TBL, the directory is changed to the path specified with the name. If name is not found in TO.TBL, the message Invalid NAME. is displayed.

The main directory for TO is C:\SOFTWARE\TO. TO is provided in three files: TO.EXE (see C:\SOFTWARE\TO\EXE) is the MS-DOS executable, TO.TBL is an example of the correspondance table between your name and the desired directory associated with this name (see C:\SOFTWARE\TO\EXE), and TO.C (see C:\SOFTWARE\TO\SOURCE) is the source code.

The format of TO.TBL is shown in Listing D.1. Note that the name *must* be separated from the desired path by a comma.

Listing D.1   Format of TO.TBL

```
name,   path
name,   path
    .       .
    .       .
name,   path
```

An example of TO.TBL is shown in Listing D.2.

Listing D.2   Example of TO.TBL

```
A,          ..\SOURCE
C,          ..\SOURCE
D,          ..\DOC
L,          ..\LST
O,          ..\OBJ
P,          ..\PROD
T,          ..\TEST
W,          ..\WORK
AIO,        \SOFTWARE\BLOCKS\AIO\SOURCE
CLK,        \SOFTWARE\BLOCKS\CLK\SOURCE
COMM,       \SOFTWARE\BLOCKS\COMM\SOURCE
DIO,        \SOFTWARE\BLOCKS\DIO\SOURCE
IX86L-FP,   \SOFTWARE\UCOS-II\IX86L-FP
KEY_MN,     \SOFTWARE\BLOCKS\KEY_MN\SOURCE
LCD,        \SOFTWARE\BLOCKS\LCD\SOURCE
LED,        \SOFTWARE\BLOCKS\LED\SOURCE
LISTC,      \SOFTWARE\BLOCKS\HPLISTC\SOURCE
TMR,        \SOFTWARE\BLOCKS\TMR\SOURCE
TO,         \SOFTWARE\TO\SOURCE
UCOS,       \SOFTWARE\UCOS\SOURCE
UCOS-II,    \SOFTWARE\UCOS-II\SOURCE
```

You may optionally add an entry by typing the path associated with a name on the command line prompt as follows:

```
TO name path
```

D

In this case, TO will append this new entry at the end of TO.TBL. This avoids having to use a text editor to add a new entry to TO.TBL. If you type:

```
TO    AIO
```

then TO will change directory to \SOFTWARE\BLOCKS\AIO\SOURCE. Similarly, if you type:

```
TO    clk
```

then TO will change directory to \SOFTWARE\BLOCKS\CLK\SOURCE. TO.TBL can be as long as needed, but each name must be unique. Note that two names can be associated with the same directory. If you add entries in TO.TBL using a text editor, all entries *must* be entered in uppercase. When you invoke TO at the DOS prompt, the name you specify is converted to uppercase before the program searches through the table. TO.TBL is searched linearly from the first entry to the last. For faster response, you may want to place your most frequently used directories at the beginning of the file.

Companion CD-ROM

R&D Books has included a companion CD-ROM to *Embedded Systems Building Blocks, Complete and Ready-to-Use Modules in C*. The CD-ROM is in MS-DOS format and contains all the source code provided in this book. The data sheets of the electronic components I have used are also on the companion CD-ROM in PDF format.

E.00 Hardware/Software Requirements

Hardware:	PC/AT compatible system
Fixed Disk Capacity:	5 Megabytes free
System Memory:	640K bytes of RAM
Operating System:	MS-DOS, Windows 95, Windows 98, or Windows NT

E.01 Installation

Use the `Install.bat` file to decompress and transfer the ESBB files from the CD to your system. `Install.bat` expects 2 arguments.

1. Load DOS or open a DOS window under Windows 95/98/NT and specify the `C:` drive as the default drive.

2. Insert the CD-ROM in your CD drive.

3. Type: `<cd-drive>:INSTALL <cd-drive> [destination]`.

where `<cd-drive>` is the drive letter of your CD-ROM and `[destination]` is the drive letter where you want ESBB installed. For example, to install ESBB on your hard disk drive `E:` from a CD drive `H:`, you would type:

```
H:INSTALL H E
```

INSTALL will create the following directory on the specified destination drive:

\SOFTWARE

INSTALL will then change the directory to \SOFTWARE and copy the file ESBB.EXE from drive <cd-drive>: to this directory. INSTALL will then execute ESBB.EXE, which will create all other directories under \SOFTWARE and transfer all source and executable files provided in this book (see Directory Structure, below). Upon completion, INSTALL will delete ESBB.EXE and change the directory to \SOFTWARE\BLOCKS\SAMPLE\TEST.

NOTE: Make sure you read the READ.ME file on the companion CD-ROM for last minute changes and notes.

E.02 Directory Structure

Once INSTALL has completed, your destination drive will contain the following subdirectories:

- \SOFTWARE
  The main directory from the root where all software-related files are placed.

- \SOFTWARE\BLOCKS
  The main directory where all building blocks are located.

- \SOFTWARE\BLOCKS\AIO\SOURCE
  This directory contains the source code for the analog I/O module (Chapter 10). The files in this directory are AIO.C and AIO.H.

- \SOFTWARE\BLOCKS\CLK\SOURCE
  This directory contains the source code for the clock/calendar module (Chapter 6). The files in this directory are CLK.C and CLK.H.

- \SOFTWARE\BLOCKS\COMM\SOURCE
  This directory contains the source code for the asynchronous serial communication modules COMM_PC, COMMBUF1, and COMMBUF2 (Chapter 11). The files in this directory are:

  COMM_PC.C, COMM_PC.H and COMM_PCA.ASM.
  COMMBGND.C and COMMBGND.H
  COMMRTOS.C and COMMRTOS.H

- \SOFTWARE\BLOCKS\DIO\SOURCE
  This directory contains the source code for the discrete I/O module (Chapter 8). The files in this directory are DIO.C and DIO.H.

- \SOFTWARE\BLOCKS\KEY_MN\SOURCE
  This directory contains the source code for the keyboard scanning module presented in Chapter 3. The source files are KEY.C and KEY.H.

- \SOFTWARE\BLOCKS\LCD\SOURCE
  This directory contains the source code for the character LCD module presented in Chapter 5. The source files are LCD.C and LCD.H.

- \SOFTWARE\BLOCKS\LED\SOURCE
  This directory contains the source code for the multiplexed LED module presented in Chapter 4. The source files are LED.C, LED_IA.ASM, and LED.H.

- \SOFTWARE\BLOCKS\PC\BC45

  This directory contains the source code for PC related services (see Chapter 1). The files in this directory are PC.C and PC.H.

- \SOFTWARE\BLOCKS\SAMPLE\SOURCE

  This directory contains the source code for the sample code (see Chapter 1). The files in this directory are: CFG.C, CFG.H, INCLUDES.H, OS_CFG.H, TEST.C, and TEST.LNK.

- \SOFTWARE\BLOCKS\SAMPLE\TEST

  This directory contains the pre-compiled DOS executable TEST.EXE. You can run this executable by opening a DOS window under either Windows 95, Windows 98, or Windows NT.

  This dicrectory also contains a 'batch' file (MAKETEST.BAT) that will rebuild the object files using the Borland 'MAKE' utility and the 'makefile' TEST.MAK. Note that the makefile assumes that the Borland C/C++ compiler is located in the E:\BC45\BIN directory but you can easily change that by editing TEST.MAK (see BORLAND and BORLAND_EXE in TEST.MAK).

- \SOFTWARE\BLOCKS\SAMPLE\OBJ

  This directory contains the compiled object files for the building blocks that are used in TEST.EXE. You will find the following files in this directory:

      AIO.OBJ
      CFG.OBJ
      CLK.OBJ
      COMMRTOS.OBJ
      COMM_PC.OBJ
      COMM_PCA.OBJ
      DIO.OBJ
      KEY.OBJ
      LCD.OBJ
      OS_CPU_A.OBJ
      OS_CPU_C.OBJ
      PC.OBJ
      TEST.OBJ
      TMR.OBJ
      UCOS_II.OBJ
      TEST.EXE
      TEST.MAP

  UCOS_II.OBJ contains the pre-compiled object code for μC/OS-II. You can obtain the source code for μC/OS-II by obtaining a copy of my other book, *MicroC/OS-II, The Real-Time Kernel*, ISBN 0-87930-543-6.

  OS_CPU_A.OBJ, OS_CPU_C.OBJ are the processor specific code for μC/OS-II for an Intel (or AMD) 80x86. The code also supports hardware floating-point.

- \SOFTWARE\BLOCKS\TMR\SOURCE

  This directory contains the source code for the timer manager module (Chapter 7). The source files are TMR.C and TMR.H.

- \SOFTWARE\HPLISTC

  This directory contains HPLISTC (Appendix D). The source file HPLISTC.C is found in \SOFTWARE\HPLISTC\SOURCE. The DOS executable file HPLISTC.EXE is found in the \SOFTWARE\HPLISTC\EXE directory.

E

- \SOFTWARE\TO

  This directory contains the files for the TO utility (Appendix D). The source file is TO.C and is found in the \SOFTWARE\TO\SOURCE directory. The DOS executable file (TO.EXE) is found in the \SOFTWARE\TO\EXE directory. Note that TO requires a file called TO.TBL which must reside on your root directory. A example of TO.TBL is also found in the .EXE directory. You will need to move TO.TBL to the root directory if you are to use TO.EXE.

- \SOFTWARE\UCOS-II\Ix86L-FP\BC45

  This directory contains the file OS_CPU.H which is the header file for the processor specific code for µC/OS-II and the 80x86 processor which supports hardware floating-point support.

- \SOFTWARE\UCOS-II\SOURCE

  This directory contains the file uCOS_II.H which is the header file for µC/OS-II. This file is used by your application code to gain access to µC/OS-II's API (Application Program Interface).

E.03 Finding Errors

I have done everything I could to test the code provided in this book. If you find errors, I would like to know about them so that I can correct them or visit my web site at www.uCOS-II.com

You can reach me through e-mail at: Jean.Labrosse@uCOS-II.com

You can also contact me through R&D Books or by sending me a letter at:

Jean J. Labrosse
949 Crestview Circle
Weston, FL 33327
U.S.A.

E.04 Licensing

Embedded Systems Building Blocks (ESBB) source code and object code can be freely distributed (to students) by accredited colleges and universities without requiring a license, as long as there is no commercial application involved. In other words, no licensing is required if ESBB is used for educational use.

You must obtain an Object Code Distribution License to embed any ESBB code (i.e., module) in a commercial product. There will be a fee for such situations, and you will need to contact me for pricing.

You must obtain an Source Code Distribution License to distribute ESBB's source code. Again, there is a fee for such a license, and you will need to contact me for pricing. You can contact me at Jean.Labrosse@uCOS-II.com or visit my web site at www.uCOS-II.com

Write me at the address provided above, or call at:

(954) 217-2036
(954) 217-2037 (fax)

Index

Symbols

Numerics

A

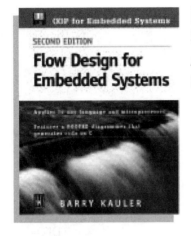

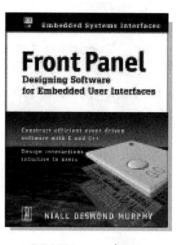

What's on the CD-ROM?

The companion CD-ROM for *Embedded Systems Building Blocks, Second Edition* is in MS-DOS format and contains all the source code provided in the book. The data sheets of the electronic components used are also on the CD in PDF format.

Hardware/Software Requirements:

Hardware:	PC/AT compatible system
Fixed Disk Capacity:	5 Megabytes free
System Memory:	640K bytes of RAM
Operating System:	MS-DOS, Windows 95, Windows 98, or Windows NT

TIP

For more information on installation and directory structure, refer to Appendix E, "Companion CD-ROM". For last minute changes or notes, see the Read.me file on the CD, or visit the web site at www.uCOS-II.com